The Search for Ultralight Bosonic Dark Matter

Derek F. Jackson Kimball • Karl van Bibber
Editors

The Search for Ultralight Bosonic Dark Matter

 Springer

Editors
Derek F. Jackson Kimball
Department of Physics
California State University, East Bay
Hayward, CA, USA

Karl van Bibber
Department of Nuclear Engineering
University of California Berkeley
Berkeley, CA, USA

This work was supported by National Science Foundation, Simons and Heising-Simons Foundatoin Gordon and Betty More Foundation

ISBN 978-3-030-95854-1   ISBN 978-3-030-95852-7   (eBook)
https://doi.org/10.1007/978-3-030-95852-7

This Springer imprint is published by the registered company Springer Nature Switzerland AG
The registered company address is: Gewerbestrasse 11, 6330 Cham, Switzerland

We dedicate this book to our mentors who have preceded us and our colleagues who have worked alongside us, who have contributed so much to our current understanding of the dark matter problem. To them we owe profound gratitude.

We further dedicate this book to our students, who are writing the next chapters of this amazing saga. To them we owe our admiration and encouragement.

Preface

The nature of dark matter, the invisible substance whose gravitational pull holds stars in their paths through galaxies and galaxies in their paths through the universe, is one of the most profound mysteries of modern physics. It is a mystery that may be at the center of a web of other mysteries, connected to physics that could explain the matter-antimatter asymmetry of the universe and why gravity is so much weaker than other fundamental forces. It is a mystery that has drawn hundreds of physicists from all over the world to propose new theories and dream up intricate experiments to search for answers. We, too, have been drawn into the community of physicists trying to understand what dark matter is, and, if you're reading this book, there's a good chance that you have been as well.

This book is intended to offer a broad introduction to one of the most intriguing hypotheses proposed to explain dark matter: the idea that perhaps dark matter is energy stored in a field composed of ultralight bosons. Such ultralight bosons are ubiquitous features of many theories extending beyond the standard model of particle physics, predicted to have nonzero masses far smaller than those of any particles we have ever observed. Depending on the particular theory predicting their existence, ultralight bosons can have many different properties, and thus require a wide range of experimental techniques to explore the vast parameter space of possibilities.

The last decade has seen a renaissance in both theoretical and experimental research into ultralight bosonic dark matter fueled by new ideas from a wide variety of disciplines: particle physics, astrophysics, cosmology, nuclear physics, atomic physics, solid state physics, optics, and electronics. Because of the rapid pace of new developments and the interdisciplinary nature of the research, the time is ripe for an introductory book to cover the basic concepts needed to understand ultralight bosonic dark matter and how to find it.

This book does not intend to be a comprehensive review covering all the work in this vibrant field of research. Neither is this book filled with technical details or complex theoretical derivations. Rather, this book is intended to be a starting point for students and researchers new to our field, focusing on clearly explaining the most important ideas and fundamental principles. Each chapter features several questions and problems (with solutions in the back of the book) that can facilitate self-study.

We sincerely hope that you find this book to be accessible and engaging, and have as much fun reading it as we have had editing it!

Hayward, CA, USA
Berkeley, CA, USA
May 2021

Derek F. Jackson Kimball
Karl van Bibber

Contents

Contributors

John W. Blanchard Helmholtz Institute Mainz, GSI Helmholtzzentrum für Schwerionenforschung, Mainz, Germany
NVision Imaging Technologies GmbH, Ulm, Germany

Dmitry Budker Helmholtz Institute Mainz, GSI Helmholtzzentrum für Schwerionenforschung, Mainz, Germany
Department of Physics, University of California, Berkeley, CA, USA

Gianpaolo Carosi Lawrence Livermore National Laboratory, Livermore, CA, USA

Andrei Derevianko Department of Physics, University of Nevada, Reno, NV, USA

Leanne D. Duffy Los Alamos National Laboratory, Los Alamos, NM, USA

Andrew A. Geraci Department of Physics and Astronomy, Northwestern University, Evanston, IL, USA

Sebastian Hoof Institut für Astrophysik, Georg-August-Universität Göttingen, Göttingen, Germany

Igor G. Irastorza Center for Astroparticle and High Energy Physics (CAPA), University of Zaragoza, Zaragoza, Spain

Derek F. Jackson Kimball Department of Physics, California State University, East Bay, Hayward, CA, USA

David J. E. Marsh Department of Physics, King's College London, London, UK

Arran Phipps Department of Physics, California State University - East Bay, Hayward, CA, USA

Szymon Pustelny Institute of Physics, Jagiellonian University, Kraków, Poland

Yun Chang Shin Center for Axion and Precision Physics Research (CAPP), IBS, Daejeon, Republic of Korea

Maria Simanovskaia Department of Physics, Stanford University, Stanford, CA, USA

Aaron D. Spector Deutsches Elektronen-Synchrotron (DESY), Hamburg, Germany

Alexander O. Sushkov Department of Physics, Boston University, Boston, MA, USA

Karl van Bibber Department of Nuclear Engineering, University of California, Berkeley, CA, USA

Julia K. Vogel Lawrence Livermore National Laboratory, Livermore, CA, USA

Arne Wickenbrock Helmholtz Institute Mainz, GSI Helmholtzzentrum für Schwerionenforschung, Mainz, Germany

Definitions of Commonly Used Acronyms and Mathematical Symbols

See Tables 1 and 2.

Table 1 General mathematical symbols used and their meanings

Symbol	Meaning
c	Speed of light
ϵ_0	Electric permittivity of vacuum
μ_0	Magnetic permeability of vacuum
G_N	Newtonian constant of gravitation
H_0	Hubble constant in the present era
$\hbar$	Planck's constant, $h = 2\pi\hbar$
e	Elementary charge
$\alpha = e^2/(\hbar c)$	Fine structure constant
k_B	Boltzmann constant
R_∞	Rydberg constant
a_0	Bohr radius
μ_B	Bohr magneton
μ_N	Nuclear magneton
G_F	Fermi constant
g	Local acceleration due to the Earth's gravity
m_e	Electron mass
m_p	Proton mass
$M_\odot$	Solar mass
σ_i	Pauli matrices, $i = 1, 2, 3$
γ_μ	Dirac matrices, $\mu = 0, 1, 2, 3$
γ_5	Dirac matrix associated with pseudoscalars
Λ_{QCD}	QCD energy scale
Q_{W}	Nuclear weak charge
ρ_{dm}	Average dark matter energy density (in the Milky Way)
$\hat{\sigma}$	Unit vector along spin

Table 2 Some acronyms and their meanings

Symbol	Meaning
ABC processes	Axio-recombination, bremsstrahlung, and Compton scattering
ADMX	Axion Dark Matter Experiment
ALP	Axionlike particle
ALPS	Any Light Particle Search
AMELIE	Axion Modulation hELIoscope Experiment
ARIADNE	Axion Resonant InterAction DetectioN Experiment
BAO	Baryon acoustic oscillations
BBN	Big Bang nucleosynthesis
CASPEr	Cosmic Axion Spin Precession Experiment
CAST	CERN Axion Solar Telescope
CCD	Charge coupled device
CDM	Cold dark matter
CMB	Cosmic microwave background radiation
DFSZ model	Dine-Fischler-Srednicki-Zhitnitsky model
DM	Dark matter
DNP	Dynamic nuclear polarization
EDM	Electric dipole moment
EMF	Electromotive force
FSR	Free spectral range
GNOME	Global Network of Optical Magnetometers for Exotic physics searches
GPS	Global Positioning System
HAYSTAC	Haloscope At Yale Sensitive To Axion Cold dark matter
HEMT	High Electron Mobility Transistor amplifier
HFET	Heterojunction Field Effect Transistor
IAXO	International AXion Observatory
JPA	Josephson parametric amplifier
KSVZ model	Kim-Shifman-Vainshtein-Zakharov model
LHC	Large Hadron Collider
LIGO	The Laser Interferometer Gravitational-Wave Observatory
LO	Local oscillator
LSW	Light-Shining-through-Wall
ly	Light year
MACHOs	Massive astrophysical compact halo objects
MAS	Magic angle spinning
MOND	MOdified Newtonian Dynamics
NMR	Nuclear magnetic resonance
PBH	Primordial black hole
PHIP	Parahydrogen-induced polarization
PQ	Peccei-Quinn
PSD	Power spectral density
QCD	Quantum chromodynamics
QFT	Quantum field theory
QUAX	QUest for AXions experiment

(continued)

Table 2 (continued)

Symbol	Meaning
RF	Radio frequency
SEOP	Spin-exchange optical pumping
SERF	Spin-exchange-relaxation free
SHM	Standard halo model
SPN	Spin-projection noise
SQUID	Superconducting Quantum Interference Device
SN	Supernova
TE	Transverse electric
TES	Transition edge sensors
TM	Transverse magnetic
UBDM	Ultralight bosonic dark matter
UCN	Ultra-cold neutron
VULF	Virialized ultralight field
WIMP	Weakly interacting massive particle
WISP	Weakly interacting sub-eV particle
ZULF NMR	Zero-to-ultralow field NMR

Units and Conversion Factors

In the search for ultralight bosonic matter, knowledge from all branches of physics is required: from particle physics to atomic physics to astrophysics. For this reason, it is common for researchers to switch between different systems of units that are best suited for description of phenomena on vastly different scales, ranging from the subatomic to the cosmological. Throughout this text, we adopt the standard approach of our community and switch between different systems of units as appropriate. In this section, we discuss common systems of units, conversion factors, and typical values in order to help the reader navigate the calculations and estimates presented throughout the text.

Three common systems of units are particularly useful in relating laboratory experiments to theoretical calculations: SI units (abbreviated from the French système international d'unités, the "International System of Units"), Gaussian or CGS units (centimeters-grams-seconds), and natural units, where

$$\hbar = c = k_B = \epsilon_0 = \mu_0 = 1. \tag{1}$$

In natural units, often used by theorists, all quantities are measured in energy to some power, and typically the unit of energy is chosen to be the electron volt (eV). Conversion factors between the key base units of the natural, SI, and Gaussian/CGS systems of units are presented in Table 3, from which the units of most other quantities can be derived. (Another common system is Lorentz-Heaviside units, discussed in Problem 5.1 of Chap. 5.)

There are many useful quantities that can be derived from the base units described in Table 3. For example, the natural unit of force can be derived from dimensional analysis:

$$[\text{force}] = \frac{[\text{mass}][\text{length}]}{[\text{time}^2]} = \frac{(\text{eV})(\text{eV}^{-1})}{(\text{eV}^{-2})} = \text{eV}^2 , \tag{2}$$

where we use $[\cdots]$ to denote the units. Using the conversion factors in Table 3, force in units of eV^2 can be converted to N or dyne. Table 4 presents approximate

Table 3 Conversion factors between key base units of natural, SI, and Gaussian/CGS systems

Quantity	Natural	SI	CGS
Length	eV^{-1}	1.9732705×10^{-7} m	1.9732705×10^{-5} cm
Mass	eV	$1.7826627 \times 10^{-36}$ kg	$1.7826627 \times 10^{-33}$ g
Time	eV^{-1}	$6.5821220 \times 10^{-16}$ s	$6.5821220 \times 10^{-16}$ s
Electric current	eV	2.8494561×10^{-3} A	8.5424545×10^{6} esu/s
Temperature	eV	1.1604518×10^{4} K	1.1604518×10^{4} K

Table 4 Approximate conversion factors between key derived units of natural, SI, and Gaussian/CGS systems.

Quantity	Natural	SI	CGS
Energy	eV	1.6×10^{-19} J	1.6×10^{-12} erg
Momentum	eV	$5.3 \times 10^{-28} \frac{\text{kg·m}}{\text{s}}$	$5.3 \times 10^{-23} \frac{\text{g·cm}}{\text{s}}$
Angular momentum	1	1.05×10^{-34} J · s	1.05×10^{-27} erg · s
Force	eV^2	8.1×10^{-13} N	8.1×10^{-8} dyne
Power	eV^2	2.4×10^{-4} W	$2.4 \times 10^{3} \frac{\text{erg}}{\text{s}}$
Charge	1	1.9×10^{-18} C	5.6×10^{-9} esu
Electric field	eV^2	$4.3 \times 10^{5} \frac{\text{V}}{\text{m}}$	$14 \frac{\text{statV}}{\text{cm}}$
Magnetic field	eV^2	1.4×10^{-3} T	14 G
Electric potential	eV	8.5×10^{-2} V	2.8×10^{-4} statV

Table 5 Approximate values of fundamental constants in natural, SI, and CGS units

Constant	Symbol	Natural	SI	CGS
Planck's constant	$\hbar$	1	1.05×10^{-34} J · s	1.05×10^{-27} erg · s
Elementary charge	e	$\sqrt{\alpha} \approx 0.085$	1.6×10^{-19} C	4.8×10^{-10} esu
Speed of light	c	1	$3.0 \times 10^{8} \frac{\text{m}}{\text{s}}$	$3.0 \times 10^{10} \frac{\text{cm}}{\text{s}}$
Gravitational constant	G_N	6.67×10^{-57} eV^{-2}	$6.67 \times 10^{-11} \frac{\text{m}^3}{\text{kg·s}^2}$	$6.67 \times 10^{-8} \frac{\text{cm}^3}{\text{g·s}^2}$
Boltzmann constant	k_B	1	$1.38 \times 10^{-23} \frac{\text{J}}{\text{K}}$	$1.38 \times 10^{-16} \frac{\text{erg}}{\text{K}}$
Fermi constant	G_F	1.17×10^{-23} eV^{-2}	$1.44 \times 10^{-36} \frac{\text{J}}{\text{m}^3}$	$1.44 \times 10^{-35} \frac{\text{erg}}{\text{cm}^3}$

conversion factors between various derived units useful for estimates; more precise conversion factors can be calculated from the values presented in Table 3.

Note the convenient and intuitive fact that electric and magnetic fields have the same units in the natural and Gaussian/CGS systems. Based on the conversion factors described in Tables 3 and 4, the values of constants in the different systems of units can also be derived (Table 5). For example, Planck's constant $\hbar$ can be converted between the various systems of units using the conversion factors for angular momentum.

Atomic units, in which

$$e = m_e = \hbar = 1 \, , \tag{3}$$

are convenient for calculating properties of atoms and molecules. The speed of light in atomic units can be derived from the fine-structure constant, which, using Gaussian units for electromagnetism, is given by

$$\alpha = \frac{e^2}{\hbar c} \approx \frac{1}{137} \, , \tag{4}$$

and so $c \approx 137$ au. Similarly, the unit of length is the Bohr radius:

$$a_0 = \frac{\hbar^2}{m_e e^2} \, , \tag{5}$$

which makes the units of electric and magnetic dipole moments ea_0. Note, however, the distinction between units and typical values: the typical size of an induced electric dipole moment in an atomic system is $\approx ea_0 = 1$ au, whereas the typical size of a magnetic dipole moment of an atom is $\approx \mu_B \approx (\alpha/2)ea_0 = \alpha/2$ au. Table 6 presents many key quantities in atomic and Gaussian/CGS units.

At the other end of the scale, Planck units are often used in cosmology and theoretical particle physics. In Planck units,

$$c = \hbar = 4\pi \epsilon_0 = k_B = G_N = 1 \, . \tag{6}$$

Table 6 Atomic units, their expressions, and approximate conversion factors to Gaussian/CGS units

Quantity	Expression	CGS
Length	$a_0 = \hbar^2 / (m_e e^2)$	5.3×10^{-9} cm
Mass	m_e	9.1×10^{-28} g
Time	$m_e a_0^2 / \hbar$	2.4×10^{-17} s
Electric current	$e\hbar / (m_e a_0^2)$	2.0×10^7 esu/s
Energy	$\hbar^2 / (m_e a_0^2) = \alpha^2 m_e c^2$	4.4×10^{-11} erg
Momentum	$\hbar / a_0$	2.0×10^{-19} g $\cdot$ m/s
Angular momentum	$\hbar$	1.05×10^{-27} erg $\cdot$ s
Force	$\hbar^2 / (m_e a_0^3)$	8.2×10^{-3} dyne
Charge	e	4.8×10^{-10} esu
Electric field	e/a_0^2	$1.7 \times 10^7 \frac{\text{statV}}{\text{cm}}$
Magnetic field	e/a_0^2	1.7×10^7 G

Table 7 Expressions for key base units of Planck units and their approximate values in natural and Gaussian/CGS systems

Quantity	Expression	Natural	CGS
Length	$\sqrt{\hbar G_N/c^3}$	8.19×10^{-29} eV^{-1}	1.62×10^{-33} cm
Mass	$\sqrt{\hbar c/G_N}$	1.22×10^{28} eV	2.18×10^{-5} g
Time	$\sqrt{\hbar G_N/c^5}$	8.19×10^{-29} eV^{-1}	6.58×10^{-16} s
Electric charge	$\sqrt{4\pi\epsilon_0 \hbar c}$	1	5.62×10^{-9} esu
Temperature	$\sqrt{\hbar c^5/(G_N k_B^2)}$	1.22×10^{28} eV	1.42×10^{32} K

The central idea of Planck units is that constants from every branch of physics are normalized to unity: c from special relativity, $\hbar$ from quantum mechanics, $4\pi\epsilon_0$ from electromagnetism, k_B from thermodynamics, and G_N from gravitational physics (general relativity). In fact, Planck's system is another class of natural units, similar to the natural units introduced at the start of this section (Tables 3, 4, and 5), where gravity determines the unit of energy rather than the eV. The fact that Planck units are another form of natural units is evident from Table 7, which shows the expressions for base units in Planck, natural (eV), and Gaussian/CGS systems: all Planck units are in terms of powers of the Planck energy ($\sqrt{\hbar c^5/G_N} \approx 1.22 \times 10^{28}$ eV).

Not only is it common to switch between different units, but frequently it is also convenient to employ a mixture of different unit systems. An example of this is the mass/energy density of dark matter in the Milky Way, ρ_{dm}. Usually in the literature, the value of ρ_{dm} is given as ≈ 0.4 GeV/cm^3, which uses a mixture of units so that the numerical value of ρ_{dm} is ~ 1. In CGS units, $\rho_{dm} \approx 7 \times 10^{-25}$ g/cm^3, and in natural units, $\rho_{dm} \approx 3 \times 10^{-6}$ eV^4. A useful representation of the dark matter density in astrophysical estimates is in units of solar masses per megaparsec ($M_\odot$/Mpc^3), where $M_\odot \approx 2 \times 10^{33}$ g and 1 Mpc $\approx 3 \times 10^{24}$ cm^3, so $\rho_{dm} \approx 10^{16}$ $M_\odot$/Mpc^3.

The values and conversion factors presented in this section are derived from values in *The NIST Reference on Constants, Units, and Uncertainty* (NIST Standard Reference Database **121**, 2019).

Chapter 1
Introduction to Dark Matter

Derek F. Jackson Kimball and Dmitry Budker

Abstract To set the stage for our study of ultralight bosonic dark matter (UBDM), we review the evidence for the existence of dark matter: galactic and stellar dynamics, gravitational lensing studies, measurements of the cosmic microwave background radiation (CMB), surveys of the large-scale structure of the universe, and the observed abundance of light elements. This diverse array of observational evidence informs what we know about dark matter: its universal abundance, its spatial and velocity distribution, and that its explanation involves physics beyond the Standard Model. But what we know about dark matter is far outweighed by what we do not know. We examine UBDM in the context of several of the most prominent alternative hypotheses for the nature of dark matter: weakly interacting massive particles (WIMPs), sterile neutrinos, massive astrophysical compact halo objects (MACHOs), and primordial black holes (PBHs). Finally we examine some of the key general characteristics of UBDM, including its wavelike nature, coherence properties, and couplings to Standard Model particles and fields.

1.1 Why Do We Think There Is Dark Matter?

Scientists have long speculated that there may be imperceptible forms of matter in the universe. Indeed, time and again forms of matter previously unknown have been discovered: Galileo used the telescope to discover the moons orbiting Jupiter, Chadwick irradiated a beryllium target to discover the neutron, Cowan and Reines used a water tank surrounded with scintillators to directly observe the neutrino flux

D. F. Jackson Kimball (✉)
Department of Physics, California State University, East Bay, Hayward, CA, USA
e-mail: derek.jacksonkimball@csueastbay.edu

D. Budker
Helmholtz Institüt Mainz, Johannes Gutenberg-Universität, Mainz, Germany

Department of Physics, University of California, Berkeley, CA, USA
e-mail: budker@uni-mainz.de

© The Author(s) 2023
D. F. Jackson Kimball, K. van Bibber (eds.), *The Search for Ultralight Bosonic Dark Matter*, https://doi.org/10.1007/978-3-030-95852-7_1

from a nuclear reactor, and so on. But the story of "dark matter" as we understand it today begins with early efforts by Kelvin, Poincaré, Öpik, Kapteyn, and Oort to use the dynamics of the stars in the Milky Way to estimate the ratio of the mass of luminous matter (stars) to the total mass of matter in the galaxy (see the historical review [1] and the references therein).

While these early estimates [2, 3] found that stars dominated the mass in our local solar neighborhood, Zwicky [4, 5] used observations of galaxy clusters to discover that on much larger scales it appeared that *dunkle Materie* (German for *dark matter*) was considerably more abundant than luminous matter. Zwicky analyzed the Coma cluster, which had roughly a thousand galaxies distributed in a sphere of radius $R \approx 10^6$ ly. Each galaxy in the Coma cluster contained, on average, stars whose total mass was $M_{tot} \sim 10^9 \, M_\odot$ (where $M_\odot$ is a solar mass), based on the mass/luminosity ratio determined from stars in our local solar neighborhood. From this information, the velocity dispersion of the galaxies can be predicted using the virial theorem

$$\langle K \rangle = -\frac{1}{2} \langle V \rangle \,, \tag{1.1}$$

where $\langle K \rangle$ is the time-averaged kinetic energy of the galaxies and $\langle V \rangle$ is the time-averaged gravitational potential energy. This estimate yields a velocity dispersion of

$$v \approx \sqrt{\frac{G_N M_{tot}}{R}} \approx 10^5 \text{ m/s} \,, \tag{1.2}$$

where G_N is the Newtonian constant of gravitation. However, the measured velocity dispersion of the galaxies in the Coma cluster was $\approx 10^6$ m/s, an order-of-magnitude discrepancy. It was later discovered that galaxy clusters contain a halo of hot gas with five times the mass of the stars [6, 7]: taking this into account reduced but did not eliminate the discrepancy between the predicted and observed velocity dispersion [8].

The next significant clues about the existence of dark matter came from observations of galactic rotation curves, the rotational velocity v of galaxies' stars as a function of their distance r from the galactic center. The mass distribution within galaxies can be inferred from the rotation curves, as discussed in Problem 1.1. Pioneering observations of numerous galaxies by astrophysicists Vera Rubin, Kent Ford, and others [9–13] showed that past the radius within which most of the luminous matter is concentrated, the rotation curves are typically flat: v is relatively independent of r (as seen in Fig. 1.1). This is in marked contrast to the expected $1/\sqrt{r}$ dependence of v on r if luminous matter alone is the source of the gravitational pull holding outer stars in their orbits (see Problem 1.1). These rotation-curve observations can be explained by the galactic masses being dominated by a spherical halo of dark matter extending far beyond the luminous matter of galaxies.

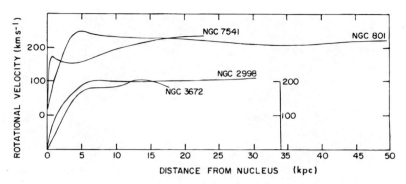

Fig. 1.1 Plot adapted from one of Rubin and Ford's papers, Ref. [13], showing the rotational velocity as a function of distance from the galactic centers (nuclei) of four different galaxies (NGC 7541, NGC 801, NGC 2998, and NGC 3672), all of which exhibit flat rotation curves

? Problem 1.1 Galactic Rotation Curves

Consider a star in a circular orbit at the periphery of a galaxy of mass M, such that most of the galaxy's mass is contained within the star's orbital radius R. How does the star's orbital velocity scale with R under these assumptions? Given that we observe flat galactic rotation curves, what can we assume is the radial dependence of the dark matter density?

Solution on page 305.

Gravitational lensing studies (see Refs. [14, 15] for reviews) considerably strengthened the case for the existence of dark matter. Because a gravitational field bends the otherwise straight-line trajectory of light, mass can distort the images of distant astrophysical objects as light travels along geodesics from those objects to the Earth. Since the images are distorted in predictable ways based on general relativity, gravitational lensing offers an independent method to investigate the distribution of mass in the universe. In 2006, Clowe et al. used gravitational lensing to study the Bullet Cluster (1E0657-558) [16], a pair of galaxy clusters that had merged $\approx$150 million years ago. Comparison of gravitational lensing to observations of stars and hot x-ray emitting gas established that the mass distribution of the Bullet Cluster does not coincide with the baryon distribution (Fig. 1.2). As seen in Fig. 1.2, the baryonic matter (dominated by hot gas, imaged by detection of x-ray emission) is clumped more closely together than the total mass of the cluster (measured by gravitational lensing), which is centered about two widely separated positions. Evidently when the two galaxy clusters merged, the baryonic matter collided and heated up while the dark matter barely interacted at all: the dark matter passed through the clusters without any observable effect, save that due to

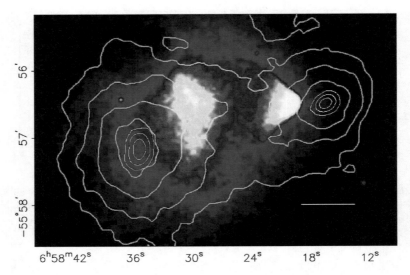

Fig. 1.2 Image of the Bullet Cluster (1E0657-558), adapted from Ref. [16], comparing x-ray emission from hot gas [the background color map with increasing x-ray intensity scaling from blue (low) to yellow/white (high)] to the mass distribution deduced from gravitational lensing (green contour plot, where the outermost contour represents low mass density and the innermost contours are highest density). The white horizontal line in the lower right represents a distance of 200 kpc at the position of the Bullet Cluster. The mass distribution is clearly different from the gas distribution

gravity. This is strong evidence that dark matter is not ordinary baryonic matter, and has been confirmed by further observations of other galaxy cluster mergers [17].

Measurements of the cosmic microwave background radiation (CMB) also point to the existence of dark matter. The CMB is a photon gas permeating the universe that essentially decoupled from baryonic matter $\approx 400,000$ years after the Big Bang. This time, known as *recombination*,[1] is when the universe had cooled to the point where the first atoms formed. From the appearance of baryons until recombination, the plasma of protons, electrons, and photons strongly interacted via Compton scattering and formed a coupled photon-baryon fluid. Thus the photons and baryons shared similar spatial patterns of density. After recombination the photons largely decoupled from baryonic matter. This is because the interaction of light with neutral atoms (integrated over all frequencies[2]) is strongly suppressed compared to the interaction of light with free charged particles. The observed CMB photons are relics

---

[1] The term *recombination* is somewhat misleading, since protons and electrons were not previously "combined"—recombination is a historical name established prior to the widespread acceptance of the Big Bang theory.

[2] Although light can strongly interact with neutral atoms at particular resonance frequencies, such resonant light-atom interactions produce spectral distortions of the CMB that are too small to detect at present [18, 19].

that provide a picture of the photon-baryon fluid at the *surface of last scattering* (the region of space a distance from which the photons from recombination have freely travelled to reach the Earth today). The CMB photons impinge on the Earth nearly uniformly from all directions and almost perfectly match a blackbody spectrum. However, there are relatively small fluctuations in the temperature and polarization of the CMB from different regions of the sky. The CMB fluctuations observed today are imprints of the photon-baryon density distribution at the time of recombination [20–22].

Based on the predictions of general relativity, cosmologists expect that the spatial fluctuations of matter density $\delta\rho_m$ should grow linearly with the expansion of the universe from the time of recombination up until the time at which $\delta\rho_m/\rho_m \gtrsim 1$. More precisely, as long as $\delta\rho_m/\rho_m \lesssim 1$, then $\delta\rho_m/\rho_m \propto a$, where a is the scale factor (see Refs. [23–25] for derivations of this relationship). The scale factor a relates the distance $d(t)$ between objects in the universe at a time t to the distance between the objects at the present time t_0:

$$d(t) = ad(t_0) \qquad (1.3)$$

and is related to the redshift z of light emitted from an object at time t by

$$a = \frac{1}{1+z} . \qquad (1.4)$$

Thus as the universe expands, the density fluctuations grow: $\delta\rho_m/\rho_m \propto a$; or, conversely, the density fluctuations observed at high redshift ($z \gg 1$) should be proportionally smaller: $\delta\rho_m/\rho_m \propto 1/z$.

An overdense region of space where $\delta\rho_m/\rho_m \gtrsim 1$ will undergo gravitational collapse, forming regions of significant overdensity: the galaxies and galactic clusters that we observe in the present epoch [26]. Recombination occurred at a redshift of $z \approx 1100$, and thus $\delta\rho_m/\rho_m$ has grown by a factor of $\approx 10^3$ since then. In order for the observed galaxy distribution throughout the universe to have grown from the density fluctuations at the moment of recombination, $\delta\rho_m/\rho_m$ at recombination should be at the level of a part per hundred[3] [28]. Measurements show, however, that the CMB is remarkably uniform throughout the sky (to more than a part in $\approx 10^5$), reflecting a similarly uniform baryon density at recombination. Thus galaxy formation could not possibly be seeded by the fluctuations in baryon density at recombination: $\delta\rho_m/\rho_m$ would be only $\lesssim 10^{-2}$ today and matter would still not have undergone gravitational collapse to form galaxies.

These apparently contradictory observations can be reconciled if the mass of baryonic matter in the universe is small compared to the mass of a gas of nonrelativistic (cold) dark matter particles that weakly interact with baryonic matter. This cold dark matter (CDM) could have begun clumping long before recombination.

---

[3] The first stars and galaxies appear at a redshift of $z \gtrsim 10$ [27].

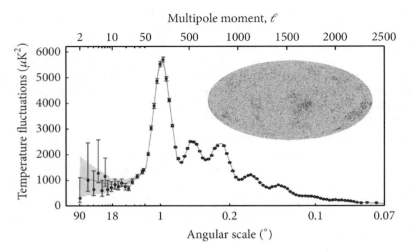

Fig. 1.3 Plot of the angular anisotropy of the square of the CMB temperature fluctuations in terms of multipole moments $\propto \langle |a_{\ell m}|^2 \rangle$ (averaged over m) and associated angular scale. Figure adapted from Ref. [35]. The black dots with red error bars are data from the Planck satellite observations and the green curve is the theoretical fit. The oval inset shows the Planck all-sky map of the CMB intensity fluctuations. The agreement between theory and data from the Planck and WMAP missions supports a flat universe whose matter density is dominated by CDM [32–34]

Thus by the time of recombination the CDM could have relatively large density fluctuations, whereas the baryons were still nearly uniformly distributed at that point in time [29]. The hot photon-baryon fluid would pass through the "clumpy" CDM with relatively little perturbation (resembling the case of the Bullet Cluster discussed above). Thus the photon-baryon fluid at recombination, and hence the CMB, exhibit small fluctuations while the large density fluctuations of the CDM could seed the formation of galaxies and give rise to the highly nonuniform distribution of matter observed today [30, 31].

This description can be made quantitative by calculating and measuring the variation of the CMB temperature $\delta T(\theta, \phi)$ across the sky, where θ and ϕ indicate the angular position. In terms of spherical harmonics $Y_{\ell m}(\theta, \phi)$:

$$\delta T(\theta, \phi) = \sum_{\ell, m} a_{\ell m} Y_{\ell m}(\theta, \phi) \, , \qquad (1.5)$$

where $a_{\ell m}$ is the expansion coefficient of the CMB temperature associated with the respective $Y_{\ell m}(\theta, \phi)$. Measurements of the angular anisotropy spectrum of the CMB by the WMAP (Wilkinson Microwave Anisotropy Probe) and Planck missions [32, 33] agree with theoretical predictions based on a cosmology with a CDM-dominated matter density [33, 34]. Data from the Planck mission are shown in Fig. 1.3.

The peaks in the plot of the angular anisotropy of the CMB temperature fluctuations appearing at different ℓ seen in Fig. 1.3 reveal both the underlying spacetime geometry of the universe and the baryon density [36], as we explain below. The peaks in Fig. 1.3 are a result of the so-called *baryon acoustic oscillations* (BAO). Baryonic matter falls into the gravitational potential wells created by concentrations of CDM, but as the baryons fall into the potential wells and the plasma density increases, the plasma heats up and generates pressure[4] that counteracts the gravitational pull. This causes the plasma to expand. These cycles of compression and expansion of the strongly coupled baryon-photon fluid after the Big Bang are the BAO.

There is an analogy between the dynamical effects of the BAO and the ripples on the surface of a pond emanating from falling droplets of rain. When baryons fall into the gravitational potential wells of the overdense regions of CDM, spherical compression (sound) waves in the photon-baryon fluid are generated and propagate outward. The speed of sound in the photon-baryon fluid is about half the speed of light. As noted above, the initial gravitational collapse that begins the BAO is seeded when $\delta \rho_m / \rho_m \gtrsim 1$ (where ρ_m is dominated by the CDM). This must happen relatively soon after the Big Bang, and so these initial compression waves propagate outward for $t_r \approx 400,000$ years until recombination, when atoms form and the light of the CMB is released. The largest peak in Fig. 1.3 occurs at an angular scale of $\Delta\theta \approx 1°$ and $\ell \approx 220$. This feature is determined by the distance these first ripples of the BAO had travelled from the time since recombination $s \approx ct_r/2$ (the sound horizon):

$$\Delta\theta = \frac{s}{d_{ls}(z)} , \qquad (1.6)$$

where $d_{ls}(z)$ is the distance to the surface of last scattering, taking into account the expansion of the universe from $z \approx 1100$.

The relationship between the angular scale of the peaks in the anisotropy of the CMB temperature fluctuations and the spatial scale of the baryon density fluctuations at recombination can be distorted by the spacetime geometry of the universe [36]. The overall spatial curvature of the universe could, in principle, be open, closed, or flat: in an open universe, initially parallel light rays would propagate along geodesics that diverge from each other; in a closed universe, initially parallel light rays would propagate along geodesics that converge; and in a flat universe (spatial curvature equals zero), initially parallel light rays would remain parallel as they propagate (geodesics are straight lines). If the spacetime geometry of the universe was open or closed, the spatial curvature would cause the observed $\Delta\theta$ and ℓ for the first peak in Fig. 1.3, corresponding to the spatial scale of the sound horizon at the surface of last scattering, to be larger or smaller than observed. The CMB measurements provide strong evidence for a flat universe (better than a part in a thousand [32–34]). The higher order peaks in Fig. 1.3 show the relative importance

---

[4] This pressure is what keeps the baryon-photon fluid density quite uniform in the early universe even in the presence of regions with significant CDM overdensities.

of the gravitational potential from the baryons themselves (which oscillates with the photon-baryon fluid density as it compresses and expands) as compared to the gravitational potential from the CDM which does not oscillate (due to negligible interactions, there is no dark matter "pressure" to counterbalance gravity). The ratio of dark matter density to baryon density derived from the CMB measurements is consistent with the ratio found from the velocity dispersion of galactic clusters [8], galactic rotation curves [13], and gravitational lensing studies [14, 15]. These topics are discussed in more detail in Chap. 3, Sect. 3.2.1.

There is yet another line of reasoning suggesting that a significant fraction of the mass of the universe is nonbaryonic: the primordial abundance of light elements produced by the Big Bang [37]. Early work by Gamow and Alpher [38–40][5] showed that light elements could be produced in the early universe via neutron capture. As the production of elements in stars and supernovae was better understood, it became apparent that, for example, deuterium (^{2}H) in the interstellar medium could not have been produced in stars but must be a relic of the Big Bang [41]. Today, this process of Big Bang nucleosynthesis (BBN) is relatively well understood based on the Standard Model of particle physics [37].[6] The basic concepts of BBN are described in the following tutorial.

Tutorial: Big Bang Nucleosynthesis (BBN)

A way to understand BBN is to follow particle reaction rates as the universe expands and cools after the Big Bang. Nucleosynthesis begins about ten seconds after the Big Bang. At this point in the evolution of the early universe, the energy density was dominated by relativistic species: photons, electrons, positrons, and neutrinos. Under these conditions the weak interaction rates were rapid compared to the expansion rate and established thermal and chemical equilibrium between neutron and proton densities via the reactions

$$n + e^+ \leftrightarrow p + \bar{\nu}_e \,,$$

$$n + \nu_e \leftrightarrow p + e^- \,,$$

$$n \leftrightarrow p + e^- + \bar{\nu}_e \,.$$

In equilibrium the ratio between the neutron and proton densities (n_n and n_p, respectively) is given by the Boltzmann factor

---

[5] This work includes the famous "$\alpha - \beta - \gamma$" paper [39] where Gamow added Hans Bethe to the author list purely for humorous purposes.

[6] Notable exceptions to the success of the BBN model are the *lithium problems*: the observed abundance of ^{7}Li is a few times smaller than the BBN predictions, and the observed abundance of ^{6}Li is about three orders of magnitude higher than the BBN predictions [42].

$$\frac{n_n}{n_p} = e^{-E_{np}/k_B T} , \tag{1.7}$$

where $E_{np} = (m_n - m_p)c^2$ is the neutron-proton mass difference and T is the temperature of the universe. As the universe continued to expand and cool, eventually the weak interaction rates fell below the expansion rate, which resulted in breaking of equilibrium: the universe was expanding too fast after this point for neutrons and protons to maintain chemical equilibrium (this is known as *freeze-out*). Specifically, freeze-out occurs when the reaction rate Γ becomes smaller than the Hubble parameter H,

$$H = \frac{\dot{a}}{a} , \tag{1.8}$$

where a is the scale factor introduced in Eqs. (1.3) and (1.4). If $\Gamma \ll H$, there is on average less than one reaction over the age of the universe ($\approx 1/H$).

The freeze-out temperature $T_f \approx 0.8$ MeV$/k_B$, for which $H = \Gamma$, is predicted by the Standard Model (which describes the rates of the aforementioned weak interactions that interconvert neutrons and protons) along with general relativity and standard cosmology (which describes the expansion rate H) and gives [37]

$$\frac{n_n}{n_p} = e^{-E_{np}/k_B T_f} \approx \frac{1}{6} . \tag{1.9}$$

After freeze-out, n_n/n_p continues to decrease because of β-decay of the neutrons. The beginning of the nucleosynthesis chain with the production of deuterium (D) is delayed because of photodissociation: the ratio of the baryon density to photon density, n_B/n_γ, is so low that photodissociation of D exceeds its production. The temperature at which nucleosynthesis begins can be found by comparing the rate of D production,

$$\Gamma_{\text{prod}} \approx n_B \sigma_{\text{nc}} v , \tag{1.10}$$

to D dissociation,

$$\Gamma_{\text{dis}} \approx n_\gamma \sigma_{\text{pd}} c e^{-E_D/k_B T} , \tag{1.11}$$

where n_B and n_γ are the baryon and photon densities, respectively, σ_{nc} and σ_{pd} are the cross-sections for neutron capture and photodissociation, respectively, v is the relative velocity between baryons, and $E_D \approx 2.23$ MeV is the deuterium binding energy. The factor $e^{-E_D/k_B T}$ must be included in Eq. (1.11) since $n_\gamma/n_B \gg 1$ and thus $\Gamma_{\text{dis}} \gg \Gamma_{\text{prod}}$ until $k_B T \ll E_D$: there is significant photodissociation due to the high-energy tail of the thermal photon distribution.

When T becomes sufficiently low (at $k_B T \approx 0.1$ MeV), Γ_{dis} drops below Γ_{prod} and deuterium is produced, starting the chain reaction that generates the light elements. From this point, most of the ratios of light element abundances can be calculated from measured nuclear reaction rates and well-known Standard Model physics [37], and agree well with observations (except for ^{6}Li and ^{7}Li as mentioned in the footnote from the previous page).

End of Tutorial

The theory of BBN outlined in the above tutorial has just one free parameter: the ratio of baryon density to photon density, n_B/n_γ, at the time of when the light elements formed. Thus the measured ratios between abundances of ^{1}H, ^{2}H, ^{3}He, and ^{4}He not only determine n_B/n_γ, but the consistency between the predicted ratios of these abundances serves as a cross-check of the theory of BBN. There is good agreement between theory and observations, validating the theory of BBN in the standard Big Bang cosmology [43, 44] and precisely measuring the baryon density produced by the Big Bang. As discussed above, from measurements of the CMB, we know that the universe has a flat spacetime geometry, which in turn implies that the total energy density measured now, $\rho_{\text{tot}}(t_0)$, is equal to the critical density, $\rho_{\text{crit}}(t_0)$, for a flat universe [45]:

$$\rho_{\text{crit}}(t_0) = \frac{3H_0^2}{8\pi G_N} , \qquad (1.12)$$

where $H_0 = H(t_0)$ is the present value of the Hubble parameter. The total energy density $\rho_{\text{tot}}(t_0)$ includes contributions from both matter and an unexplained form of energy known as *dark energy*[7] (described in the standard Big Bang cosmology by a cosmological constant Λ). The value of the dark energy density ρ_Λ can be determined from surveys of distant type Ia supernovae [51–53]. With the baryon mass density $\rho_B(t_0)$ given by measurements and calculations of the relic density of light elements [54], the density of nonbaryonic CDM can be determined:

$$\rho_{\text{CDM}}(t_0) = \frac{3H_0^2}{8\pi G_N} - \rho_B(t_0) - \rho_\Lambda(t_0) . \qquad (1.13)$$

The amount of dark matter found from these considerations is consistent with that found from other lines of reasoning: over 80% of the matter content of the universe is dark.

Given the diversity of evidence for dark matter outlined above, not to mention additional evidence from detailed modeling of the cosmological evolution of the universe and galaxy formation [55], is there any possibility that dark matter does

---

[7] The problem of the nature of dark energy is in some ways even more perplexing than the problem of dark matter, and there may even be connections between explanations of the two phenomena [46, 47]. The interested reader is referred to Refs. [48–50] for reviews.

not exist? Historically, when the primary evidence for dark matter was derived from the rotation curves of galaxies, a plausible alternative hypothesis to explain the data was proposed by Milgrom [56, 57]: Modified Newtonian Dynamics (MOND). The main idea of MOND is that rather than introducing new particles, the laws of physics should be modified: if the nonrelativistic force due to gravity behaved as

$$F = \frac{ma^2}{a_0} \qquad (1.14)$$

in the limit of very small accelerations $a \ll a_0 \approx 10^{-10}$ m/s^2, then the motion of stars in galaxies could be understood without postulating the existence of dark matter. MOND remained a viable alternative to dark matter for quite some time, but in spite of valiant attempts to extend the theory [58], MOND struggles to explain the combined observational evidence for dark matter derived from galactic clusters, gravitational lensing, CMB measurements, and BBN without, ultimately, introducing new particles [1, 59]. This is not to rule out the possibility that MOND or variants on these ideas could account for *some* of the observations described above. However, based on the multiple and distinct observations and calculations supporting the dark matter hypothesis, it is difficult to envision a scenario without some form of dark matter.

Nonetheless, one should keep in mind the complexity of the Standard Model when imagining that but a single type of particle makes up all of the dark matter: the plethora of known particles and fields in the Standard Model apparently constitute less than a fifth of the matter in the universe. Furthermore, there is always the possibility of discovering new physics that could significantly alter our understanding of the case for dark matter. For instance, a nonzero mass of the photon could partially explain the flat galactic rotation curves [60]. So while the evidence for dark matter is compelling, one should not turn a blind eye to alternative theories.

1.2   What Do (We Think) We Know About Dark Matter?

In this section we consider in turn several crucial characteristics of dark matter established by the observational evidence discussed in Sect. 1.1. Already we have seen that multiple, independent observations provide a good understanding of the total amount of mass in the form of dark matter in the universe. We also know that the dark matter must either be stable or long-lived, since the evidence shows that dark matter has been present and played a crucial role throughout the cosmological history of the universe. Furthermore, dark matter:

1. is not predominantly any of the known Standard Model particles (without the introduction of some new physics beyond the Standard Model),
2. is predominantly nonrelativistic (cold), and
3. is distributed in halos that extend well beyond the luminous matter of galaxies.

The fundamental Standard Model constituents of matter are leptons and quarks. The known stable, long-lived form of quarks are baryons: protons and bound neutrons. The preponderance of observational evidence establishes that dark matter is not made of such baryons. The baryonic content of the universe, as noted in Sect. 1.1, is determined from measurements of the CMB and the abundance of light elements produced by BBN, and establishes that dark matter cannot be ordinary baryons. The only known stable charged lepton is the electron, which when free interacts strongly with light: electrons can contribute significantly to dark matter only if they are bound to nuclei in the form of atoms, in which case the constraint on baryon density rules them out as a candidate. That leaves neutrinos.

At first glance, neutrinos appear to be an intriguing dark matter candidate: they only interact via the weak interaction (so they are indeed dark) and they are produced as a thermal relic of the Big Bang [61–63]. However, Standard Model neutrinos cannot be a substantial fraction of the dark matter for a reason related to the second item in the above list of dark matter characteristics: dark matter must be nonrelativistic (cold) rather than relativistic (hot) during the formation of structure in the early universe. This point was alluded to in the discussion of the CMB fluctuations in Sect. 1.1: only the cold dark matter (CDM) scenario can connect the measured scale of density fluctuations at recombination seen in the CMB to the observed large-scale structure of the matter in the universe in the present epoch. The random thermal motion of hot dark matter would wash out the small-scale density fluctuations needed to seed galaxy formation. When detailed cosmological models and simulations are compared to extensive surveys of the distribution of galaxies in the universe, it is clear that the observed universe matches the CDM scenario (see, for example, Refs. [64–66]).[8]

It turns out, for this reason, that Standard Model neutrinos cannot be CDM. Measurements of neutrino oscillations determine the differences between the squares of the masses of neutrino flavors: the largest square of the mass difference between neutrino flavors is $\Delta\left(mc^2\right)^2 \lesssim 2.5 \times 10^{-3}$ eV^2 [67]. Direct measurements of the electron neutrino mass from beta-decay experiments set an upper limit of $m_{\nu_e} c^2 \lesssim 2$ eV [68–70], proving that in fact all the Standard Model neutrinos have masses <10 eV. Neutrinos with masses <10 eV decouple from thermal equilibrium in the early universe at a temperature where they are highly relativistic and thus cannot be CDM [63].

Furthermore, the contribution of neutrinos to the overall mass-energy density of the universe can be determined from BBN [71] and CMB measurements [32–34] and turns out to be far too small to be the dominant component of dark matter. Yet

---

[8] However, it should be noted that warm dark matter, something which is relativistic but not highly relativistic, may make up some substantial fraction of the dark matter density [66].

another argument against neutrinos being the dominant contribution to dark matter[9] (and in fact any fermion with mass below ≈ 10 eV) is considered in Problem 1.2.

? Problem 1.2 Minimum Mass of Fermionic Dark Matter

Derive a lower limit on the mass of a spin-1/2 fermionic dark matter candidate based on the facts that (a) the average mass density of dark matter in the Milky Way is $\rho_{dm} \approx 0.4$ GeV/cm^3 [73] and (b) the escape velocity of the Milky Way galaxy is $v_{esc} \approx 2 \times 10^{-3}c$ [74].

Solution on page 306.

The third item on our list of dark matter characteristics concerns the distribution of dark matter in galaxies. The distribution in our own Milky Way galaxy is of particular interest for many of the experiments discussed in this text that seek to directly measure nongravitational interactions of dark matter using Earthbound detectors. As noted in Problem 1.1, dark matter must be distributed in a halo that extends far beyond the luminous matter in galaxies (about 6–8 times the distance from the galactic center as compared to luminous matter [75]). Presently, most researchers assume that the galactic dark matter distribution is described by what is known as the *standard halo model* (SHM) [76–78]. While there are certainly some notable discrepancies between the SHM's predictions and observations [79–81], the SHM generally accounts well for galactic rotation curves within present uncertainties. Using the SHM along with observations of stars' rotation curves in the Milky Way, a number of groups have estimated the dark matter energy density in the vicinity of our solar system to be $\rho_{dm} \approx 0.3$–0.4 GeV/cm^3, with a model-dependent uncertainty of about a factor of two [73, 82–84]. This corresponds to a mass density equivalent to one hydrogen atom per a few cm^3.

Dark matter particles are trapped within the gravitational potential well of the Milky Way galaxy and in the SHM are assumed to be virialized[10] but not thermalized (since the absence of significant nongravitational interactions is assumed). The SHM assumes that in the galactic rest frame the velocity distribution of dark matter is isotropic with a dispersion $\Delta v \approx 290$ km/s. The distribution of gravitationally bound dark matter in the galaxy (Fig. 1.4) naturally has a cutoff above the galactic escape velocity of $v_{esc} \approx 544$ km/s [74]; however, it should be noted that the speed of dark matter particles can exceed the cutoff velocity in the local vicinity of massive bodies due to gravitational acceleration, and there can also be a small fraction of

---

[9] It should be noted that the argument presented in Problem 1.2 does not apply if somehow neutrinos violate the spin-statistics theorem [72], in which case they may yet be a viable dark matter candidate.

[10] Although it should be recognized that there is evidence that fairly recently (within 1–2 billion years) a number of smaller galaxies have merged with the Milky Way, and the stars and dark matter from these galaxies have not had sufficient time to completely virialize [85, 86].

Fig. 1.4 Probability distribution function describing the speed of dark matter particles in the galactic frame of the Milky Way according to the SHM. There is a cutoff at the escape velocity of the galaxy ($v_{esc} \approx 544$ km/s). Figure courtesy of G. Blewitt

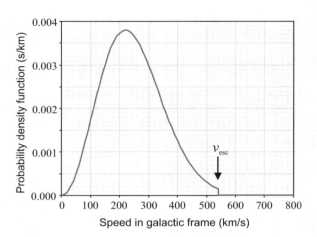

unbound dark matter passing through the galaxy at velocities above v_{esc}. Our solar system moves through the dark matter halo with relative velocity with respect to the galactic rest frame of ≈ 220 km/s $\approx 10^{-3}c$ toward the Cygnus constellation. It is important to note that the relative velocity of an Earthbound dark matter sensor also has both daily and seasonal modulations due to Earth's rotation about its axis and orbit around the Sun: the Earth's orbit creates a 10% modulation of the velocity and the Earth's rotation can create up to a 0.2% modulation [74, 87, 88].

A final characteristic, of keen interest for the experiments discussed in this text, is the degree to which dark matter interacts nongravitationally. Some generic upper limits on the strength of interactions between dark matter and Standard Model particles and fields can be obtained from observations of the Bullet Cluster and similar galaxy cluster mergers [17], as well as measurements of galaxies and satellites of galaxies moving through dark matter halos [89] and constraints on dissipation and thermalization within dark matter halos [90]. Based on this evidence, nongravitational interactions (long-range and contact) between dark matter particles are constrained to have an average scattering-cross-section-to-mass ratio $\sigma_{dm}/m_{dm} \lesssim 0.5$ cm^2/g ≈ 1 barn/GeV. This turns out to be similar to the ratio of scattering-cross-section-to-mass ratios for nuclei. Thus, generically, from astrophysical evidence it is difficult to say that the interaction strength between dark and ordinary matter is "small." Direct experimental searches for particular classes of dark matter candidates, however, significantly constrain the interactions of such particles with Standard Model constituents [91]. It is also relevant to note that dark matter particles must be neutral (or have infinitesimal charge [92]) so that they do not interact electromagnetically (otherwise dark matter would not be dark!).

1.3   What Could Dark Matter Be?

There are a plethora of hypotheses about the nature of dark matter that span an enormous range of parameter space. For example, the masses of dark matter particle candidates range from 10^{-22} eV (fuzzy dark matter [93, 94]) up to 10^{21} eV (WIMPzillas [95]); if dark matter particles have significant self-interactions, then they can coalesce into composite objects with masses up to 10^{50} eV [96]. Several review articles explore in detail many of these hypotheses (see Refs. [83, 97–100], and, for amusement, Fig. 1.5). For brevity, here we highlight general principles and a few of the most popular hypotheses and their current experimental status.

Dark matter hypotheses regarded as "theoretically well-motivated" usually share several key attributes. The first is a plausible production mechanism that generates an abundance matching the observed dark matter density in the universe. Of course, as mentioned in Sect. 1.2, in order to match the observed density the dark matter particles must be stable: long-lived compared to the age of the universe so that they persist to the modern epoch. Another key attribute is that dark matter particles proposed in well-motivated theories also solve some other mystery of modern physics: multiple puzzles hint of their existence.

These attributes are exemplified by the hypothesis that has attracted the most attention over the last several decades: the idea that dark matter consists of *weakly interacting massive particles* (WIMPs). The WIMP hypothesis developed from the observation that particles interacting via the weak interaction would be created at just the right abundance to match the observed dark matter density [62, 102]. This is the so-called WIMP miracle. If the dark matter particles were thermally produced in the early universe, meaning that they were created in equilibrium with Standard Model particles via collisions at sufficiently high temperature, then the interaction cross-section can be estimated from arguments similar to those used to understand BBN (see the tutorial in Sect. 1.1). In the case of BBN, the

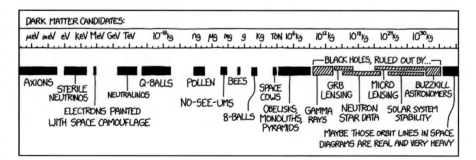

Fig. 1.5 Comical portrayal of the wide range of possible dark matter candidates and their masses from the xkcd comic strip (*https://xkcd.com/*), not too far off from the actual state of affairs at present. Actually, in some cases the cartoonist is a bit too conservative: for example, axions can have masses as small as 10^{-12} eV [100] and axionlike particles (ALPs) could have masses $\lesssim 10^{-22}$ eV [94, 101]

weak-interaction-maintained equilibrium between neutrons and protons until the universe cooled below the freeze-out temperature; analogously, there could be an interaction that maintained equilibrium between Standard Model particles (SM) and dark matter particles (χ) through a process $\chi\chi \leftrightarrow$ SM in the early universe.[11] As the universe continued to cool after the Big Bang, $k_B T$ would become smaller than $m_\chi c^2$, where m_χ is the dark matter particle mass, and the density n_χ would scale as $e^{-m_\chi c^2/(k_B T)}$. The decline in the dark matter density as T decreased would halt at a freeze-out temperature, leaving a relic density of dark matter—just like the relic density of baryons in the BBN scenario. This process is described by the Boltzmann equation [99]:

$$\frac{dn_\chi}{dt} = -3H_0 n_\chi - \langle \sigma_\chi v \rangle \left(n_\chi^2 - n_\chi(\text{eq})^2 \right), \qquad (1.15)$$

where σ_χ is the cross-section for $\chi\chi \leftrightarrow$ SM, v is the relative velocity between particles, $\langle \cdots \rangle$ indicates the thermal average, and $n_\chi(\text{eq})$ is the dark matter density in equilibrium. The first term on the right-hand side of Eq. (1.15) describes the decrease in dark matter density due to the expansion of the universe while the second term describes the creation and annihilation of dark matter from Standard Model particles. The solution of Eq. (1.15) yields [97–99]

$$\langle \sigma_\chi v \rangle \approx \frac{6 \times 10^{-27} \text{ cm}^3/\text{s}}{\Omega_{\text{dm}}}, \qquad (1.16)$$

where $\Omega_{\text{dm}} = \rho_{\text{dm}}/\rho_{\text{crit}} \approx 0.22$ is the ratio of the dark matter density to the critical density for a flat universe. The estimate of $\langle \sigma_\chi v \rangle$ from Eq. (1.16) turns out to equal the characteristic scale of the weak interaction if $10 \text{ GeV} \lesssim m_\chi c^2 \lesssim 1 \text{ TeV}$ [99]: hence the "WIMP miracle"—weakly interacting particles can be thermally produced with a relic abundance matching the dark matter density. Furthermore, WIMPs with such masses would be nonrelativistic at the freeze-out temperature and thus would fit the CDM scenario.

It also turns out that many leading theories of physics beyond the Standard Model predict new physics at the weak interaction scale. The key motivation for these theories is the *hierarchy problem*: the mystery of why gravity is so feeble compared to the other fundamental forces of nature, the strong and electroweak interactions. In the framework of quantum field theory, the hierarchy problem can be reframed in terms of the puzzle of the smallness of the Higgs-boson mass. The Higgs mass is $m_H c^2 \approx 125$ GeV, which can be compared to the natural mass scale of the gravitational interaction, the Planck scale:

---

[11] Notably, one of the intriguing facts about dark matter is that its density is actually quite similar to the baryon density: there is only about five times more dark matter than ordinary matter as opposed to orders of magnitude more or less, suggesting that perhaps baryons and dark matter were produced by similar processes that equilibrate their densities in the early universe.

Table 1.1 Examples of theories proposing WIMP dark matter candidates and related references

WIMP candidate	Description	References
Neutralino	Lightest superpartner in many supersymmetric models, a linear combinations of the photino, higgsino, and Z-ino.	[103, 105]
Gravitino	Superpartner of the graviton in supersymmetric models, in many scenarios only interacts gravitationally.	[106]
Little Higgs	A $\approx$ TeV scalar WIMP predicted by an alternative to supersymmetry's solution of the hierarchy problem.	[107, 108]
Kaluza-Klein excitation	Compactified extra spatial dimensions, an ingredient of string theory, have excited modes that correspond to an infinite number of partners to standard model particles; the lightest one is a WIMP candidate.	[109, 110]

$$M_{\mathrm{Pl}}c^2 = \sqrt{\frac{\hbar c}{G_N}} \approx 10^{19} \text{ GeV}. \qquad (1.17)$$

Quantum field theory predicts that the measured (physical) Higgs mass is given by

$$m_H^2 \approx m_H(0)^2 + \Delta m_H^2, \qquad (1.18)$$

where Δm_H is from radiative corrections to the "bare" mass $m_H(0)$. The natural scale of Δm_H is the energy scale at which beyond-standard-model physics appears: if there were no new physics until the Planck scale, $\Delta m_H^2 \approx M_{\mathrm{Pl}}^2$. Unless there is a coincidental cancellation at a level of a part in 10^{34} between contributions to the radiative correction term Δm_H, the Higgs mass should be close to M_{Pl}. Since m_H is measured to be close to the weak scale, there should be beyond-standard-model physics at the weak scale in order to set $\Delta m_H c^2 \approx 100$ GeV.

Thus, many theories proposing WIMPs share both key attributes of a well-motivated dark matter hypothesis: they give the correct dark matter abundance and also solve another mystery of modern physics, in this case the hierarchy problem.[12] Table 1.1 presents a list of some WIMP candidates and associated references.

Experiments have shown, however, that if the WIMP hypothesis is correct, the story must not be so simple. If all of dark matter consisted of particles with masses 10 GeV $\lesssim m_\chi c^2 \lesssim 1$ TeV that interacted with nuclei via the weak force with unsuppressed couplings, they would have been experimentally observed decades

---

[12] Supersymmetry [103] at the $\approx$ TeV scale, one of the leading theories of WIMP dark matter, also predicts a unification of the electromagnetic, strong, and weak coupling constants at the "Grand Unification Theory" (GUT) scale of $\approx 10^{16}$ GeV [104]. This is widely viewed as another tantalizing theoretical hint of WIMP dark matter.

ago. Cryogenic experiments searching for energy deposition from collisions of WIMPs with nuclei, first proposed in the 1980s [111, 112], have been pursued by a number of collaborations over the past decades. Despite several tantalizing hints of detections,[13] ultimately none of the experiments searching for WIMPs has found evidence of WIMP dark matter. The resulting constraints from these null experiments have become increasingly stringent, ruling out many of the most attractive WIMP theories [119]. Similarly, searches for WIMP candidates at the Large Hadron Collider (LHC) have placed tight constraints on many WIMP models [120]. The situation has become increasingly dire for the WIMP hypothesis, and the motivation to explore other explanations for the nature of dark matter has become correspondingly stronger.

A hypothesis closely related to the WIMP paradigm is the suggestion that dark matter might be sterile neutrinos. Perhaps there is a heavy neutrino species that does not interact via the weak interaction but could be generated by mixing with standard model neutrinos. The sterile neutrino hypothesis possesses the key attributes of theoretically well-motivated dark matter candidate: there is a production mechanism that can give a reasonable abundance (mixing with standard model neutrinos [121]) and sterile neutrinos can also solve a number of puzzles in neutrino physics, for example, as a mechanism to generate the nonzero standard model neutrino mass [122]. Because of the mixing with standard model neutrinos, sterile neutrinos can decay into a photon and a lighter neutrino. Thus searches for x-rays from sterile neutrino decay in nearby galaxies have been able to rule out a wide region of sterile neutrino parameter space [123, 124]. Most of the rest of the sterile neutrino parameter space is ruled out by its effect on small-scale structure in the universe [125], although loopholes remain [126].

Another dark matter hypothesis that received considerable attention in the past was the possibility that dark matter consists of *massive astrophysical compact halo objects* (MACHOs): composite baryonic objects that are non-luminous, such as planets, brown dwarfs, white dwarfs, neutron stars, and black holes. The term MACHO was coined to contrast with the term WIMP, and MACHOs had the notable advantage in that they were known to exist.[14] However, it turns out that MACHOs do not exist in sufficient abundance: today there is consensus that MACHOs do not constitute a large fraction of the dark matter in the universe. One of the main

---

[13] The most well-known, persistent, and controversial hint of a WIMP dark matter signal comes from the DAMA/LIBRA collaboration's reports of an annually modulated rate of scattering events on top of a background [113]. WIMP scattering rates should exhibit annual modulation due to the relative motion of the Earth with respect to the dark matter halo [88], and the DAMA/LIBRA uses this annual modulation to identify possible WIMP signals. However, the measured WIMP mass and coupling constants corresponding to the DAMA/LIBRA signals have been ruled out by a number of other experiments [114, 115]. Independent experiments undertaken specifically to resolve this controversy have recently ruled out the possibility that the DAMA/LIBRA results are evidence of dark matter [116–118].

[14] Along these lines, an alternative meaning of MACHO was suggested by astrophysicist Chris Stubbs: *maybe astronomy can help out!*

arguments against MACHOs as dark matter is the evidence discussed in Sects. 1.1 and 1.2 from CMB measurements and BBN that dark matter is nonbaryonic. A second argument against MACHOs as dark matter comes from gravitational microlensing studies [127]. If the dark matter halo consisted primarily of MACHOs in the mass range of $10^{-7} M_\odot \lesssim M \lesssim 10^2 M_\odot$, gravitational lensing of light from visible stars by the MACHOs would cause a significant fraction of those stars (one in a million) to exhibit transient variation of their apparent brightness. Large-scale microlensing surveys have been able to constrain the contribution of MACHOs to the dark matter mass content at $\lesssim 8\%$ [127]. Importantly, these constraints apply not only to MACHOs, but also to compact objects composed of nonbaryonic matter.

It should be noted that there are special, possibly baryonic, MACHO dark matter candidates that evade the CMB and BBN bounds: primordial black holes (PBHs). In the early universe, prior to BBN, there might be regions of space with energy so dense that they gravitationally collapse into black holes [128]. This is in contrast to black holes that are later produced as the end state of stellar evolution, and hence subject to the CMB limits on baryon density at recombination and BBN limits at the time of light element formation. The PBH mass is constrained to be $\gtrsim 10^{-19} M_\odot$, otherwise the PBHs would have evaporated via Hawking radiation prior to the present epoch [129]. Gravitational microlensing surveys constrain the PBH mass to be $\lesssim 10^{-7} M_\odot$ [127].

This brings us, at last, to the dark matter hypothesis that is the subject of this book: the idea that dark matter consists primarily of ultralight bosons.

1.4   Ultralight Bosonic Dark Matter

Ultralight bosonic dark matter (UBDM) is qualitatively quite different from the dark matter particles considered in Sect. 1.3. WIMPs and sterile neutrinos are particles with masses $\gg 10$ eV and the search methods are aimed at detecting individual interactions of dark matter particles. In contrast, UBDM consists of bosons with masses $\ll 10$ eV (hence *ultralight*) and the search methods are aimed at detecting coherent effects of UBDM waves. This difference in search methodologies arises from the fact that in order to match the observed dark matter density, the mode occupation number of the ultralight bosons can be quite high (Problem 1.3). In this case it is natural to treat UBDM as a classical field and take advantage of its coherent wavelike properties. A useful analogy can be made with radio waves: an efficient method of detection is to measure the electron current coherently driven by the radio waves using an antenna, as opposed to detecting single photons.

?   Problem 1.3   Ultralight Bosonic Dark Matter Waves

Suppose that dark matter consists mostly of bosons with mass $m_b c^2 = 10^{-6}$ eV. What are the Compton frequency and Compton wavelength of such bosons? Recalling that the virialized velocity of dark matter in the Milky Way is $\approx 10^{-3} c$, what is the de Broglie wavelength λ_{dB} of such bosons? Given that the local dark matter density is $\rho_{dm} \approx 0.4 \, \text{GeV/cm}^3$, estimate how many bosons occupy a volume corresponding to $\lambda_{dB}{}^3$? Repeat these estimates for dark matter bosons with mass $m_b c^2 = 10^{-12} \, \text{eV}$.

Solution on page 307.

In the rest frame of the UBDM, the oscillation frequency of the UBDM field is given by the Compton frequency,

$$\omega_c = \frac{mc^2}{\hbar} \, . \tag{1.19}$$

Of course, as noted in Sect. 1.2, in the SHM the dark matter particles are assumed to be virialized in the gravitational potential well of the galaxy. This leads to a random distribution of boson velocities. In the Milky Way, the characteristic width of the distribution is $\Delta v \approx 10^{-3} c$, about equal to the velocity of our solar system relative to the galactic rest frame. The spread in boson velocities gives rise to frequency dispersion, since an observable UBDM field arises from interference between a multitude of bosons with different velocities. Therefore an UBDM field has a characteristic coherence time τ_{coh} and coherence length L_{coh}, as considered in Problem 1.4.

? Problem 1.4 Coherence of Ultralight Bosonic Dark Matter Fields

Given that the characteristic width of the UBDM velocity distribution in the Milky Way is $\Delta v \approx 10^{-3} c$, derive τ_{coh} and L_{coh} for the UBDM field. Carry out numerical estimates of τ_{coh} and L_{coh} for the boson masses considered in Problem 1.3 ($m_b c^2 = 10^{-6} \, \text{eV}$ and $m_b c^2 = 10^{-12} \, \text{eV}$). What would be the corresponding Q-factor for the UBDM in the Milky Way, $Q = \omega / \Delta \omega$?

Solution on page 308.

Since, if we assume UBDM is described by the SHM, the observable UBDM field is the result of the interference of bosons with random velocities, its properties undergo stochastic variation with characteristic time scale τ_{coh} and length scale L_{coh}. Figure 1.6 shows a simulated virialized UBDM field over several coherence times. The amplitude of the UBDM field, while relatively constant over time durations $\Delta t \ll \tau_{coh}$, varies randomly on longer time scales. In fact, the stochastically varying amplitude of a virialized UBDM field is described by the Rayleigh distribution, which also describes the statistical properties of thermal (chaotic) light. As long as an experiment measures the UBDM field for a time $\Delta t \gg \tau_{coh}$, the experimental results can be interpreted based on the average dark matter properties. However, for

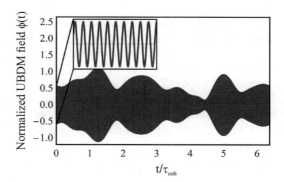

Fig. 1.6 Simulated virialized UBDM field $\phi(t)$. The inset shows the coherent oscillations of the UBDM field over a time scale $\ll \tau_{coh}$

extremely low-mass bosons it is impractical to measure for a time longer than τ_{coh}. For example, fuzzy dark matter [93, 94] with boson mass $m_b c^2 \approx 10^{-22}$ eV would have $\tau_{coh} \approx 4 \times 10^{13}$ s (roughly a million years!). In such cases, the interpretation of experiments must take into account the stochastic nature of UBDM [130].

It should be noted that the distribution of UBDM in the Milky Way may deviate from the predictions of the SHM in various ways. There can be enhancement (or suppression) of the local dark matter density due to formation of "clumps" or streams [131]. A related possibility is that self-interactions or topological properties of the UBDM field could lead to the formation of large composite structures such as condensates [132], clusters [133], boson stars [134], or domain walls [135, 136]. A reasonable assumption is that the motion and distribution of such composite structures are described by the SHM. On the other hand, some fraction of the UBDM could become trapped in the local gravitational potential of the Earth or Sun [137], creating a local halo where the UBDM density is enhanced. The fact that much is unknown about the local dark matter density should be taken into account when interpreting terrestrial experiments searching for UBDM.

One of the most well-motivated UBDM candidates from the perspective of theory, according to the criteria developed in Sect. 1.3, is the *axion* [138, 139]. The existence of axions is predicted by a proposal to solve the so-called *strong-CP problem*. CP refers to the combined symmetry with respect to charge-conjugation (C), transformation between matter and anti-matter, and spatial inversion, i.e., parity (P).[15] The strong CP problem arises from a CP-violating term appearing in the Lagrangian describing quantum chromodynamics (QCD) [140, 141]. The magnitude of CP violation in the strong interaction caused by this term is governed by a phase $\bar{\theta}_{QCD}$. Experimentally, $\bar{\theta}_{QCD}$ is found to be vanishingly small: constraints on the neutron electric dipole moment (EDM) imply that $\bar{\theta}_{QCD} \lesssim 10^{-10}$ [142]. This creates a so-called *fine-tuning problem*, since $\bar{\theta}_{QCD}$ is an arbitrary phase in QCD that could, in principle, take on any value from zero to 2π: the fact that $\bar{\theta}_{QCD}$ is near zero seems to be an unlikely coincidence. A solution to the strong CP problem

---

[15] A P-invariant interaction is said to possess *chiral symmetry*.

was proposed by Roberto Peccei and Helen Quinn [143, 144]: perhaps $\bar{\theta}_{QCD}$ does not possess a constant value, but rather evolves dynamically and naturally tends to a value near zero due to spontaneous symmetry breaking (see Ref. [145] for an intuitive explanation).[16] In this model, the CP-violating $\bar{\theta}_{QCD}$ term is replaced by a term in the QCD Lagrangian representing a dynamical field, and the quantum of this field is a spin-0 particle known as the axion. Furthermore, there are a number of plausible mechanisms to generate axions matching the observed abundance of dark matter [146–151], and such axions naturally fit the CDM paradigm [100, 152] (although, it is important to note as discussed in Chap. 3, Sect. 3.2, the CDM and UBDM scenarios are not entirely equivalent and can, in principle, be distinguished). The axion mass m_a is quite small: upper limits based on astrophysical observations are $m_a c^2 \lesssim 10$ meV [153], and in principle m_a can be smaller than 10^{-12} eV [154].

Independent of the strong CP problem, ultralight spin-0 bosons are ubiquitous features of many theories of physics beyond the Standard Model. For example, axionlike particles (ALPs) appear in theories with spontaneous breaking of flavor symmetry (familons [155, 156]), models with spontaneous breaking of chiral lepton symmetry (arions [157]), and versions of quantum gravity (spin-0 gravitons [158–161]). Axions and ALPs also generically arise in string theory as excitations of quantum fields that extend into extra compactified spacetime dimensions [162], with masses ranging all the way to $m_a c^2 \approx 10^{-33}$ eV [101]. Another ALP, known as the relaxion, has been proposed to solve the hierarchy problem [163]. Axions and ALPs have also been shown to offer a plausible mechanism to generate the matter-antimatter asymmetry of the universe [164, 165].

The characteristic amplitude of the axion dark matter field is estimated in Problem 1.5.

? Problem 1.5 Axion Dark Matter Field Amplitude

Based on the fact that the axion field φ is described (ignoring self-interactions) by the Lagrangian

$$\mathcal{L} = \frac{1}{2}(\partial_\mu \varphi)\partial^\mu \varphi - \frac{1}{2}\left(\frac{m_a c}{\hbar}\right)^2 \varphi^2 , \qquad (1.20)$$

where ∂_μ denotes the four-derivative and $\mathcal{L}$ has units of energy density, what are the units of φ? What is the relationship between the time-averaged value of the square of the axion field $\langle \varphi^2 \rangle$ and the average dark matter density in the galaxy ρ_{dm}?

Solution on page 308.

---

[16] The underlying physics of the Peccei-Quinn solution to the strong CP problem is closely related to the physics behind the Higgs mechanism endowing particles with mass in the Standard Model.

Axions are also involved in a rather different CDM theoretical framework (see [166, 167] and the references therein) that appears to be able to account for the origin of dark matter and also explain a number of other puzzles, including the baryon asymmetry of the Universe, the roughly similar abundance of luminous and dark matter, the lithium anomalies in the BBN [168], etc. In this model, dark matter consists of "nuggets" of some 10^{25} quarks at roughly the nuclear density held together by an "axion domain wall." The axion-quark-nugget model assumes the existence of both nuggets containing quarks and "anti-nuggets" containing antiquarks, such that the total number of quarks and antiquarks in the universe is roughly the same, thus resolving the mystery of the matter-antimatter asymmetry. The axion-quark nugget radius is on the order of 10^{-5} cm and, in contrast to most other dark matter scenarios, the interactions of such a nugget with normal matter are not feeble. For example, the cross-section for proton annihilation is on the order of the geometrical cross-section of 3×10^{-10} cm^2. The reason such nuggets are "dark" is that they have an unusually small cross-section-to-mass ratio.

Spin-1 bosons form another class of UBDM candidates. There are twelve fundamental spin-1 bosons in the Standard Model: the photon, the $W^{\pm}$ and Z bosons, and the eight gluons. Generally speaking, a massless spin-1 boson appears for any unbroken $\mathbb{U}(1)$ *gauge symmetry.*[17] New massless spin-1 bosons are referred to as paraphotons γ' [170] in analogy with photons, the quanta arising from the $\mathbb{U}(1)$ gauge symmetry of electromagnetism. Of interest as dark matter candidates are exotic spin-1 bosons that possess nonzero mass, as does the Z boson in the Standard Model. A nonzero mass for such a hypothetical Z' boson could arise from the breaking of a new $\mathbb{U}(1)$ gauge symmetry. There are a plethora of theoretical models predicting new Z' bosons and theoretically motivated masses and couplings to quarks and leptons extend over a broad range [171]. Z' bosons that do not directly interact with Standard Model particles (and therefore reside in the so-called hidden sector) are commonly referred to as *hidden photons* [170]. Like axions and ALPs, ultralight spin-1 bosons could plausibly be produced with the correct abundance to be the dark matter [172–174]. The characteristic magnitudes of the hidden electric and magnetic fields are estimated in Problem 1.6.

? Problem 1.6 Dark Electromagnetic Fields

Equating the average dark matter density ρ_{dm} to the energy density in the hidden electric field $\mathcal{E}'$ (given by an analog to standard electromagnetism) yields:

$$\rho_{\mathrm{dm}} \approx \frac{1}{8\pi}\left(\mathcal{E}'\right)^2 . \tag{1.21}$$

---

[17] Such symmetries arise quite naturally, for example, in string theory [169] and other Standard Model extensions. $\mathbb{U}(1)$ refers to the unitary group of degree 1, the collection of all complex numbers with absolute value 1 under multiplication.

Because the hidden photons have nonzero mass, there is a rest frame of the hidden photons for which there is only a hidden electric field oscillating at the Compton frequency. The hidden magnetic field is given by the relativistic transform of $\mathcal{E}'$ when there is relative motion between an observer and the hidden photons

$$\mathcal{B}' \approx \frac{v}{c}\mathcal{E}' . \tag{1.22}$$

Using Eqs. (1.21) and (1.22), the local dark matter density $\rho_{dm} \approx 0.4$ GeV/cm^3, and the characteristic relative velocity of Earth with respect to the dark matter halo of $v \approx 10^{-3}c$, estimate $\mathcal{E}'$ and $\mathcal{B}'$.

Solution on page 309.

Ultralight bosons can couple to Standard Model particles and fields through a number of distinct *portals* [175] as discussed in Chap. 2. A spin-0 bosonic field φ can directly couple to fermions in two possible ways: through a *scalar* vertex or through a *pseudoscalar* vertex [176–178]. In the nonrelativistic limit (small fermion velocity and momentum transfer), a fermion coupling to φ via a scalar vertex acts as a monopole and a fermion coupling to φ via a pseudoscalar vertex acts as a dipole. This can be understood from the fact that in the particle's center of mass frame, there are only two vectors from which to form a scalar/pseudoscalar quantity: the spin **s** and the momentum **p** (since the field φ is a scalar), so either the vertex does not involve **s** (monopole coupling) or if it does, it depends on **s** · **p**, which is a P-odd, pseudoscalar term. Hence the pseudoscalar interaction of φ is the source of new dipole interactions that are manifest as spin-dependent energy shifts. The scalar interaction gives rise to apparent variations of fundamental constants [175]. Spin-0 fields can also couple to the electromagnetic field:[18] a number of experiments exploit this coupling to search for conversion of axions into photons in strong magnetic fields. As suggested by the original theoretical motivation for the axion, the Peccei-Quinn solution of the strong CP problem [143, 144], axions couple to the gluon field and can generate EDMs along the spin direction [182]. Analogously to photons, spin-1 bosons can generate spin-dependent energy shifts and can also mix with the electromagnetic field [175]. These distinct portals for observing the effects of UBDM offer a variety of possibilities for direct detection, discussed in detail in the subsequent chapters of this book.

---

[18] In general, pseudoscalar particles such as axions can be produced by the interaction of two photons via a process known as the Primakoff effect [179] (discussed in Chaps. 2–5), and consequently an axion interacting with an electromagnetic field can produce a photon via the inverse Primakoff effect [180]; see also the review [181].

1.5   Conclusion

There is a strong case for the existence of dark matter: multiple independent astrophysical observations point to a consistent model where over 80% of the matter in the universe is dark. But the fundamental nature of dark matter is a complete mystery. A wide range of theories of physics beyond the Standard Model suggest there may exist heretofore undiscovered ultralight bosons with the right characteristics to explain the mystery of dark matter. In the following chapters, the rich and interesting physics of UBDM and the diverse array of experiments searching for evidence of its existence are explored.

Acknowledgments We are grateful to Alex Sushkov, Arne Wickenbrock, Alex Gramolin, Gary Centers, Peter Graham, Surjeet Rajendran, Maxim Pospelov, Andrei Derevianko, Victor Flambaum, Mikhail Kozlov, Gilad Perez, and Ariel Zhitnitsky for helpful discussions.

References

1. G. Bertone, D. Hooper, Rev. Mod. Phys. **90**, 045002 (2018)
2. E. Öpik, Bull. de la Soc. Astr. de Russie **21**, 5 (1915)
3. J.H. Oort et al., Bull. Astron. Inst. Neth. **6**, 249 (1932)
4. F. Zwicky, Helv. Phys. Acta **6**, 138 (1933)
5. F. Zwicky, Astrophys. J. **86**, 217 (1937)
6. A. Vikhlinin, A. Kravtsov, W. Forman, C. Jones, M. Markevitch, S. Murray, L. Van Speybroeck, Astrophys. J. **640**, 691 (2006)
7. L.P. David, C. Jones, W. Forman, Astrophys. J. **748**, 120 (2012)
8. R.W. Schmidt, S. Allen, Mon. Not. Roy. Astron. Soc. **379**, 209 (2007)
9. V.C. Rubin, W.K. Ford Jr, Astrophys. J. **159**, 379 (1970)
10. D. Rogstad, G. Shostak, Astrophys. J. **176**, 315 (1972)
11. R.N. Whitehurst, M.S. Roberts, Astrophys. J. **175**, 347 (1972)
12. M. Roberts, A. Rots, Astron. Astrophys. **26**, 483 (1973)
13. V.C. Rubin, W.K. Ford Jr, N. Thonnard, Astrophys. J. **225**, L107 (1978)
14. M. Bartelmann, Classical Quantum Gravity **27**, 233001 (2010)
15. H. Hoekstra, M. Bartelmann, H. Dahle, H. Israel, M. Limousin, M. Meneghetti, Space Sci. Rev. **177**, 75 (2013)
16. D. Clowe, M. Bradač, A.H. Gonzalez, M. Markevitch, S.W. Randall, C. Jones, D. Zaritsky, Astrophys. J. Lett. **648**, L109 (2006)
17. D. Harvey, R. Massey, T. Kitching, A. Taylor, E. Tittley, Science **347**, 1462 (2015)
18. C. Hernández-Monteagudo, J. Rubiño-Martín, R. Sunyaev, Mon. Not. Roy. Astron. Soc. **380**, 1656 (2007)
19. A. Lewis, J. Cosm. Astropart. Phys. **2013**, 053 (2013)
20. P.J. Peebles, J. Yu, Astrophys. J. **162**, 815 (1970)
21. R.A. Sunyaev, Y.B. Zeldovich, Astrophys. Space Sci. **7**, 3 (1970)
22. A. Doroshkevich, Y.B. Zel'dovich, R. Syunyaev, Sov. Astron. **22**, 523 (1978)
23. T. Padmanabhan, *Structure Formation in the Universe* (Cambridge University Press, Cambridge, 1993)

24. P.J.E. Peebles, *Principles of Physical Cosmology* (Princeton University Press, Princeton, 1993)
25. S. Weinberg, *Cosmology* (Oxford University Press, Oxford, 2008)
26. L. Anderson, E. Aubourg, S. Bailey, D. Bizyaev, M. Blanton, A.S. Bolton, J. Brinkmann, J.R. Brownstein, A. Burden, A.J. Cuesta et al., Mon. Not. Roy. Astron. Soc. **427**, 3435 (2012)
27. R.S. Ellis, R.J. McLure, J.S. Dunlop, B.E. Robertson, Y. Ono, M.A. Schenker, A. Koekemoer, R.A. Bowler, M. Ouchi, A.B. Rogers et al., Astrophys. J. Lett. **763**, L7 (2012)
28. G.R. Blumenthal, S. Faber, J.R. Primack, M.J. Rees, Nature **311**, 517 (1984)
29. P. Peebles, Astrophys. J. **263**, L1 (1982)
30. D.J. Eisenstein, W. Hu, Astrophys. J. **496**, 605 (1998)
31. A. Meiksin, M. White, J. Peacock, Mon. Not. Roy. Astron. Soc. **304**, 851 (1999)
32. C.L. Bennett, D. Larson, J. Weiland, N. Jarosik, G. Hinshaw, N. Odegard, K. Smith, R. Hill, B. Gold, M. Halpern et al., Astrophys. J. Suppl. **208**, 20 (2013)
33. P.A. Ade, N. Aghanim, M. Arnaud, M. Ashdown, J. Aumont, C. Baccigalupi, A. Banday, R. Barreiro, J. Bartlett, N. Bartolo et al., Astron. Astrophys. **594**, A13 (2016)
34. G. Hinshaw, D. Larson, E. Komatsu, D.N. Spergel, C. Bennett, J. Dunkley, M. Nolta, M. Halpern, R. Hill, N. Odegard et al., Astrophys. J. Suppl. **208**, 19 (2013)
35. P. Gorenstein, W. Tucker, Adv. High Energy Phys. **2014** (2014)
36. W. Hu, S. Dodelson, Ann. Rev. Astron. Astrophys. **40**, 171 (2002)
37. B.D. Fields, K.A. Olive, Nucl. Phys. A **777**, 208 (2006)
38. G. Gamow, Phys. Rev. **70**, 572 (1946)
39. R.A. Alpher, H. Bethe, G. Gamow, Phys. Rev. **73**, 803 (1948)
40. R.A. Alpher, J.W. Follin Jr, R.C. Herman, Phys. Rev. **92**, 1347 (1953)
41. H. Reeves, J. Audouze, W.A. Fowler, D.N. Schramm, *The Big Bang and Other Explosions in Nuclear and Particle Astrophysics*, vol. 179 (World Scientific, Singapore, 1996), pp. 65–86
42. B.D. Fields, Ann. Rev. Nucl. Part. Sci. **61**, 47 (2011)
43. T.P. Walker, G. Steigman, D.N. Schramm, K.A. Olive, H.S. Kang, *The Big Bang and Other Explosions in Nuclear and Particle Astrophysics* (World Scientific, Singapore, 1991), pp. 43–61
44. K.A. Olive, G. Steigman, T.P. Walker, Phys. Rep. **333**, 389 (2000)
45. L. Bergström, Phys. Scr. **2013**, 014014 (2013)
46. T. Padmanabhan, T.R. Choudhury, Phys. Rev. D **66**, 081301 (2002)
47. E.P. Verlinde, SciPost Phys. **2**, 016 (2017)
48. M. Tanabashi, K. Hagiwara, K. Hikasa, K. Nakamura, Y. Sumino, F. Takahashi, J. Tanaka, K. Agashe, G. Aielli, C. Amsler et al., Phys. Rev. D **98**, 030001 (2018)
49. P.J.E. Peebles, B. Ratra, Rev. Mod. Phys. **75**, 559 (2003)
50. L. Amendola, S. Tsujikawa, *Dark Energy: Theory and Observations* (Cambridge University Press, Cambridge, 2010)
51. A.G. Riess, A.V. Filippenko, P. Challis, A. Clocchiatti, A. Diercks, P.M. Garnavich, R.L. Gilliland, C.J. Hogan, S. Jha, R.P. Kirshner et al., Astron. J. **116**, 1009 (1998)
52. S. Perlmutter, G. Aldering, G. Goldhaber, R. Knop, P. Nugent, P. Castro, S. Deustua, S. Fabbro, A. Goobar, D. Groom et al., Astrophys. J. **517**, 565 (1999)
53. D. Scolnic, D. Jones, A. Rest, Y. Pan, R. Chornock, R. Foley, M. Huber, R. Kessler, G. Narayan, A. Riess et al., Astrophys. J. **859**, 101 (2018)
54. S. Burles, K.M. Nollett, M.S. Turner, Astrophys. J. Lett. **552**, L1 (2001)
55. J. Schaye, R.A. Crain, R.G. Bower, M. Furlong, M. Schaller, T. Theuns, C. Dalla Vecchia, C.S. Frenk, I. McCarthy, J.C. Helly et al., Mon. Not. Roy. Astron. Soc. **446**, 521 (2014)
56. M. Milgrom, Astrophys. J. **270**, 365 (1983)
57. M. Milgrom, Astrophys. J. **302**, 617 (1986)
58. J.D. Bekenstein, Phys. Rev. D **70**, 083509 (2004)
59. B. Famaey, S.S. McGaugh, Living Rev. Relativ. **15**, 10 (2012)
60. D.D. Ryutov, D. Budker, V.V. Flambaum, Astrophys. J. **871**, 218 (2019)

61. S. Gershtein, Y.B. Zel'dovich, JETP Lett. **4**, 1 (1966)
62. Y.B. Zel'dovich, A. Klypin, M.Y. Khlopov, V. Chechetkin, Sov. J. Nucl. Phys. **31**, 664 (1980)
63. S.D. White, C. Frenk, M. Davis, Astrophys. J. **274**, L1 (1983)
64. U. Seljak, A. Makarov, P. McDonald, H. Trac, Phys. Rev. Lett. **97**, 191303 (2006)
65. M. Vogelsberger, S. Genel, V. Springel, P. Torrey, D. Sijacki, D. Xu, G. Snyder, S. Bird, D. Nelson, L. Hernquist, Nature **509**, 177 (2014)
66. P. Bode, J.P. Ostriker, N. Turok, Astrophys. J. **556**, 93 (2001)
67. D. Forero, M. Tortola, J. Valle, Phys. Rev. D **90**, 093006 (2014)
68. C. Kraus, B. Bornschein, L. Bornschein, J. Bonn, B. Flatt, A. Kovalik, B. Ostrick, E. Otten, J. Schall, T. Thümmler et al., Eur. Phys. J. C **40**, 447 (2005)
69. E. Otten, C. Weinheimer, Rep. Prog. Phys. **71**, 086201 (2008)
70. V. Aseev, A. Belesev, A. Berlev, E. Geraskin, A. Golubev, N. Likhovid, V. Lobashev, A. Nozik, V. Pantuev, V. Parfenov et al., Phys. Rev. D **84**, 112003 (2011)
71. R.H. Cyburt, B.D. Fields, K.A. Olive, E. Skillman, Astropart. Phys. **23**, 313 (2005)
72. A. Dolgov, A.Y. Smirnov, Phys. Lett. B **621**, 1 (2005)
73. J. Bovy, S. Tremaine, Astrophys. J. **756**, 89 (2012)
74. K. Freese, M. Lisanti, C. Savage, Rev. Mod. Phys. **85**, 1561 (2013)
75. J. Diemand, M. Kuhlen, P. Madau, Astrophys. J. **657**, 262 (2007)
76. J. Dubinski, R. Carlberg, Astrophys. J. **378**, 496 (1991)
77. J.F. Navarro, in *Symposium: International Astronomical Union*, vol. 171 (Cambridge University Press, Cambridge, 1996), pp. 255–258
78. P. Salucci, A. Borriello, in *Particle Physics in the New Millennium* (Springer, Berlin, 2003), pp. 66–77
79. A. Klypin, A.V. Kravtsov, O. Valenzuela, F. Prada, Astrophys. J. **522**, 82 (1999)
80. B. Moore, S. Ghigna, F. Governato, G. Lake, T. Quinn, J. Stadel, P. Tozzi, Astrophys. J. Lett. **524**, L19 (1999)
81. M. Boylan-Kolchin, J.S. Bullock, M. Kaplinghat, Mon. Not. Roy. Astron. Soc.: Lett. **415**, L40 (2011)
82. L. Bergström, P. Ullio, J.H. Buckley, Astropart. Phys. **9**, 137 (1998)
83. G. Bertone, D. Hooper, J. Silk, Phys. Rep. **405**, 279 (2005)
84. M. Kamionkowski, S.M. Koushiappas, Phys. Rev. D **77**, 103509 (2008)
85. M. Lisanti, D.N. Spergel, Phys. Dark Universe **1**, 155 (2012)
86. C.A.J. O'Hare, C. McCabe, N.W. Evans, G. Myeong, V. Belokurov, Phys. Rev. D **98**, 103006 (2018)
87. A. Bandyopadhyay, D. Majumdar, Astrophys. J. **746**, 107 (2012)
88. A.K. Drukier, K. Freese, D.N. Spergel, Phys. Rev. D **33**, 3495 (1986)
89. F. Kahlhoefer, K. Schmidt-Hoberg, M.T. Frandsen, S. Sarkar, Mon. Not. Roy. Astron. Soc. **437**, 2865 (2013)
90. O.Y. Gnedin, J.P. Ostriker, Astrophys. J. **561**, 61 (2001)
91. D.S. Akerib, H. Araújo, X. Bai, A. Bailey, J. Balajthy, S. Bedikian, E. Bernard, A. Bernstein, A. Bolozdynya, A. Bradley et al., Phys. Rev. Lett. **112**, 091303 (2014)
92. J.M. Cline, Z. Liu, W. Xue, Phys. Rev. D **85**, 101302 (2012)
93. W. Hu, R. Barkana, A. Gruzinov, Phys. Rev. Lett. **85**, 1158 (2000)
94. L. Hui, J.P. Ostriker, S. Tremaine, E. Witten, Phys. Rev. D **95**, 043541 (2017)
95. D.J. Chung, E.W. Kolb, A. Riotto, Phys. Rev. D **59**, 023501 (1998)
96. D.M. Grabowska, T. Melia, S. Rajendran, Phys. Rev. D **98**, 115020 (2018)
97. G. Bertone, *Particle Dark Matter: Observations, Models and Searches* (Cambridge University Press, Cambridge, 2010)
98. L.E. Strigari, Phys. Rep. **531**, 1 (2013)
99. J.L. Feng, Ann. Rev. Astron. Astrophys. **48**, 495 (2010)
100. P.W. Graham, I.G. Irastorza, S.K. Lamoreaux, A. Lindner, K.A. van Bibber, Ann. Rev. Nucl. Part. Sci. **65**, 485 (2015)

101. A. Arvanitaki, S. Dimopoulos, S. Dubovsky, N. Kaloper, J. March-Russell, Phys. Rev. D **81**, 123530 (2010)
102. B.W. Lee, S. Weinberg, Phys. Rev. Lett. **39**, 165 (1977)
103. G. Jungman, M. Kamionkowski, K. Griest, Phys. Rep. **267**, 195 (1996)
104. U. Amaldi, W. de Boer, H. Fürstenau, Phys. Lett. B **260**, 447 (1991)
105. H. Goldberg, Phys. Rev. Lett. **50**, 1419 (1983)
106. J.L. Feng, A. Rajaraman, F. Takayama, Phys. Rev. Lett. **91**, 011302 (2003)
107. A. Birkedal, A. Noble, M. Perelstein, A. Spray, Phys. Rev. D **74**, 035002 (2006)
108. H.C. Cheng, I. Low, J. High Energy Phys. **2003**, 051 (2003)
109. H.C. Cheng, J.L. Feng, K.T. Matchev, Phys. Rev. Lett. **89**, 211301 (2002)
110. K. Kong, K.T. Matchev, J. High Energy Phys. **2006**, 038 (2006)
111. A. Drukier, L. Stodolsky, Phys. Rev. D **30**, 2295 (1984)
112. M.W. Goodman, E. Witten, Phys. Rev. D **31**, 3059 (1985)
113. R. Bernabei, P. Belli, F. Cappella, R. Cerulli, C. Dai, A. d'Angelo, H. He, A. Incicchitti, H. Kuang, X. Ma et al., Eur. Phys. J. C **67**, 39 (2010)
114. R. Agnese, T. Aramaki, I. Arnquist, W. Baker, D. Balakishiyeva, S. Banik, D. Barker, R.B. Thakur, D. Bauer, T. Binder et al., Phys. Rev. Lett. **120**, 061802 (2018)
115. A. Tan, X. Xiao, X. Cui, X. Chen, Y. Chen, D. Fang, C. Fu, K. Giboni, F. Giuliani, H. Gong et al., Phys. Rev. D **93**, 122009 (2016)
116. G. Adhikari et al., Nature **564**, 83 (2018)
117. G. Adhikari, P. Adhikari, E.B. de Souza, N. Carlin, S. Choi, M. Djamal, A. Ezeribe, C. Ha, I. Hahn, E. Jeon et al., Phys. Rev. Lett. **123**, 031302 (2019)
118. G. Adhikari, E.B. de Souza, N. Carlin, J. Choi, S. Choi, M. Djamal, A. Ezeribe, L. França, C. Ha, I. Hahn et al., Sci. Adv. **7**, eabk2699 (2021)
119. L. Roszkowski, E.M. Sessolo, S. Trojanowski, Rep. Prog. Phys. **81**, 066201 (2018)
120. G. Bertone, N. Bozorgnia, J.S. Kim, S. Liem, C. McCabe, S. Otten, R.R. de Austri, J. Cosm. Astropart. Phys. **2018**, 026 (2018)
121. S. Dodelson, L.M. Widrow, Phys. Rev. Lett. **72**, 17 (1994)
122. L. Canetti, M. Drewes, M. Shaposhnikov, Phys. Rev. Lett. **110**, 061801 (2013)
123. M. Viel, J. Lesgourgues, M.G. Haehnelt, S. Matarrese, A. Riotto, Phys. Rev. D **71**, 063534 (2005)
124. U. Seljak, A. Slosar, P. McDonald, J. Cosm. Astropart. Phys. **2006**, 014 (2006)
125. K. Abazajian, S.M. Koushiappas, Phys. Rev. D **74**, 023527 (2006)
126. X. Shi, G.M. Fuller, Phys. Rev. Lett. **82**, 2832 (1999)
127. P. Tisserand, L. Le Guillou, C. Afonso, J. Albert, J. Andersen, R. Ansari, É. Aubourg, P. Bareyre, J. Beaulieu, X. Charlot et al., Astron. Astrophys. **469**, 387 (2007)
128. B.J. Carr, S.W. Hawking, Mon. Not. Roy. Astron. Soc. **168**, 399 (1974)
129. A.S. Josan, A.M. Green, K.A. Malik, Phys. Rev. D **79**, 103520 (2009)
130. G.P. Centers, J.W. Blanchard, J. Conrad, N.L. Figueroa, A. Garcon, A.V. Gramolin, D.F. Jackson Kimball, M. Lawson, B. Pelssers, J.A. Smiga et al., Nature Comm. **12**, 7321 (2021)
131. J. Diemand, M. Kuhlen, P. Madau, M. Zemp, B. Moore, D. Potter, J. Stadel, Nature **454**, 735 (2008)
132. P. Sikivie, Q. Yang, Phys. Rev. Lett. **103**, 111301 (2009)
133. E.W. Kolb, I.I. Tkachev, Phys. Rev. Lett. **71**, 3051 (1993)
134. D. Jackson Kimball, D. Budker, J. Eby, M. Pospelov, S. Pustelny, T. Scholtes, Y. Stadnik, A. Weis, A. Wickenbrock, Phys. Rev. D **97**, 043002 (2018)
135. M. Pospelov, S. Pustelny, M.P. Ledbetter, D.F. Jackson Kimball, W. Gawlik, D. Budker, Phys. Rev. Lett. **110**, 021803 (2013)
136. A. Derevianko, M. Pospelov, Nature Phys. **10**, 933 (2014)
137. A. Banerjee, D. Budker, J. Eby, H. Kim, G. Perez, Commun. Phys. **3**, 1 (2020)
138. K. van Bibber, L.J. Rosenberg, Phys. Today **59**, 30 (2006)

139. K. van Bibber, K. Lehnert, A. Chou, Phys. Today **72**, 48 (2019)
140. R. Peccei, Lect. Notes Phys. **741**, 3 (2008)
141. J.E. Kim, G. Carosi, Rev. Mod. Phys. **82**, 557 (2010)
142. C.A. Baker, D.D. Doyle, P. Geltenbort, K. Green, M.G.D. van der Grinten, P.G. Harris, P. Iaydjiev, S.N. Ivanov, D.J.R. May, J.M. Pendlebury, J.D. Richardson, D. Shiers, K.F. Smith, Phys. Rev. Lett. **97**, 131801 (2006)
143. R. Peccei, H. Quinn, Phys. Rev. Lett. **38**, 1440 (1977)
144. R. Peccei, H. Quinn, Phys. Rev. D **16**, 1791 (1977)
145. P. Sikivie, Phys. Today **49**, 22 (1996)
146. L.F. Abbott, P. Sikivie, Phys. Lett. B **120**, 133 (1983)
147. J. Preskill, M.B. Wise, F. Wilczek, Phys. Lett. B **120**, 127 (1983)
148. M. Dine, W. Fischler, Phys. Lett. B **120**, 137 (1983)
149. R.L. Davis, Phys. Rev. D **32**, 3172 (1985)
150. S. Chang, C. Hagmann, P. Sikivie, Phys. Rev. D **59**, 023505 (1998)
151. M. Nagasawa, M. Kawasaki, Phys. Rev. D **50**, 4821 (1994)
152. L.D. Duffy, K. van Bibber, New J. Phys. **11**, 105008 (2009)
153. G.G. Raffelt, Annu. Rev. Nucl. Part. Sci. **49**, 163 (1999)
154. P.W. Graham, A. Scherlis, Phys. Rev. D **98**(3), 035017 (2018)
155. F. Wilczek, Phys. Rev. Lett. **49**, 1549 (1982)
156. G. Gelmini, S. Nussinov, T. Yanagida, Nucl. Phys. B **219**, 31 (1983)
157. A. Ansel'm, Pis'ma Zh. Eksp. Teor. Fiz. **36**, 46 (1982)
158. J. Scherk, Phys. Lett. B **88**, 265 (1979)
159. D.E. Neville, Phys. Rev. D **21**, 2075 (1980)
160. D.E. Neville, Phys. Rev. D **25**, 573 (1982)
161. S.M. Carroll, G.B. Field, Phys. Rev. D **50**, 3867 (1994)
162. P. Svrček, E. Witten, J. High Energy Phys. **06**, 051 (2006)
163. P.W. Graham, D.E. Kaplan, S. Rajendran, Phys. Rev. Lett. **115**, 221801 (2015)
164. K. Harigaya et al., Phys. Rev. Lett. **124**, 111602 (2020)
165. R.T. Co, L.J. Hall, K. Harigaya, J. High Energ. Phys. **2021**, 172 (2021)
166. S. Ge, K. Lawson, A. Zhitnitsky, Phys. Rev. D **99**, 116017 (2019)
167. D. Budker, V.V. Flambaum, X. Liang, A. Zhitnitsky, Phys. Rev. D **101**, 043012 (2020)
168. V.V. Flambaum, A.R. Zhitnitsky, Phys. Rev. D **99**, 023517 (2019)
169. M. Cvetic, P. Langacker, Phys. Rev. D **54**, 3570 (1996)
170. B. Holdom, Phys. Lett. B **166**, 196 (1986)
171. P. Langacker, Rev. Mod. Phys. **81**, 1199 (2009)
172. N. Arkani-Hamed, D.P. Finkbeiner, T.R. Slatyer, N. Weiner, Phys. Rev. D **79**, 015014 (2009)
173. P. Arias, D. Cadamuro, M. Goodsell, J. Jaeckel, J. Redondo, A. Ringwald, J. Cosm. Astropart. Phys. **2012**, 013 (2012)
174. A.E. Nelson, J. Scholtz, Phys. Rev. D **84**, 103501 (2011)
175. M.S. Safronova, D. Budker, D. DeMille, D.F. Jackson Kimball, A. Derevianko, C.W. Clark, Rev. Mod. Phys. **90**, 025008 (2018)
176. J.E. Moody, F. Wilczek, Phys. Rev. D **30**, 130 (1984)
177. B.A. Dobrescu, I. Mocioiu, J. High Energy Phys. **11**, 005 (2006)
178. P. Fadeev, Y.V. Stadnik, F. Ficek, M.G. Kozlov, V.V. Flambaum, D. Budker, Phys. Rev. A **99**, 022113 (2019)
179. H. Primakoff, Phys. Rev. **81**, 899 (1951)
180. G. Raffelt, D. Seckel, Phys. Rev. Lett. **60**, 1793 (1988)
181. R. Battesti, J. Beard, S. Böser, N. Bruyant, D. Budker, S.A. Crooker, E.J. Daw, V.V. Flambaum, T. Inada, I.G. Irastorza et al., Phys. Rep. **765**, 1 (2018)
182. D. Budker, P.W. Graham, M. Ledbetter, S. Rajendran, A.O. Sushkov, Phys. Rev. X **4**, 021030 (2014)

Chapter 2
Ultralight Bosonic Dark Matter Theory

Derek F. Jackson Kimball, Leanne D. Duffy, and David J. E. Marsh

Abstract The basic theoretical concepts motivating the hypothesis that dark matter may consist of ultralight spin-0 or spin-1 bosons are explored. The origin of bosons with masses $\ll 1$ eV from spontaneous and explicit symmetry breaking is illustrated with examples. The origins and characteristics of nongravitational couplings or "portals" between ultralight bosons and Standard Model particles and fields are considered, with particular attention paid to the cases of the axion-photon and axion-fermion interactions. Theoretical motivations for the existence of ultralight bosons, besides as an explanation of dark matter, are examined, with particular focus on the Peccei-Quinn solution to the strong CP problem (resulting in the QCD axion) and a dynamical solution to the hierarchy problem (the "relaxion" hypothesis, based on a particular axion-Higgs coupling in the early universe). Mechanisms for non-thermal production of ultralight bosonic dark matter are examined.

2.1 Introduction

This book explores the hypothesis that dark matter consists predominantly of ultralight bosons. In this chapter we discuss the theoretical motivation for the ultralight bosonic dark matter (UBDM) hypothesis and the testable predictions

---

The original version of the chapter has been revised. A correction to this chapter can be found at https://doi.org/10.1007/978-3-030-95852-7_11

D. F. Jackson Kimball (✉)
Department of Physics, California State University, East Bay, Hayward, CA, USA
e-mail: derek.jacksonkimball@csueastbay.edu

L. D. Duffy
Los Alamos National Laboratory, Los Alamos, NM, USA
e-mail: ldd@lanl.gov

D. J. E. Marsh
Department of Physics, King's College London, London, UK
e-mail: david.j.marsh@kcl.ac.uk

derived from it, considering a number of relevant examples along the way. At the outset several questions naturally arise:

- If we suppose that dark matter is a bosonic field, how do we describe that from a theoretical perspective?
- Why would such bosons be "ultralight"—with masses $\ll 1$ eV/c^2?
- How could such ultralight bosonic matter interact with known Standard Model particles and fields?
- Why should one expect that there exist bosons beyond those already discovered (e.g., photons, gluons, W and Z-bosons, and the Higgs boson)?
- How could ultralight bosons be created in the early universe in sufficient abundance to match the dark matter density observed today?

2.2 Bosonic Field Lagrangians

From the perspective of both classical and quantum field theory (QFT), a common place to begin trying to understand the physics of a new particle is to write down the Lagrangian (or more specifically, the Lagrangian density $\mathcal{L}$) of the corresponding field. The following several sections draw heavily from textbooks on QFT, such as Refs. [1–5], which offer more detail and further explanation of many of the key points addressed. Let us start by assuming we are dealing with a scalar field $\phi(r, t)$; the quantum excitations of the scalar field $\hat{\phi}(r, t)$ are spin-0 bosons.[1] This choice is motivated both by simplicity and because axions and axionlike particles (ALPs), some of the most prominent dark matter candidates, are spin-0 bosons. Further motivation for considering scalar fields is derived from the discovery of the Higgs boson [6, 7], proving that elementary spin-0 bosons do indeed exist in nature [8].

The Lagrangian $\mathcal{L}$ describing the scalar field will naturally depend on the rate of change of ϕ in time, $\partial_0\phi = \partial\phi/\partial t$, and the derivative of ϕ with respect to the spatial coordinates, $\nabla\phi$. (In this chapter we will use natural units where $\hbar = c = 1$, see the discussion of units and conversion factors in the prefatory material at the beginning of this text.) We require that $\mathcal{L}$ be Lorentz invariant, so we will build our Lagrangian from the four-derivative of ϕ,

$$\partial_\mu\phi = \frac{\partial\phi}{\partial x^\mu} = \left(\frac{\partial}{\partial t}, \nabla\right)\phi \,, \tag{2.1}$$

$$= \left(\frac{\partial\phi}{\partial t}, \frac{\partial\phi}{\partial x}, \frac{\partial\phi}{\partial y}, \frac{\partial\phi}{\partial z}\right), \tag{2.2}$$

which is manifestly Lorentz invariant. In the following we use the Einstein summation convention for repeated indices, with Greek indices such as μ running from $0 \rightarrow 3$, where 0 indicates the time-like component and 1, 2, and 3 are the spacelike components. The metric tensor describing flat spacetime is

---

[1] The "hat" on the scalar field denotes that we treat $\hat{\phi}$ as an operator.

$$g_{\mu\nu} = \begin{pmatrix} 1 & 0 & 0 & 0 \\ 0 & -1 & 0 & 0 \\ 0 & 0 & -1 & 0 \\ 0 & 0 & 0 & -1 \end{pmatrix} = \mathrm{diag}\,[1, -1, -1, -1]\,, \tag{2.3}$$

which takes contravariant (upper) indices to covariant (lower) indices: $x_\mu = g_{\mu\nu}x^\nu$.

For simplicity, motivated at least in part by the principle of Occam's razor, we will also want to choose a form of $\mathcal{L}$ that depends on the lowest order of derivatives possible.[2] Since $\mathcal{L}$ is a scalar and $\partial_\mu\phi$ is a four-vector, at a minimum we must use the inner product of the four-derivatives of ϕ, and so our first guess at $\mathcal{L}$ is

$$\mathcal{L} = \frac{1}{2}\partial^\mu\phi\,\partial_\mu\phi = \frac{1}{2}\big(\partial_\mu\phi\big)^2\,, \tag{2.4}$$

$$= \frac{1}{2}\frac{\partial^2\phi}{\partial t^2} - \frac{1}{2}(\nabla\phi)^2\,, \tag{2.5}$$

where the factor of $1/2$ is included to simplify future results, and we use the metric for flat spacetime. In analogy with the Lagrangian from classical mechanics describing particles, the term $(1/2)\big(\partial_\mu\phi\big)^2$ is often associated with a "kinetic" energy of the field.

So what can we learn from our guess for $\mathcal{L}$ about the properties of ϕ? By using Eq. (2.4) in the Euler–Lagrange equation,

$$\frac{\partial\mathcal{L}}{\partial\phi} - \partial_\mu\left(\frac{\partial\mathcal{L}}{\partial\big(\partial_\mu\phi\big)}\right) = 0\,, \tag{2.6}$$

noting that

$$\frac{\partial\mathcal{L}}{\partial\phi} = 0 \tag{2.7}$$

and

$$\frac{\partial\mathcal{L}}{\partial\big(\partial_\mu\phi\big)} = \partial^\mu\phi\,, \tag{2.8}$$

we find from Eq. (2.6) that

$$\partial_\mu\partial^\mu\phi = \frac{\partial^2\phi}{\partial t^2} - \nabla^2\phi = 0\,. \tag{2.9}$$

---

[2] In principle, theories with higher-order derivatives are possible, but are associated with non-local effects and causality violation. Models involving such higher-order derivatives include, for example, theories of modified gravity (see, e.g., Ref. [9]).

Note that Eq. (2.9) shows that $j^\mu = \partial^\mu \phi$ is a conserved current, since $\partial_\mu j^\mu = 0$. The conservation of the current j^μ is a consequence of the continuous *shift symmetry* of the Lagrangian under the transformation $\phi \to \phi +$ constant, a result of Noether's theorem [10].

Equation (2.9) is a wave equation for ϕ and thus has solutions of the form

$$\phi(\boldsymbol{r}, t) = \varphi_0 e^{i(\boldsymbol{k}\cdot\boldsymbol{r} - \omega t)} , \qquad (2.10)$$

where φ_0 is the amplitude of this particular mode of the scalar field, ω is the frequency, and $\boldsymbol{k}$ is the wave vector. In natural units, the frequency ω is equivalent to the energy E of ϕ, as can be derived by applying the energy operator $\hat{E} = i(\partial/\partial t)$ to $\phi(\boldsymbol{r}, t)$. Similarly, the wave vector $\boldsymbol{k}$ is equivalent to the momentum $\boldsymbol{p}$ of ϕ, as can be derived by applying the momentum operator $\hat{\boldsymbol{p}} = -i\nabla$ to $\phi(\boldsymbol{r}, t)$.

Substituting Eq. (2.10) into Eq. (2.9), we obtain the dispersion relation

$$\omega^2 = |\boldsymbol{k}|^2 , \qquad (2.11)$$

or, equivalently,

$$E = |\boldsymbol{p}| . \qquad (2.12)$$

What does Eq. (2.12) imply about our scalar field ϕ? One of the key ideas of QFT is that particles can be interpreted as quantum excitations of fields. The dispersion relation (2.12) thus implies that if the field ϕ has zero momentum, $|\boldsymbol{p}| = 0$, then it has zero energy, $E = 0$, which means the particles associated with ϕ have zero rest mass ($m = 0$). Note that, in fact, these considerations also apply to classical fields. The dispersion relation for a classical field defines a "mass" based on the curvature of the dispersion around $|\boldsymbol{k}| = 0$.

But in order to match the astrophysical observations discussed in Chaps. 1 and 3, the particles associated with ϕ must behave as cold dark matter and thus cannot be massless. To get a theory of particles with mass, we need to modify the Lagrangian density (2.5) so that there is some energy "cost" to having a non-vacuum field value. This can be done by adding to our Lagrangian a potential energy term that depends on ϕ such that

$$\mathcal{L} = \frac{1}{2}(\partial_\mu \phi)^2 - \frac{1}{2}m^2 \phi^2 , \qquad (2.13)$$

where, again, the factor $m^2/2$ is chosen to obtain the correct units and with future results in mind, and the potential energy term has a minus sign since the Lagrangian is the kinetic minus the potential energy (thus the larger the field ϕ, the larger the potential energy). To show that our theory describes massive particles, we can re-derive the dispersion relation using $\mathcal{L}$ from Eq. (2.13). Since now

Fig. 2.1 Plot comparing the dispersion relation for a massless boson (red line) based on Eq. (2.12) with that for a massive boson (blue curve) based on Eq. (2.17). A key feature of the massive boson is the energy cost for zero-momentum excitations of the field, shown by the nonzero intercept of the dispersion curve on the energy axis

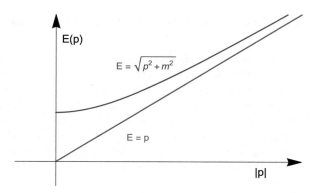

$$\frac{\partial \mathcal{L}}{\partial \phi} = -m^2 \phi \,, \tag{2.14}$$

the Euler–Lagrange equation (2.6) gives

$$\left(\partial_\mu \partial^\mu + m^2 \right) \phi = 0 \,, \tag{2.15}$$

which is the Klein–Gordon equation. The solutions of the Klein–Gordon equation (2.15) are also of the form

$$\phi(\boldsymbol{r}, t) = \varphi_0 e^{-i(Et - \boldsymbol{p} \cdot \boldsymbol{r})} \,, \tag{2.16}$$

but with the dispersion relation

$$E^2 = |\boldsymbol{p}|^2 + m^2 \,, \tag{2.17}$$

which shows that if the field ϕ has zero momentum, $|\boldsymbol{p}| = 0$, then it has energy equal to the rest mass of the associated particle $E = m$. Thus the Lagrangian in Eq. (2.13) describes a relatively simple model for massive particles that could be dark matter.

Figure 2.1 compares the dispersion relation for massless particles derived from Eq. (2.4) to that for massive particles derived from Eq. (2.13). Already we can note an interesting feature of the scalar field that will be repeatedly referenced throughout this text, namely that a nonrelativistic bosonic field, for which $|\boldsymbol{p}| \ll m$, oscillates at the Compton frequency: $\omega \approx m$.

2.3   Why New Bosons Might Be Ultralight

So far, from the considerations in Sect. 2.2, we have from Eq. (2.13) a model of a
scalar field whose particles have mass m. But, as discussed in Chap. 1, the UBDM
hypothesis suggests that the dark matter particles have masses $\lesssim 0.1$ eV (and even
perhaps as small as $m \sim 10^{-22}$ eV!), a small value compared to most known
Standard Model particles with nonzero masses.[3] This invites the question: from a
theoretical perspective, why might we expect new bosons to be ultralight? One of the
main motivations for postulating the existence of new bosons with ultralight masses
comes from the physics of *spontaneous symmetry breaking*, which we explore in
this section.

Let us reconsider our model Lagrangian for the scalar field,

$$\mathcal{L} = \frac{1}{2}\left(\partial_\mu \phi\right)^2 - V(\phi) \,, \tag{2.18}$$

where we designate $V(\phi)$ as the potential energy density term. In Eq. (2.13), we
chose $V(\phi) = m^2\phi^2/2$, but in principle we could try other potentials and investigate
the consequences. In fact, this is a familiar approach used throughout physics: one
might imagine that the true potential describing nature is some complicated function
of ϕ, but one can always Taylor expand such a function:

$$V(\phi) = \sum_{n=0}^{\infty} \frac{c_n}{n!}\phi^n \,, \tag{2.19}$$

where c_n are constants. As long as the series converges, the first few terms of the
Taylor expansion (2.19) may offer a reasonable approximation for $V(\phi)$. With this
in mind, let us consider the following potential:

$$V(\phi) = \frac{\mu^2}{2}\phi^2 + \frac{\lambda}{4!}\phi^4 \,, \tag{2.20}$$

where λ is a constant. In Eq. (2.20) we take only the first two terms with even powers
of ϕ from the expansion (2.19) to keep $V(\phi)$ symmetric about $\phi = 0$, so that $V(\phi)$
is invariant under the transformation $\phi_0 \rightarrow -\phi_0$. Also note that truncating the series
at the ϕ^4 term is convenient as it makes the theory renormalizable (see, for example,
chapter 31 of Ref. [1]). The potential described by Eq. (2.20) is shown in the plot
on the top in Fig. 2.2. The minimum of this potential at $\phi = 0$ corresponds to the
vacuum state of the field and the quantum excitations of ϕ are bosons with mass
$m = \mu$, as can be seen in the limit where $\phi \ll 1$, in which case $V(\phi) \rightarrow \mu^2\phi^2/2$
and thus matches the potential from Eq. (2.13).

---

[3] Neutrinos, of course, have nonzero but comparatively small masses: the sum of the three different
mass eigenstates for neutrinos is $\lesssim 0.1$ eV [11].

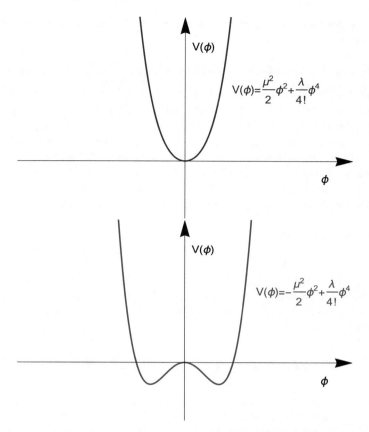

Fig. 2.2 The purple plot on the top shows a quartic scalar field potential $V(\phi)$ with a positive quadratic term [Eq. (2.20)], the blue plot on the bottom shows $V(\phi)$ with a negative quadratic term [Eq. (2.21)]

What if instead we construct a potential

$$V(\phi) = -\frac{\mu^2}{2}\phi^2 + \frac{\lambda}{4!}\phi^4 \,, \tag{2.21}$$

where the quadratic term is negative instead of positive? Then we get a shape of the potential as shown in the plot on the bottom in Fig. 2.2. Now there are two minima of the field at $\phi \neq 0$ (Problem 2.1). This means that the ground state of the field, which will be one of the two minima, *breaks reflection symmetry* and is not invariant under the transformation $\phi_0 \rightarrow -\phi_0$ (whereas, crucially, $V(\phi)$ still possesses reflection symmetry). This illustrates the essence of *spontaneous symmetry breaking* and shows how the vacuum expectation value of the field acquires a nonzero amplitude (see Problem 2.1).

? Problem 2.1 Vacuum Field and Boson Mass in Spontaneous Symmetry Breaking

Solve for the minima of the potential $V(\phi)$ described by Eq. (2.21). These are the two possible vacua or "vacuum expectation values" of the field ϕ, both of which are nonzero. Thus the field ϕ has the property that even when there are no bosons present, the field is nonzero, possibly with relatively large amplitude. This is in contrast to the more familiar case of the electromagnetic field whose vacuum expectation value is zero, so that where there are no photons present the average electromagnetic field is zero. Also find the new mass of the boson.

Solution on page 309.

Still we have not yet seen why bosons associated with the field ϕ might be ultralight. Let us introduce a new model, this time with two different scalar fields, $\alpha(r, t)$ and $\beta(r, t)$. We construct a potential for these two scalar fields similar to that from Eq. (2.21):

$$V(\alpha, \beta) = -\frac{\mu^2}{2}\left(\alpha^2 + \beta^2\right) + \frac{\lambda}{4!}\left(\alpha^2 + \beta^2\right)^2 . \qquad (2.22)$$

The potential $V(\alpha, \beta)$, plotted in Fig. 2.3, possesses what is known as a global $\mathbb{SO}(2)$ symmetry: it is invariant with respect to rotations in the α-β plane. It is a *global symmetry* because in order to maintain invariance with respect to the transformation, the fields at all points in spacetime must be rotated in the same way in the α-β plane. The label $\mathbb{SO}(2)$ for the symmetry originates from group theory: "$\mathbb{SO}$" refers to the special orthogonal group, namely the group of all orthogonal matrices[4] whose determinants $= 1$ (this condition is what makes this subgroup of all orthogonal matrices "special"). $\mathbb{SO}(2)$ is the special orthogonal group of 2×2 matrices, equivalent to the group of rotations about a point in two dimensions.

Now, instead of two potential minima as in the case of $V(\phi)$ from Eq. (2.21) (see Problem 2.1), there are an infinite number of minima on a ring of radius $\rho_0 = \sqrt{\alpha^2 + \beta^2} = \sqrt{6\mu^2/\lambda}$ centered at ($\alpha = 0$, $\beta = 0$). This is seen by writing Eq. (2.22) in terms of $u = \alpha^2 + \beta^2$,

$$V(u) = -\frac{\mu^2}{2}u + \frac{\lambda}{4!}u^2 , \qquad (2.23)$$

---

[4] An orthogonal matrix is a matrix whose inverse equals its transpose.

Fig. 2.3 Plot of the potential
$V(\alpha, \beta)$ from Eq. (2.22)

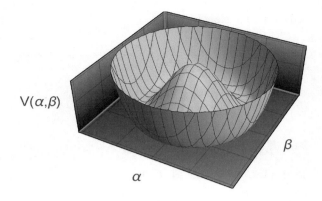

$V(\alpha,\beta)$

β

α

and then finding the minimum with respect to u:

$$\left[\frac{\partial V}{\partial u}\right]_{u=u_{\min}} = -\frac{\mu^2}{2} + \frac{\lambda}{12}u_{\min} = 0 \qquad (2.24)$$

$$\Rightarrow u_{\min} = \frac{6\mu^2}{\lambda} . \qquad (2.25)$$

What happens when this system undergoes spontaneous symmetry breaking?
Suppose the system "falls into" a particular ground state of the system. Without loss
of generality, let us choose the ground state ($\alpha = \alpha_0 = \sqrt{6\mu^2/\lambda}$, $\beta = \beta_0 = 0$).
In order to investigate small perturbations around this particular field minimum, we
can re-write the Lagrangian in terms of the variables

$$\bar{\alpha} \equiv \alpha - \alpha_0 = \alpha - \sqrt{\frac{6\mu^2}{\lambda}} , \qquad (2.26)$$

$$\bar{\beta} \equiv \beta - \beta_0 = \beta , \qquad (2.27)$$

noting that

$$\partial_\mu \bar{\alpha} = \partial_\mu \alpha , \qquad (2.28)$$

$$\partial_\mu \bar{\beta} = \partial_\mu \beta . \qquad (2.29)$$

The Lagrangian with the potential from Eq. (2.22), written in terms of $\bar{\alpha}$ and $\bar{\beta}$, is
given by

$$\mathcal{L} = \frac{1}{2}(\partial_\mu \bar{\alpha})^2 + \frac{1}{2}(\partial_\mu \bar{\beta})^2 + \frac{\mu^2}{2}\Big[(\bar{\alpha} + \alpha_0)^2 + \bar{\beta}^2\Big] - \frac{\lambda}{4!}\Big[(\bar{\alpha} + \alpha_0)^2 + \bar{\beta}^2\Big]^2 ,$$
$$(2.30)$$

which is equivalent to

$$\mathcal{L} = \frac{1}{2}\left(\partial_\mu \bar{\alpha}\right)^2 + \frac{1}{2}\left(\partial_\mu \bar{\beta}\right)^2 + \frac{3}{2}\frac{\mu^4}{\lambda} - \mu^2\bar{\alpha}^2 - \mu\sqrt{\frac{\lambda}{6}}\bar{\alpha}^3 - \frac{\lambda}{4!}\bar{\alpha}^4$$
$$- \frac{\lambda}{4!}\bar{\beta}^4 - \mu\sqrt{\frac{\lambda}{6}}\bar{\alpha}\bar{\beta}^3 - \frac{\lambda}{12}\bar{\alpha}^2\bar{\beta}^2 \ . \tag{2.31}$$

? Problem 2.2 Lagrangian for Two Scalar Fields

Derive Eq. (2.31) from Eq. (2.30).

Solution on page 310.

The physics described by $\mathcal{L}$ is unchanged by resetting the zero of the potential, so the constant term in Eq. (2.31), $3\mu^4/(2\lambda^2)$, can be subtracted. As a first approximation, let us consider only small amplitude field excitations and therefore neglect terms higher than second order in the fields $\bar{\alpha}$, $\bar{\beta}$:

$$\mathcal{L} \approx \frac{1}{2}\left(\partial_\mu \bar{\alpha}\right)^2 + \frac{1}{2}\left(\partial_\mu \bar{\beta}\right)^2 - \mu^2\bar{\alpha}^2 \ . \tag{2.32}$$

The part of the Lagrangian describing the $\bar{\alpha}$ field has a form analogous to Eq. (2.13) and thus represents a field whose quantum excitations are bosons of mass $m = \sqrt{2}\mu$. The part of the Lagrangian describing the $\bar{\beta}$ field has a form analogous to Eq. (2.4) and thus represents a field whose quantum excitations are massless bosons. The $m = 0$ excitations of the $\bar{\beta}$ field are known as *Goldstone bosons* [12], massless bosons appearing whenever a continuous symmetry, in this case $\mathbb{SO}(2)$, is spontaneously broken (a consequence of Goldstone's theorem [13]). The reason that the $\bar{\beta}$ bosons are massless can be intuited from the shape of the potential plotted in Fig. 2.3, shown in an "overhead" view in Fig. 2.4. Small excitations of the $\bar{\beta}$ field (indicated by the double-headed purple arrow in Fig. 2.4) around the ground state (indicated by the purple dot in Fig. 2.4) occur essentially without any increase in potential energy, as they are along the ring of minima in the "trough" of the potential $V(\alpha, \beta)$. In contrast, excitations of the $\bar{\alpha}$ field are perpendicular to the double-headed purple arrow in Fig. 2.4, where the potential resembles that of a simple harmonic oscillator, corresponding to the massive bosons associated with the potential of Eq. (2.13).

So far, our model based on the potential from Eq. (2.22) shows no indication of an ultralight field: rather we have one field ($\bar{\alpha}$) that has an arbitrary mass and another field ($\bar{\beta}$) that is massless. The appearance of an ultralight field requires one more ingredient in our model: *explicit symmetry breaking* on top of the spontaneous symmetry breaking. By explicit symmetry breaking we mean that the global $\mathbb{SO}(2)$ symmetry of the potential $V(\alpha, \beta)$ of Eq. (2.22) is itself broken, so that $V(\alpha, \beta)$ is no longer symmetric with respect to rotations in the α-β plane. In theories proposing ultralight bosons, such explicit symmetry breaking occurs due to, for example, non-perturbative effects in quantum chromodynamics (QCD), leading to so-called "soft"

Fig. 2.4 Overhead view of the potential $V(\alpha, \beta)$ from Eq. (2.22). The purple dot indicates the (arbitrary) ground state after spontaneous symmetry breaking at $(\alpha = \alpha_0 = \sqrt{6\mu^2/\lambda}, \ \beta = \beta_0 = 0)$. The double-headed purple arrow indicates small perturbations of the β field around $\beta = \bar{\beta} = 0$, requiring approximately zero energy as seen from Eq. (2.32). Thus the quantum excitations of the β field are massless bosons, a consequence of Goldstone's theorem

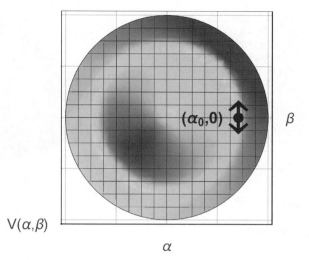

explicit breaking of the symmetry (where "soft" refers to the fact that the symmetry is restored at high energy scales), or even effects associated with quantum gravity (which is generically expected to violate global symmetries), see the reviews [14–19] for further discussion. For the purposes of our present investigations, let us invoke explicit symmetry breaking of the potential by "tilting" $V(\alpha, \beta)$ toward the original vacuum state from the spontaneously broken symmetry (α_0, β_0) by adding the term

$$V_\epsilon = -\epsilon \lambda \alpha_0^3 \alpha \tag{2.33}$$

to the potential of Eq. (2.22), so the Lagrangian is now

$$\mathcal{L} = \frac{1}{2}(\partial_\mu \alpha)^2 + \frac{1}{2}(\partial_\mu \beta)^2 + \frac{\mu^2}{2}(\alpha^2 + \beta^2) - \frac{\lambda}{4!}(\alpha^2 + \beta^2)^2 + \epsilon \lambda \alpha_0^3 \alpha . \tag{2.34}$$

In Eqs. (2.33) and (2.34), $\epsilon \ll 1$ is a small parameter characterizing the symmetry breaking. Figure 2.5 shows a plot of the potential (the tilt is greatly exaggerated so as to be clearly visible).

The explicit symmetry breaking due to V_ϵ shifts the minimum of the potential with respect to α, as seen in Problem 2.3.

? Problem 2.3 Explicit and Spontaneous Symmetry Breaking

Keeping only terms to first order in ϵ, verify that the minimum of the potential in Eq. (2.34), namely

$$V(\alpha, \beta) = -\frac{\mu^2}{2}(\alpha^2 + \beta^2) + \frac{\lambda}{4!}(\alpha^2 + \beta^2)^2 - \epsilon \lambda \alpha_0^3 \alpha , \tag{2.35}$$

Fig. 2.5 Plot of the potential
from Eq. (2.34), showing
explicit symmetry breaking.
The potential is tilted toward
the original vacuum state
(α_0, β_0) identified in Fig. 2.4

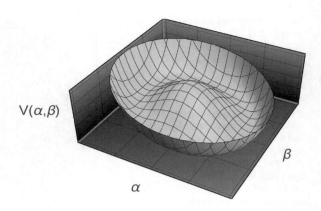

$V(\alpha, \beta)$

β

α

occurs at

$$\alpha = \alpha_0(1 + 3\epsilon) \, , \qquad (2.36)$$

$$\beta = 0 \, , \qquad (2.37)$$

where, as before, $\alpha_0 = \sqrt{6\mu^2/\lambda}$. Thus in order to investigate small perturbations around this particular field minimum, the Lagrangian (2.34) can be re-written in terms of the variable

$$\bar{a} = \alpha - \alpha_0(1 + 3\epsilon) \, . \qquad (2.38)$$

By writing $\mathcal{L}$ in terms of $\bar{a}$, keeping only first order terms in ϵ and second order or smaller terms in the fields $\bar{a}$ and β, and also appropriately resetting the zero of the potential (allowing all constant terms to be subtracted), show that the potential (2.35) can be approximated as

$$V(\bar{a}, \beta) \approx \mu^2 \bar{a}^2 + 3\epsilon \mu^2 \beta^2 \, . \qquad (2.39)$$

Solution on page 311.

Based on Eq. (2.39), the Lagrangian for the fields resulting from both spontaneous and explicit symmetry breaking can be approximately described as

$$\mathcal{L} = \frac{1}{2}(\partial_\mu \bar{a})^2 + \frac{1}{2}(\partial_\mu \beta)^2 - \mu^2 \bar{a}^2 - 3\epsilon \mu^2 \beta^2 \, , \qquad (2.40)$$

which shows that due to the explicit symmetry breaking, the β field has acquired a small mass $\propto \sqrt{\epsilon}$,

$$m_\beta^2 \approx 6\epsilon \mu^2 \, . \qquad (2.41)$$

Thus β represents the sought-after ultralight bosonic field: the quantum excitations of the β field are commonly known as a *pseudo-Goldstone bosons* or *pseudo-Nambu-Goldstone bosons* (pNGBs in the literature, see, for example, Ref. [12]).[5]

In order to connect our somewhat simplistic model to more realistic UBDM scenarios, it is useful to re-parameterize the descriptions of the explicit and spontaneous symmetry breaking. We can associate a characteristic energy scale f with the spontaneous symmetry breaking based on the depth of the potential [see, e.g., Eqs. (2.23) and (2.25)],

$$|V_{\min}| \sim \frac{\mu^4}{\lambda^2} \sim f^4 , \tag{2.42}$$

where we note that $V(\alpha, \beta)$ represents an energy density and thus, in natural units, is proportional to the fourth power of energy. The energy scale Λ describing the explicit symmetry breaking can be characterized by the associated part of the potential [Eq. (2.33)], namely

$$|V_\epsilon| \approx \epsilon \lambda \alpha_0^4 \sim \epsilon \frac{\mu^4}{\lambda} \sim \Lambda^4 . \tag{2.43}$$

The mass of the β boson [Eq. (2.41)] can now be re-written in terms of f and Λ:

$$m_\beta^2 \sim \epsilon \mu^2 \sim \left(\epsilon \frac{\mu^4}{\lambda} \right) \times \left(\frac{\lambda}{\mu^2} \right) , \tag{2.44}$$

$$\sim \frac{\Lambda^4}{f^2} . \tag{2.45}$$

Since the mass of the β boson scales as $m \sim \Lambda^2/f$, if $f \gg \Lambda$ (which corresponds to ϵ being small), as is the case in many beyond-the-Standard-Model theories incorporating such effects, then indeed the new boson can be ultralight. Note that we have an additional symmetry restored in the limit where $\epsilon \to 0$, namely the $\mathbb{SO}(2)$ symmetry, and thus we say that the ultralight mass of the pseudo-Goldstone boson is "technically natural."

Specific models of ultralight bosons suggest particular values for the spontaneous symmetry breaking scale f and the explicit symmetry breaking scale Λ. For example, the spontaneous symmetry breaking might occur at the Planck scale, in which case $f \sim 10^{28}$ eV. A possible source of (soft) explicit symmetry breaking arises from the strong interaction, in which case the explicit symmetry breaking scale is given by the QCD confinement scale (the energy scale above which calculations of the strong coupling constant diverge), i.e., $\Lambda \sim 10^8$ eV. Employing

---

[5] Goldstone bosons resulting from spontaneous symmetry breaking are massless, while pseudo-Goldstone bosons, possessing relatively small but nonzero masses, result from the combination of spontaneous and explicit symmetry breaking as considered here.

these energy scales in Eq. (2.45) gives a boson mass of $m \sim 10^{-12}$ eV, which is much, much lighter than any Standard Model boson with nonzero mass.

Tutorial: Spontaneous and Explicit Breaking of the $\mathbb{U}(1)$ Symmetry of a Complex Scalar Field

In this tutorial, we offer another example elucidating the origin of an ultralight bosonic field from the combination of spontaneous and explicit symmetry breaking. Instead of the two real scalar fields α and β considered above, let us consider a single complex scalar field φ, where we can make the correspondence:

$$\varphi = \alpha + i\beta \,. \tag{2.46}$$

Then the Lagrangian corresponding to the potential in Eq. (2.22) can be written as

$$\mathcal{L} = \frac{1}{2}\left(\partial^{\mu}\varphi\right)^{\dagger}\left(\partial_{\mu}\varphi\right) + \frac{\mu^2}{2}\varphi^{\dagger}\varphi - \frac{\lambda}{4!}\left(\varphi^{\dagger}\varphi\right)^2 \,. \tag{2.47}$$

Next we can re-parametrize the complex field using polar coordinates:

$$\varphi = \rho e^{i\theta} \,, \tag{2.48}$$

which yields a new form for the Lagrangian (2.47):

$$\mathcal{L} = \frac{1}{2}\left(\partial_{\mu}\rho\right)^2 + \frac{1}{2}\rho^2\left(\partial_{\mu}\theta\right)^2 + \frac{\mu^2}{2}\rho^2 - \frac{\lambda}{4!}\rho^4 \,. \tag{2.49}$$

Note that the Lagrangian described by Eqs. (2.47) and (2.49) exhibits a global $\mathbb{U}(1)$ symmetry for φ, namely that a global transformation $\varphi \to \varphi e^{i\theta'}$ has no effect on $\mathcal{L}$. $\mathbb{U}(1)$ refers to the one-dimensional unitary group, i.e., complex numbers with magnitude $= 1$, and so the $\mathbb{U}(1)$ symmetry is a symmetry with respect to rotations in the complex plane. The correspondence between rotations in the complex plane for φ and rotations in the real α-β plane is a consequence of the fact that $\mathbb{U}(1)$ is isomorphic to $\mathbb{SO}(2)$.

Similarly to the case of the two real-valued fields α and β, minima of the potential occur in a ring with radius $\rho = \rho_0 = \sqrt{6\mu^2/\lambda}$. Let us assume that the $\mathbb{U}(1)$ symmetry is spontaneously broken such that $\rho \to \rho_0$ and $\theta \to 0$. Then we can re-write the Lagrangian in terms of $\bar{\rho} = \rho - \rho_0$, which, after some algebra, yields

$$\mathcal{L} = \frac{1}{2}\left(\partial_{\mu}\bar{\rho}\right)^2 + \frac{1}{2}\rho_0^2\left(\partial_{\mu}\theta\right)^2 - \mu^2\bar{\rho}^2 - \frac{\lambda}{6}\rho_0\bar{\rho}^3 - \frac{\lambda}{24}\bar{\rho}^4 + \left(\frac{\bar{\rho}^2}{2} + \rho_0\bar{\rho}\right)\left(\partial_{\mu}\theta\right)^2 \,, \tag{2.50}$$

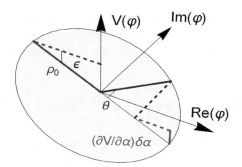

Fig. 2.6 Schematic diagram showing the effect of explicit symmetry breaking due to a tilt by an angle ϵ of the quartic potential for the complex scalar field φ (see Figs. 2.3 and 2.5 for illustrations of the analogous case for two real scalar fields, with and without tilt, respectively). The edge of the disk represents the ring of minima with respect to ρ at $\rho \approx \rho_0$ (radius ρ_0 shown by the solid blue line). If the potential is tilted by an angle ϵ, the potential acquires a θ-dependence given by $(\partial V/\partial \alpha)\delta\alpha$ (illustrated by the solid vertical red line). Here the solid purple radial line indicates a particular value of $\varphi = \rho_0 e^{i\theta}$, $\alpha = \mathrm{Re}(\varphi)$, $\delta\alpha \approx \rho_0(\cos\theta - 1)$ (illustrated by the dashed red line), and $\partial V/\partial \alpha = -\epsilon\mu^2\rho_0$

where in Eq. (2.50) we have dropped all constant terms, since they have no effect on the physics. Note that in Eq. (2.50), the terms independent of θ and linear in $\bar{\rho}$ have cancelled out, similarly to the derivation of Eq. (2.31) discussed in Problem 2.2. Retaining only second order or lower terms in $\bar{\rho}$ and θ, we obtain

$$\mathcal{L} \approx \frac{1}{2}\left(\partial_\mu\bar{\rho}\right)^2 + \frac{1}{2}\rho_0^2\left(\partial_\mu\theta\right)^2 - \mu^2\bar{\rho}^2 \,, \tag{2.51}$$

which is analogous to Eq. (2.32). Note that $\theta = \beta/\rho_0 \sim \beta/f$, where f is the spontaneous symmetry breaking scale defined in Eq. (2.43).

Next, we introduce explicit symmetry breaking by tilting the potential appearing in the Lagrangian (2.49) by an angle ϵ toward $\theta = 0$. Figure 2.6 illustrates the parametrization of the explicit symmetry breaking. The tilt by ϵ causes the potential to acquire a θ-dependence. For $\varphi = \rho_0 e^{i\theta}$, the real part of the field is $\mathrm{Re}(\varphi) = \alpha = \rho_0 \cos\theta$. The minimum of the tilted potential is at $\theta = 0$, and so the change in the potential with respect to the minimum is given by

$$\delta V(\theta) = \frac{\partial V}{\partial \alpha}\delta\alpha = \epsilon\mu^2\rho_0^2(1 - \cos\theta) \,, \tag{2.52}$$

where $\delta\alpha = -\rho_0(1 - \cos\theta)$. Including this term in the Lagrangian (2.51), we have

$$\mathcal{L} \approx \frac{1}{2}\left(\partial_\mu\bar{\rho}\right)^2 + \frac{1}{2}\rho_0^2\left(\partial_\mu\theta\right)^2 - \mu^2\bar{\rho}^2 - \epsilon\mu^2\rho_0^2(1 - \cos\theta) \,. \tag{2.53}$$

As a final step, to connect this result to the form of the potential most commonly encountered in the literature on UBDM, we use the relationships outlined in Eqs. (2.42), (2.43), and (2.45), along with the correspondence noted earlier, $\theta \sim \beta/f$, to write:

$$V(\beta) = m_b^2 f^2 \left[1 - \cos\left(\frac{\beta}{f}\right) \right] = \Lambda^4 \left[1 - \cos\left(\frac{\beta}{f}\right) \right]. \tag{2.54}$$

$V(\beta)$ can be expanded about $\beta = 0$ to give

$$V(\beta) \approx \frac{1}{2} m_b^2 \beta^2 \approx \frac{1}{2} \frac{\Lambda^4}{f^2} \beta^2 , \tag{2.55}$$

which can be compared to the β^2 term in Eq. (2.40).

End of Tutorial

2.4 Portals Between the Dark Sector and the Standard Model

The next major question we will address is how ultralight bosonic fields can interact nongravitationally with Standard Model particles and fields. To develop some intuition about such interactions, let us begin by continuing to work with our simple model of an ultralight bosonic field developed in Sect. 2.3. From a QFT perspective, interactions between two different fields arise when terms appear in the Lagrangian involving both fields as factors. In this way we can investigate interactions between the α (or $\bar{a}$) and β fields analyzed in Sect. 2.3. While the approximate Lagrangian of Eq. (2.40) has no such terms, they appear if we expand the Lagrangian of Eq. (2.34) to third order in the products of the fields, as shown in Problem 2.4.

? Problem 2.4 Interactions Between Two Scalar Fields

Using the results from the solution to Problem 2.3, expand the potential of Eq. (2.34) to third order in the products of the fields, thereby deriving two new "interaction" terms:

$$V_{\text{int}}(\bar{a}, \beta) = \frac{\lambda}{6} \alpha_0 \bar{a}^3 + \frac{\lambda}{6} \alpha_0 \beta^2 \bar{a} . \tag{2.56}$$

Solution on page 313.

The constant factor in front of the terms in Eq. (2.56) represents the coupling constant g characterizing the strength of the interaction between the fields (or the self-interaction in the case of the $\bar{a}^3$ term). The coupling constant can be re-written in terms of the spontaneous symmetry breaking scale f:

$$g = \frac{\lambda \alpha_0}{6} = \frac{1}{\sqrt{6}} \frac{\mu^2}{f} \sim \frac{\mu^2}{f}. \tag{2.57}$$

Accounting for the interaction terms gives a new approximate Lagrangian,

$$\mathcal{L} \approx \frac{1}{2}(\partial_\mu \bar{a})^2 + \frac{1}{2}(\partial_\mu \beta)^2 - \mu^2 \bar{a}^2 - 3\epsilon \mu^2 \beta^2 + \frac{1}{\sqrt{6}} \frac{\mu^2}{f} \bar{a}^3 + \frac{1}{\sqrt{6}} \frac{\mu^2}{f} \beta^2 \bar{a}. \tag{2.58}$$

Equations (2.57) and (2.58) highlight another important generic feature of ultralight bosonic fields that makes them good candidates to be dark matter: the coupling to other particles and fields generally scales as $1/f$, so if the symmetry breaking scale is at a very large energy, such as the grand unified theory (GUT) scale ($f \sim 10^{25}$ eV $= 10^{16}$ GeV) or Planck scale ($f \sim 10^{28}$ eV $= 10^{19}$ GeV), nongravitational interactions of the ultralight bosons are strongly suppressed, consistent with astrophysical observations as discussed in Chaps. 1 and 3, and also consistent with the results of the many null experiments described throughout this book.

2.4.1 Interactions Between Ultralight Bosonic Fields and Standard Model Particles

If terms describing the Standard Model particles and fields and their interactions are incorporated into the Lagrangian, along with terms describing ultralight bosonic fields, a variety of interaction terms are possible [20]. Many of the couplings studied both in the experiments discussed in this book, as well as in numerous theories of beyond-the-Standard-Model physics, are listed in Table 2.1 (note that the list of couplings is not exhaustive[6]). If dark matter consists primarily of ultralight bosonic fields, these possible nongravitational interactions can be classified into a

---

[6] The couplings listed in Table 2.1 only include operators up to a certain dimension (see discussion in Ref. [20]). Also, Table 2.1 is compiled assuming a particular basis for the fermions, other bases permit different forms of the couplings. For axions, in particular, it is significant that the fermion interactions generate the other axion interactions via the *chiral anomaly*, called an "anomaly" because it is a case where a classical symmetry of the Lagrangian does not map to a quantum symmetry for the corresponding Lagrangian. In the low temperature limit (where T is well below the QCD phase transition temperature $\sim$200 MeV), the gluon interaction generates the axion mass via soft explicit breaking of the chiral symmetry due to mixing with pions as described in the tutorial at the end of Sect. 2.5.1. In the high temperature limit, the axion mass is generated via

Table 2.1 Couplings of ultralight bosonic fields to Standard Model particles and fields. Examples of ultralight bosons include scalars ϕ, axions (or axionlike particles, ALPs) a, and dark/hidden photons, described by a vector potential X_μ and field strength $\mathcal{F}_{\mu\nu}$. Standard Model particles include Higgs bosons h, gluons $G^{\mu\nu}$, photons $F^{\mu\nu}$, and fermions ψ. The dual gluon field tensor is denoted $\tilde{G}^{\mu\nu}$ and the dual electromagnetic tensor is denoted $\tilde{F}^{\mu\nu}$, and A_μ is the photon vector potential. General terms from the Standard Model are denoted by O_{sm}. Note that because the Lagrangian is real-valued, the operators must take the appropriate form depending on whether the considered fields are real or complex. The usual Dirac matrices are denoted γ_μ and $\gamma_5 = -i\gamma_0\gamma_1\gamma_2\gamma_3$, and $\sigma^{\mu\nu} = (i/2)[\gamma^\mu, \gamma^\nu]$. The rightmost column list the chapters of the present book in which experiments probing such effects are discussed. Table adapted from Refs. [20] and [24]

Spin	Type	Operator	Interaction	Chapters
0	Scalar	$\phi h^\dagger h$	Higgs portal	8, 10
0	Scalar	$\phi^n O_{\mathrm{sm}}$ $(n = 1, 2)$	Dilaton	8, 10
0	Scalar	$\phi^\dagger \partial_\mu \phi \psi^\dagger \gamma^\mu \psi$	Current-current	8, 10
0	Pseudoscalar	$a G^{\mu\nu} \tilde{G}_{\mu\nu}$	Axion-gluon	6
0	Pseudoscalar	$a F^{\mu\nu} \tilde{F}_{\mu\nu}$	Axion-photon	4, 5, 7, 9
0	Pseudoscalar	$(\partial_\mu a)\psi^\dagger \gamma^\mu \gamma_5 \psi$	Axion-fermion	6, 8, 10
1	Vector	$X_\mu \psi^\dagger \gamma^\mu \psi$	Minimally coupled	8
1	Vector	$F_{\mu\nu} \mathcal{F}^{\mu\nu}, A_\mu X^\mu$	Photon-hidden-photon mixing	7
1	Vector	$\mathcal{F}_{\mu\nu} \psi^\dagger \sigma^{\mu\nu} \psi$	Dipole interaction	6, 8, 10
1	Axial vector	$X_\mu \psi^\dagger \gamma^\mu \gamma^5 \psi$	Minimally coupled	6, 8, 10

few different phenomenological "portals" between the Standard Model and the dark sector [24], where the portals can be classified by the physical effects the UBDM generates in experiments. In this section, for illustrative purposes, we analyze a few of these different interactions and portals.

Before analyzing particular cases, though, let us consider some general features of the interactions listed in Table 2.1. The first column of Table 2.1 lists the spin of the boson. Here we consider spin-0 (as discussed in Sects. 2.2 and 2.3) and spin-1 bosons, encompassing the majority of presently studied beyond-the-Standard-Model theories.[7] The second column considers the parity symmetry (P) of the interaction. Parity is the symmetry with respect to spatial inversion (reflection of coordinate axes through the origin): under spatial inversion, P-odd quantities change sign (pseudoscalars and vectors) and P-even quantities are invariant (scalars and axial vectors). Parity symmetry is among the key discrete symmetries characterizing interactions, others include time-reversal (T) and charge-

---

instantons [21, 22]. For further discussion of the chiral anomaly and instantons, see, e.g., Ref. [2]. For the dilaton, the interactions are defined in the Einstein conformal frame [23].

[7] The limitation to bosons with spin ≤ 1 is due in part to the fact that at present there are unresolved theoretical questions concerning the validity, naturalness, and allowed interactions for spin-2 fields with nonzero mass [20]. Presently there is no known effective field theory for bosons with spin ≥ 3 that is valid above the boson mass [20].

conjugation (C).[8] The discrete symmetry properties of an interaction inform the nature of the experiment necessary to observe signatures of particular classes of UBDM candidates.

2.4.2 Axion-Photon Interaction

Let us begin by considering one of the most widely studied UBDM interactions, the axion-photon coupling. The axion-photon coupling is used to convert axions or ALPs into photons in the presence of strong magnetic fields. This is the technique at the heart of the microwave cavity haloscopes described in Chap. 4, the axion helioscopes searching for axion/ALP emission from the Sun described in Chap. 5, axion/ALP searches with "dark matter radios" using lumped-element resonators described in Chap. 7, and light-shining-through-walls experiments discussed in Chap. 9. The fourth row of Table 2.1 describes an operator involving factors of both a spin-0 pseudoscalar axion (ALP) field a and the product of the electromagnetic field tensor (Faraday tensor) $F^{\mu\nu}$ with the dual field tensor $\tilde{F}_{\mu\nu}$. The Faraday tensor $F^{\mu\nu}$ is given by [30]

$$F^{\mu\nu} = \partial^\mu A^\nu - \partial^\nu A^\mu \tag{2.59}$$

$$= \begin{pmatrix} 0 & -E_x & -E_y & -E_z \\ E_x & 0 & -B_z & B_y \\ E_y & B_z & 0 & -B_x \\ E_z & -B_y & B_x & 0 \end{pmatrix}, \tag{2.60}$$

where A^μ is the four-potential and E_i and B_i are the electric and magnetic field components in the Cartesian basis. The dual field tensor is given by

$$\tilde{F}_{\alpha\beta} = \frac{1}{2}\varepsilon_{\alpha\beta\mu\nu}F^{\mu\nu}, \tag{2.61}$$

where $\varepsilon_{\alpha\beta\mu\nu}$ is the Levi-Civita totally antisymmetric tensor. We note the general structure of the operator for the axion-photon interaction, one factor of the ultralight bosonic field a, and two factors of the photon field. This structure is similar to the interaction terms studied in Problem 2.4, seen perhaps most clearly by writing the operator in terms of the four-potential A^μ:

$$aF^{\mu\nu}\tilde{F}_{\mu\nu} = a\epsilon^{\mu\nu\alpha\beta}\left(\partial_\mu A_\nu \partial_\alpha A_\beta\right), \tag{2.62}$$

---

[8] Famously, Wu et al. [25] discovered that the weak interaction violated parity conservation in 1957, and later in 1964 Christenson, Cronin, Fitch, and Turlay [26] discovered violation of the combined CP symmetry. Observations of atomic parity violation [27–29] were crucial in establishing the existence of parity-violating neutral weak currents mediated by the Z-boson.

showing that indeed this term represents an interaction between an axion and two photons.

The term in the Lagrangian describing the axion-photon interaction is

$$\mathcal{L}_{a\gamma\gamma} = \frac{g_\gamma}{4}\frac{\alpha}{\pi}\frac{a}{f_a}F^{\mu\nu}\tilde{F}_{\mu\nu} = \frac{g_{a\gamma\gamma}}{4}aF^{\mu\nu}\tilde{F}_{\mu\nu}\,, \tag{2.63}$$

where g_γ is a dimensionless model-dependent coupling factor, α is the fine structure constant, f_a is the spontaneous symmetry breaking scale for the axion/ALP field, and $g_{a\gamma\gamma} = g_\gamma\alpha/(\pi f_a)$ is the axion-photon coupling constant. Note that the axion-photon coupling is proportional to $1/f_a$, exhibiting the characteristic suppression derived in Eqs. (2.57) and (2.58). The form of the Lagrangian in terms of the electric field E and magnetic field B is

$$\mathcal{L}_{a\gamma\gamma} = g_\gamma\frac{\alpha}{\pi}\frac{a}{f_a}E\cdot B \approx g_{a\gamma\gamma}aE\cdot B\,. \tag{2.64}$$

? Problem 2.5 Axion-Photon Interaction

Derive Eq. (2.64).

Solution on page 314.

In experiments, the magnetic field B appearing in Eq. (2.64) is generated in the laboratory by, for example, current circulating in a superconducting coil, and the electric field E represents the field of the resultant photon generated from the axion. The conversion of axions into photons in a magnetic field is known as the *inverse Primakoff effect* [31–33], illustrated by the Feynman diagram in Fig. 2.7.

One method to calculate the observable physical consequences resulting from the axion-photon interaction is to apply the Euler–Lagrange equation to the Lagrangian describing electromagnetism plus the axion-photon Lagrangian of Eq. (2.63), namely

$$\mathcal{L} = -\frac{1}{4}F^{\mu\nu}F_{\mu\nu} - J^\mu A_\mu + \frac{g_{a\gamma\gamma}}{4}aF^{\mu\nu}\tilde{F}_{\mu\nu}\,, \tag{2.65}$$

where J^μ is the electromagnetic current and A^μ is the gauge potential. The Euler–Lagrange equation in this case produces a version of Maxwell's equations that includes the effects of an axion field, as discussed in Refs. [32–35]:

$$\nabla\cdot E = \rho + g_{a\gamma\gamma}B\cdot\nabla a\,, \tag{2.66}$$

$$\nabla\cdot B = 0\,, \tag{2.67}$$

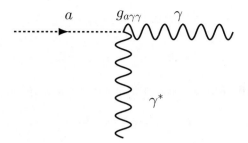

Fig. 2.7 Feynman diagram illustrating the inverse Primakoff effect, where an axion a is converted into a real photon γ by interacting with a virtual photon γ^* sourced by a magnetic field (a virtual photon is one that does not need to satisfy the energy-momentum relationship or "on-shell" dispersion equation, see discussion in Refs. [1–5]). The axion-photon interaction is parameterized by the axion-photon coupling constant $g_{a\gamma\gamma}$, see Eqs. (2.63) and (2.64)

$$\nabla \times \boldsymbol{E} = -\frac{\partial \boldsymbol{B}}{\partial t} \,, \tag{2.68}$$

$$\nabla \times \boldsymbol{B} = \frac{\partial \boldsymbol{E}}{\partial t} + \boldsymbol{J} + g_{a\gamma\gamma}\left(\boldsymbol{E} \times \nabla a - \frac{\partial a}{\partial t}\boldsymbol{B}\right), \tag{2.69}$$

where ρ is the charge density and $\boldsymbol{J}$ is the electric current density.

Physical observables that can be searched for in experiments can be derived from these modified Maxwell's equations. (A similar approach for understanding hidden photon experiments is described in detail in Chap. 7.) Consider, for example, a region of vacuum ($\rho = 0$ and $\boldsymbol{J} = 0$) bounded by a perfect conductor in the shape of an infinite cylinder with radius R. Inside this cylindrical region, a magnetic field $\boldsymbol{B}_0$ is applied along the cylinder axis ($\hat{z}$), such that $\boldsymbol{B} = B_0 z$ for $r \leq R$ and $\boldsymbol{B} = 0$ for $r > R$. Further, let us assume that the Compton wavelength of the axion (equal to $1/m_a$ in natural units, where m_a is the axion mass) is large compared to the cylinder dimensions, $m_a R \ll 1$. This is the case for "dark matter radio" experiments (discussed in Chap. 7) that search for UBDM candidates whose Compton wavelengths are so large that construction of resonant cavities is impractical. In such cases the axion de Broglie wavelength is also large compared to the cylinder dimensions, meaning that the spatial gradient of the axion field can be neglected in this treatment ($\nabla a \approx 0$). To analyze this system, we differentiate between the total magnetic field $\boldsymbol{B}$ and the induced fields $\mathcal{E}$ and $\mathcal{B}$ from the axion-photon interaction, such that $\boldsymbol{B} = \boldsymbol{B}_0 + \mathcal{B}$. With these assumptions, noting that $\boldsymbol{B}_0 \gg \mathcal{B}$, the modified Maxwell's equations become

$$\nabla \cdot \mathcal{E} = 0 \,, \tag{2.70}$$

$$\nabla \cdot \mathcal{B} = 0 \,, \tag{2.71}$$

$$\nabla \times \mathcal{E} = -\frac{\partial \mathcal{B}}{\partial t} , \tag{2.72}$$

$$\nabla \times \mathcal{B} = \frac{\partial \mathcal{E}}{\partial t} - g_{a\gamma\gamma}\frac{\partial a}{\partial t} \mathbf{B}_0 . \tag{2.73}$$

Taking the curl of Eq. (2.73), and making use of the identity

$$\nabla \times (\nabla \times \mathcal{B}) = \nabla(\nabla \cdot \mathcal{B}) - \nabla^2\mathcal{B} , \tag{2.74}$$

as well as Eqs. (2.71) and (2.72), we find

$$-\nabla^2\mathcal{B} = \frac{\partial}{\partial t}\left(-\frac{\partial \mathcal{B}}{\partial t}\right) - g_{a\gamma\gamma}\frac{\partial a}{\partial t}(\nabla \times \mathbf{B}_0) = -\frac{\partial^2\mathcal{B}}{\partial t^2} , \tag{2.75}$$

where we used the fact that $\nabla \times \mathbf{B}_0 = 0$. A similar approach yields

$$-\nabla^2\mathcal{E} = -\frac{\partial^2\mathcal{E}}{\partial t^2} + g_{a\gamma\gamma}\frac{\partial^2 a}{\partial t^2}\mathbf{B}_0 , \tag{2.76}$$

and so we arrive at the wave equations

$$\nabla^2\mathcal{B} - \frac{\partial^2\mathcal{B}}{\partial t^2} = 0 , \tag{2.77}$$

$$\nabla^2\mathcal{E} - \frac{\partial^2\mathcal{E}}{\partial t^2} = -g_{a\gamma\gamma}\frac{\partial^2 a}{\partial t^2}\mathbf{B}_0 . \tag{2.78}$$

As discussed in Sect. 2.2, if axions are the dark matter, they are nonrelativistic and thus manifest as a field oscillating at the Compton frequency m_a. As noted in Chap. 1, the axion field has a relatively long coherence time, so a good initial model for the axion field is

$$a(\mathbf{r}, t) = a_0 e^{i(\mathbf{k}\cdot\mathbf{r} - m_a t)} , \tag{2.79}$$

where $\mathbf{k}$ is the wave vector. Taking into account the cylindrical symmetry of the cavity and the boundary condition that the electric field parallel to the conducting surface at $r = R$ is zero, the wave equations (2.77) and (2.78) are solved by

$$\mathcal{E}(\mathbf{r}, t) = g_{a\gamma\gamma}a_0 e^{-im_a t}\mathbf{B}_0\left(1 - \frac{J_0(m_a r)}{J_0(m_a R)}\right) , \tag{2.80}$$

$$\mathcal{B}(\mathbf{r}, t) = i g_{a\gamma\gamma}a_0 e^{-im_a t}\mathbf{B}_0\hat{\boldsymbol{\phi}}\left(\frac{J_1(m_a r)}{J_1(m_a R)}\right) , \tag{2.81}$$

where $J_n(x)$ is the is the nth order Bessel function of the first kind [36, 37], and where we have used the fact that $e^{i\mathbf{k}\cdot\mathbf{r}} \approx 1$. For $m_a r \leq m_a R \ll 1$, the Bessel functions can be approximated by the lowest order terms in their Taylor expansion, and so

$$\mathcal{E}(\boldsymbol{r}, t) \approx g_{a\gamma\gamma} a_0 e^{-im_a t} \boldsymbol{B}_0 \left(m_a^2 R^2 - m_a^2 r^2 \right) , \tag{2.82}$$

$$\mathcal{B}(\boldsymbol{r}, t) \approx i g_{a\gamma\gamma} a_0 e^{-im_a t} B_0 \hat{\boldsymbol{\phi}}(m_a r) . \tag{2.83}$$

Note that in this case, the magnitude of the induced electric field is suppressed compared to that of the magnetic field by a factor of $\approx m_a R \ll 1$.

Based on the above analysis, it is evident that the axion field is a source term that can, in principle, generate measurable electromagnetic energy via the inverse Primakoff effect. Experiments searching for axion and ALP dark matter using the axion-photon coupling are discussed in detail in Chaps. 4, 5, and 9, and the closely related case of dark matter radio searches for hidden photons is discussed in Chap. 7.

2.4.3 Axion-Fermion Interaction

A number of experiments search for couplings between axions/ALPs and fermions, for example, the Cosmic Axion Spin Precession Experiment (CASPEr, see Ref. [38]) and the QUest for AXions experiment (QUAX, see Ref. [39]) discussed in Chap. 6 and the Global Network of Optical Magnetometers to search for Exotic physics (GNOME, see Refs. [40, 41]) described in Chap. 10, as well as experiments searching for long-range interactions between fermions mediated by axions or ALPs (such as the Axion Resonant InterAction Detection Experiment, ARIADNE, see Ref. [42]), discussed in Chap. 8.

One possible axion-fermion interaction is described by the Lagrangian term

$$\mathcal{L}_{\text{aff}} = \frac{g_{\text{f}}}{f_a} \left(\partial_\mu a \right) \psi^\dagger \gamma^\mu \gamma_5 \psi , \tag{2.84}$$

where g_{f} is a dimensionless model-dependent coupling factor and $\psi^\dagger \gamma^\mu \gamma_5 \psi$ is the axial-vector current for a Standard Model fermion f. The Hamiltonian $\mathcal{H}_{\text{af}}$ describing this interaction can be calculated from the Euler–Lagrange equations according to

$$\mathcal{H}_{\text{af}} \psi = -\gamma_0 \left[\frac{\partial \mathcal{L}_{\text{aff}}}{\partial \psi^\dagger} - \partial_\mu \left(\frac{\partial \mathcal{L}_{\text{aff}}}{\partial \left(\partial_\mu \psi^\dagger \right)} \right) \right] , \tag{2.85}$$

$$= -\frac{g_{\text{f}}}{f_a} \gamma_0 \gamma^\mu \gamma_5 \psi \left(\partial_\mu a \right) . \tag{2.86}$$

The Dirac matrices can be evaluated according to

$$\gamma_0 \gamma^\mu \gamma_5 = (\gamma_0\gamma_0\gamma_5, \ -\gamma_0\gamma_1\gamma_5, \ -\gamma_0\gamma_2\gamma_5, \ -\gamma_0\gamma_3\gamma_5) \,, \tag{2.87}$$

$$= (\gamma_5, \ -\Sigma_1, \ -\Sigma_2, \ -\Sigma_3) \,, \tag{2.88}$$

$$= (\gamma_5, \ -\boldsymbol{\Sigma}) \,, \tag{2.89}$$

where the parentheses enclose a list of the individual components of $\gamma_0\gamma^\mu\gamma_5$, evident from the definition $\gamma^\mu = (\gamma_0, \ \gamma_1, \ \gamma_2, \ \gamma_3)$, and where

$$\boldsymbol{\Sigma} = \begin{pmatrix} \sigma & 0 \\ 0 & \sigma \end{pmatrix} \,, \tag{2.90}$$

with σ being the Pauli spin matrices, and where we have employed the identities

$$\gamma_0\gamma_0 = 1 \tag{2.91}$$

and

$$\gamma_0\gamma_i\gamma_5 = \Sigma_i \,. \tag{2.92}$$

Thus the Hamiltonian $\mathcal{H}_{af}$ appearing in Eq. (2.86) can be written as

$$\mathcal{H}_{af}\psi = -\frac{g_f}{f_a}(\gamma_5, \ -\boldsymbol{\Sigma})\psi \partial_\mu a \,, \tag{2.93}$$

$$= -\frac{g_f}{f_a}\left(\gamma_5\psi \frac{\partial a}{\partial t} + (\boldsymbol{\Sigma}\psi)\cdot\nabla a\right) \,, \tag{2.94}$$

and taking the nonrelativistic limit, in which the spacelike component is much larger than the time-like component, Eq. (2.94) becomes

$$\mathcal{H}_{af} \approx -\frac{g_f}{f_a}\frac{\boldsymbol{S}}{|S|}\cdot\nabla a \,, \tag{2.95}$$

where S is the fermion spin and $|S|$ is the spin magnitude. It is important to note that not only does $\mathcal{H}_{af}$ generate an interaction between spins and the spatial gradient of the axion field but also an interaction between spins who are moving with respect to an axion field, since the momentum operator $p = -i\nabla$. This effect is known as the "axion wind" interaction and is a consequence of the fact that the field gradient is frame-dependent. The axion gradient interaction (encompassing the effects of spatial gradients and the wind interaction) is searched for in experiments such as CASPEr (the Cosmic Axion Spin Precession Experiment [43–45]) and GNOME (the Global Network of Optical Magnetometers for Exotic physics searches [40, 41, 46]) as discussed in Chaps. 6 and 10, respectively.

2.5   Theoretical Motivations for Ultralight Bosons

As noted in Chap. 1, theoretically well-motivated dark matter candidates have additional hints of their existence beyond just the evidence for dark matter. In other words, well-motivated dark matter candidates also solve other mysteries of physics. One of the most prominent examples of such a UBDM candidate is the axion, which originally emerged from an elegant solution to the strong CP problem [47, 48], the mystery of why CP-violating nuclear electric dipole moments are many orders of magnitude smaller than nominally predicted by quantum chromodynamics (QCD). As a consequence this particular ultralight boson is known as the *QCD axion*. A variety of other theories have emerged predicting similar axionlike particles (ALPs) [19]. One example such is the *relaxion*, proposed to solve the hierarchy problem [49], the question of why the Higgs boson mass is so much lighter than the Planck mass (or, in other words, why the electroweak interaction so much stronger than gravity). Axions and ALPs also offer a mechanism to explain the asymmetry between matter and antimatter in the universe [50, 51]. Attempts to unify general relativity and quantum field theory, such as string theories, generically predict the existence of axions, ALPs and other spin-0 bosons [52, 53] as well as spin-1 bosons such as dark or hidden photons [54, 55]. The key takeaway is that ultralight bosons are well-motivated from a wide variety of theoretical perspectives. In this section we explore the basic ideas behind some illustrative examples of ultralight bosons, the QCD axion, the relaxion, and axions arising from the extra dimensions appearing in string theory.

2.5.1   *Peccei-Quinn Solution to the Strong CP Problem and the QCD Axion*

The QCD axion is a natural consequence of the solution to the strong CP problem first proposed by Peccei and Quinn [47, 48, 56, 57]. The strong CP problem is related to the non-observation of a permanent electric dipole moment (EDM) of the neutron [58] and various nuclei [59] (such as ^{199}Hg, which gives the best constraint at present [60]). The magnitude of the neutron EDM d_n is predicted by the Standard Model to be [61–64]

$$|d_n| \sim 10^{-16} \bar{\theta}_{\text{QCD}} \, e \cdot \text{cm} \,, \tag{2.96}$$

where $\bar{\theta}_{\text{QCD}}$ is a CP-violating parameter appearing in the Lagrangian for the strong interaction. $\bar{\theta}_{\text{QCD}}$ is a phase angle that, in principle, can take on any value, so, based on "naturalness" its value (modulo 2π) should nominally be $\bar{\theta}_{\text{QCD}} \sim 1$. Thus the Standard Model nominally predicts a neutron EDM of $|d_n| \sim 10^{-16} \, e \cdot$cm. However, the current experimental limit on the neutron EDM is [58]

$$|d_n| < 1.8 \times 10^{-26} \, e \cdot \text{cm} \,, \tag{2.97}$$

which leads to the conclusion that $|\bar{\theta}_{QCD}| \lesssim 2 \times 10^{-10}$ (the ^{199}Hg EDM constraint [60] suggests a similar limit [65, 66]). One may wonder if $\bar{\theta}_{QCD}$ is simply a very small number by accident. However, the observable $\bar{\theta}_{QCD}$ actually arises from two contributions to the Standard Model. For these two contributions to cancel with such precision would be unnatural.

The first of these contributions is the θ parameter, which appears in a term in the QCD Lagrangian:

$$\mathcal{L}_\theta = \theta \frac{\alpha_s}{8\pi} G^{(a)}_{\mu\nu} \tilde{G}^{(a)\mu\nu} , \tag{2.98}$$

where $\alpha_s \sim 1$ is the coupling constant for the gluon field, $G^{(a)}_{\mu\nu}$ is the gluon field strength tensor (where $a = 1, 2, \ldots, 8$ indicate the eight gluon color charges), and $\tilde{G}^{(a)\mu\nu} = (1/2)\varepsilon^{\mu\nu\alpha\beta} G^{(a)}_{\alpha\beta}$ is the dual gluon field strength tensor (the gluon field strength tensor is analogous to the Faraday tensor for electromagnetism, see, for example, Refs. [2, 3]). Note that $G^{(a)}_{\mu\nu}\tilde{G}^{(a)\mu\nu}$ violates CP symmetry, just as $F^{\mu\nu}\tilde{F}_{\mu\nu} \propto \boldsymbol{E} \cdot \boldsymbol{B}$ does for electromagnetism (as seen from the fact that $\boldsymbol{E} \cdot \boldsymbol{B}$ is P- and T-odd). The θ parameter is associated with the QCD vacuum state $|\theta\rangle$ parametrized by the angle $0 \le \theta < 2\pi$ (see Refs. [14, 16] for further discussion). However, it turns out that the angle θ is not invariant with respect to *chiral transformation* (i.e., parity transformation or helicity exchange) for nonzero quark masses.

While in the limit of massless quarks, QCD would possess a chiral symmetry, such a symmetry is broken by the Adler-Bell-Jackiw anomaly [67, 68] if the quark masses are nonzero. For massive quarks, QCD physics is invariant under the following transformation of the quark fields and masses, q_i and m_i, respectively, and the vacuum parameter θ:

$$q_i \to e^{i\alpha_i \gamma_5/2} , \tag{2.99}$$

$$m_i \to e^{-i\alpha_i} m_i , \tag{2.100}$$

$$\theta \to \theta - \sum_{i=1}^{N} \alpha_i , \tag{2.101}$$

where α_i are the phases of the N quark fields.[9] While θ is thus not an invariant of QCD, the combination

$$\bar{\theta}_{QCD} \equiv \theta - \arg(\det\mathcal{M}_q) = \theta - \arg\left(\prod_{i=1}^{N} m_i\right) \tag{2.102}$$

---

[9] Note that Eq. (2.101), a rotation of the fermion determinant, is highly nontrivial: for more detailed discussion see Refs. [14–17] and for a pedagogical treatment see Ref. [5].

is invariant and thus observable ($\mathcal{M}_q$ is the quark mass matrix, see Refs. [2, 69] for definition and discussion). The strong CP problem is the question of why $\bar{\theta}_{\text{QCD}}$ is so small. Given that θ describes the QCD vacuum and that quark masses are due to the Higgs mechanism, a naive estimate for such a phase parameter is that it is of order one. Therefore the observed exceedingly small $\bar{\theta}_{\text{QCD}}$ is unnatural.

The Peccei-Quinn solution to the strong CP problem allows $\bar{\theta}_{\text{QCD}}$ to be small in a natural way, by promoting it to a dynamical variable that naturally relaxes to zero, at the minimum of a potential. To do this, the Standard Model must be extended with the introduction of additional degrees of freedom, while preserving the existing symmetries of the Standard Model. To achieve this, Peccei and Quinn [47, 48] introduced a global, chiral $\mathbb{U}(1)$ symmetry, now known as the Peccei-Quinn (PQ) symmetry, $\mathbb{U}(1)_{PQ}$ (see the tutorial at the end of Sect. 2.3 involving the $\mathbb{U}(1)$ symmetry). This symmetry is spontaneously broken at some scale, f_a, a parameter of the model, and the resulting pseudo-Nambu-Goldstone boson is the axion.

The way in which the required additional degree of freedom is introduced is model-dependent. Peccei and Quinn originally tied the symmetry breaking scale to the electroweak scale, but this resulted in an axion with a mass and couplings that would have been observed in experiments, and thus this original axion model was rapidly ruled out. Other axion models were quickly proposed that resulted in a much lighter axion with small couplings to Standard Model particles. The nature of these couplings made these axions difficult to detect, and thus they are sometimes called "invisible" axion models.

Here, we will review the Peccei-Quinn-Weinberg-Wilczek (PQWW) axion model [47, 48, 56, 57]. The original Peccei-Quinn proposal was implemented using two Higgs doublets h_u and h_d, which, respectively, couple to the up-type quarks with isospin $+1/2$, and the down-type quarks with isospin $-1/2$. The quark masses are then generated from the following Yukawa couplings to the neutral components of the Higgs fields

$$\mathcal{L}_m = y_i^u u_{Li}^\dagger h_u^0 u_{Ri} + y_i^d d_{Li}^\dagger h_d^0 d_{Ri} + \text{h.c.}, \tag{2.103}$$

where for N total quarks, there are $N/2$ up-type quarks, u_i, and $N/2$ down-type quarks, d_i, subscripts L and R denote left and right quark chirality, respectively, and the y_i are the Yukawa couplings to the quark type denoted by the superscript. Peccei and Quinn chose the Higgs potential to be

$$V(h_u, h_d) = -\mu_u^2 h_u^\dagger h_u + \mu_d^2 h_d^\dagger h_d + \sum_{i,j} \left(A_{ij} h_i^\dagger h_i h_j^\dagger h_j + B_{ij} h_i^\dagger h_j h_j^\dagger h_i \right), \tag{2.104}$$

where the coefficient matrices (A_{ij}) and (B_{ij}) are real and symmetric, and the sum is over the two types of Higgs fields. The $\mathbb{U}_{PQ}(1)$ invariance is manifested as the Lagrangian, $\mathcal{L} \equiv \mathcal{L}_m + V$, is invariant under the following transformations:

$$u_i \to e^{-i\alpha_u \gamma_5} u_i \tag{2.105}$$

$$d_i \rightarrow e^{-i\alpha_d \gamma_5} d_i \tag{2.106}$$

$$h_u \rightarrow e^{i2\alpha_u} h_u \tag{2.107}$$

$$h_d \rightarrow e^{i2\alpha_d} h_d \ . \tag{2.108}$$

Under the transformations (2.105)–(2.108), by applying Eqs. (2.101) and (2.102) one finds that $\bar{\theta}_{\mathrm{QCD}}$ is also transformed according to

$$\bar{\theta}_{\mathrm{QCD}} \rightarrow \bar{\theta}_{\mathrm{QCD}} - N(\alpha_u + \alpha_d) \ . \tag{2.109}$$

In this model, when the Universe cools to the electroweak symmetry breaking scale, the neutral Higgs acquire vacuum expectation values,

$$\langle h_u^0 \rangle = v_u e^{i P_u / v_u} \tag{2.110}$$

$$\langle h_d^0 \rangle = v_d e^{i P_d / v_d} \ , \tag{2.111}$$

where P_u and P_d are the Nambu-Goldstone fields. One linear combination of these fields becomes the longitudinal component of the Z-boson, $\mathcal{Z}$, as per standard electroweak symmetry breaking, and the other combination is the axion field, a:

$$\mathcal{Z} = P_u \cos \beta_v - P_d \sin \beta_v \tag{2.112}$$

$$a = P_u \sin \beta_v + P_d \cos \beta_v \ . \tag{2.113}$$

Using Eqs. (2.110) through (2.113) gives the following for the quark masses in Eq. (2.103):

$$-\mathcal{L}_m = m_i^u u_{Li}^\dagger e^{(i \sin \beta_v / v_u) a} u_{Ri} + m_i^d d_{Li}^\dagger e^{(i \cos \beta_v / v_d) a} d_{Ri} + \mathrm{h.c.} \ , \tag{2.114}$$

where the quark masses are $m_i^u = y^{u_i} v_u$ and $m_i^d = y_i^d v_d$.

Using the quark transformations of Eqs. (2.105) and (2.106) with Eq. (2.109), the axion dependence can be removed from the quark mass terms. The change in $\bar{\theta}_{\mathrm{QCD}}$ due to the transformation of Eq. (2.109) can be absorbed by a redefinition of the axion field. This end result is that the $\bar{\theta}_{\mathrm{QCD}}$ parameter of QCD is replaced by the axion field a. That is, a static parameter required to have a single value, which is not necessarily small, is replaced by a dynamical field. When given a potential, this dynamical field will relax to the minimum of the potential, providing a natural explanation for CP conservation in the Standard Model.

Tutorial: Mass of the QCD Axion

The axion mass, m_a, depends on the value of the axion decay constant, f_a, via

$$m_a \simeq 6 \times 10^{-6} \text{ eV} \left(\frac{10^{12} \text{ GeV}}{f_a} \right) . \tag{2.115}$$

This was first derived using the methods of current algebra by Weinberg [56], and by Bardeen and Tye [70], although Bardeen and Tye used the name "higglet" for the axion at this early stage of its study. Note that as in Eq. (2.45), the mass of the axion is $\propto 1/f_a$.

The axion mass can be determined by considering the chiral effective Lagrangian at low energies for axions and pions. This may be written as

$$
\begin{aligned}
\mathcal{L}_{\pi a} = {} & \frac{1}{2} \partial_\mu a' \partial^\mu a' + \frac{f_\pi^2}{4} \operatorname{Tr} \left[\partial_\mu U^\dagger(\boldsymbol{\pi}) \partial^\mu U(\boldsymbol{\pi}) \right] \\
& + \Lambda_{\text{QCD}}^3 \operatorname{Tr} \left[\mathcal{M}_q U(\boldsymbol{\pi}) e^{-ia'/(2f_a)} + \text{h.c.} \right] ,
\end{aligned} \tag{2.116}
$$

where $\Lambda_{\text{QCD}} \sim 200$ MeV is the QCD confinement scale (which gives rise to the explicit symmetry breaking for the QCD axion, and is thus roughly equivalent to the Λ discussed in Sect. 2.3), the pion triplet is represented by the field $\boldsymbol{\pi}$, and

$$U(\boldsymbol{\pi}) = \exp \left(\frac{i \boldsymbol{\pi} \cdot \boldsymbol{\sigma}}{f_\pi} \right) , \tag{2.117}$$

with f_π as the pion decay constant, 93 MeV, and $\boldsymbol{\sigma}$ are the Pauli matrices. The third term in Eq. (2.116) describes the explicit breaking of chiral symmetry for pions and axions and works in much the same way as the breaking of the $\mathbb{U}(1)$ symmetry discussed in the tutorial at the end of Sect. 2.3 and illustrated in Fig. 2.6. Therefore Λ_{QCD} plays a role analogous to the Λ discussed in Sect. 2.3. The origin of this symmetry breaking term is discussed in further detail in Refs. [71, 72] and can also be understood in analogy with the theory of antiferromagnetism [12].

The physical axion and neutral pion fields can be evaluated by expanding around the minimum of the potential arising from explicit symmetry breaking, assuming two light quarks [56, 71, 72], to give

$$\pi_{\text{phys}}^0 = \pi^0 + \frac{m_d - m_u}{m_d + m_u} \frac{f_\pi}{2 f_a} a' + O \left(\frac{f_\pi^2}{f_a^2} \right) \tag{2.118}$$

$$a_{\text{phys}} = a' - \frac{m_d - m_u}{m_d + m_u} \frac{f_\pi}{2 f_a} \pi^0 + O \left(\frac{f_\pi^2}{f_a^2} \right) , \tag{2.119}$$

and the corresponding masses for these fields are then

$$m_{\pi^0}^2 = \Lambda_{\text{QCD}}^3 \frac{m_u + m_d}{f_\pi^2} + O \left(\frac{f_\pi^2}{f_a^2} \right) \tag{2.120}$$

$$m_a^2 = \Lambda_{QCD}^3 \frac{m_u m_d}{f_a^2 (m_u + m_d)} + O\left(\frac{f_\pi^2}{f_a^2}\right) \tag{2.121}$$

$$\approx \frac{f_\pi^2 m_\pi^2}{f_a^2} \frac{m_u m_d}{(m_u + m_d)^2} \; . \tag{2.122}$$

With the accepted values of m_π, f_π, m_u, and m_d, the axion mass is as given in Eq. (2.115).

This tutorial computed the axion mass using *chiral perturbation theory* in QCD, which is valid for temperatures far below the QCD phase transition (technically, a cross over), $T \ll \Lambda_{QCD} \approx 200$ MeV. At high temperatures, the axion mass becomes temperature dependent, i.e., $m_a = m_a(T)$. The temperature dependence can be estimated using the so-called instanton methods, where the canonical "dilute instanton gas approximation" leads to [73, 74]:

$$m_a \propto T^{-4} \; . \tag{2.123}$$

Non-perturbative lattice QCD methods can interpolate through the QCD phase transition between the two regimes, see Ref. [75]. As we will see, the temperature dependence of the axion mass plays an important role in determining the UBDM relic density in this model.

The power of temperature in the relation Eq. (2.123) depends on the particle content of the Standard Model. The power T^{-4} is valid in a limited regime, and changes at higher temperatures where there are more effectively massless particles. A generic ALP does not obtain its mass from QCD. If the ALP mass comes, for example, from a strongly coupled "hidden sector" based on, but not equivalent to, the Standard Model, then the temperature dependence can be found via methods described in, for example, Ref. [76].

End of Tutorial

2.5.2   The Hierarchy Problem and the Relaxion

One of the greatest mysteries of theoretical physics is the hierarchy problem: why is gravity is so much weaker than all other forces? At the heart of this problem is the question of why the observed Higgs mass ($m_h \approx 125$ GeV) is so much lighter than the Planck mass ($M_{Pl} \sim 10^{19}$ GeV), for one would expect that quantum corrections would cause the effective Higgs mass to be closer to the Planck scale [77–79]. Attempts to solve the hierarchy problem include, for example, supersymmetry [80] and large (sub-mm) extra dimensions [81, 82]. Graham et al. [49] propose that instead the hierarchy problem can be solved by dynamic relaxation of the effective Higgs mass from the Planck scale to the electroweak scale in the early universe.

The dynamics are driven by inflation and a coupling of the Higgs boson to a spin-0 particle dubbed the relaxion. The relaxion could, in principle, be the QCD axion or an ALP [49] and could also constitute the dark matter [83–85]. (although it should be noted that there are issues with fine-tuning in some models [86].)

The basic idea is that inflation in the early universe causes the relaxion field to evolve in time, and because of the coupling between the relaxion and the Higgs, the effective Higgs mass evolves as well. The coupling between the relaxion and the Higgs generates a periodic potential for the relaxion once the Higgs' vacuum expectation value (VEV) becomes nonzero. When the periodic potential barriers become large enough, the time evolution of the relaxion halts and the effective mass of the Higgs settles at its observed value. The electroweak symmetry breaking scale is a special point in the evolution of the Higgs mass. This explains why the Higgs mass eventually settles at the observed value: relatively close to the electroweak scale and far from the Planck scale.

Following the discussion of Refs. [49, 87], let us suppose that the dynamics of the Higgs h and a relaxion φ are governed by a potential of the form

$$V_r(\varphi, h) = \Lambda^3 g\varphi - \frac{1}{2}\left(\Lambda^2 - g\Lambda\varphi\right)|h|^2 + \epsilon\Lambda_c^3 h \cos\left(\varphi/f\right), \qquad (2.124)$$

where Λ is the "ultraviolet cutoff" of the effective field theory (the energy scale beyond which the theory is no longer valid), g is a coupling parameter, Λ_c is the energy scale at which soft explicit symmetry breaking for the relaxion occurs ($\Lambda_c \sim \Lambda_{\rm QCD}$ for the QCD axion), and f is the spontaneous symmetry breaking scale for the relaxion. The first term in Eq. (2.124), $\Lambda^3 g\varphi$, is the leading order term of a Taylor expansion of the relaxion potential arising due to the g-coupling. The second term in Eq. (2.124) gives the effective mass m_h of the Higgs since it is of the form $m_h^2|h|^2/2$ [see, for example, the discussion surrounding Eq. (2.13)], so

$$m_h^2 \approx g\Lambda\varphi - \Lambda^2 . \qquad (2.125)$$

The third term in Eq. (2.124), $\epsilon\Lambda_c^3 h \cos\left(\varphi/f\right)$, describes the periodic potential for the relaxion arising from explicit symmetry breaking (for example, due to QCD effects). A sketch of the potential $V_r(\varphi, h)$ is shown in Fig. 2.8.

Now suppose that in the very early universe during inflation, the relaxion field starts with a large value, $\varphi \gtrsim \Lambda/g$ (indicated by the rightmost faded red dot in Fig. 2.8). It is energetically favorable for φ to decrease, and so, under certain conditions, the relaxion field will "slowly roll" down the potential (as indicated by the dashed green arrow and subsequent less faded red dots appearing to the left in Fig. 2.8). The rolling can be slow due to *Hubble friction*, which arises from the term $3H(t)\partial\varphi/\partial t$ appearing in the equation of motion for a scalar field in an expanding universe, where $H(t)$ is the Hubble parameter (as discussed in Sect. 2.6.1). As long as the Hubble friction is sufficiently large so that the dynamics are in the overdamped regime, then φ reaches a "terminal velocity" and the dynamics are independent of the initial conditions. When the evolution of φ reaches the critical

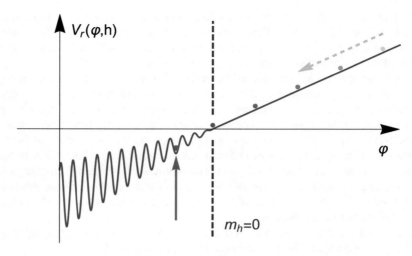

Fig. 2.8 Plot of the relaxion potential $V_r(\varphi, h)$ and illustration of the dynamics. The relaxion field φ starts at a relatively large value (shown by the faded red dots) and then "slowly rolls" down the potential (as indicated by the green dashed arrow), decreasing in amplitude, which in turn decreases m_h according to Eq. (2.125). When the Higgs' vacuum expectation value becomes nonzero at the onset of spontaneous symmetry breaking at $m_h = 0$ (marked by the dashed purple line), the amplitude of the periodic potential for φ increases. Shortly after spontaneous symmetry breaking occurs the potential wells become too deep and φ becomes trapped in a local minimum (shown by the leftmost red dot marked by the red arrow), which sets the scale of m_h at a value $\ll \Lambda$, far from the Planck scale

point $\varphi = \Lambda/g$ where $m_h = 0$, spontaneous symmetry breaking occurs (via mechanisms analogous to those discussed in Sect. 2.3), and the Higgs develops a nonzero vacuum expectation value $\langle h \rangle$. As φ decreases further, $\langle h \rangle$ grows and the amplitude of the periodic potential for φ grows as well. When the periodic potential barriers become sufficiently large, the relaxion will settle into a local minimum (as indicated by the leftmost red dot marked with a red arrow in Fig. 2.8). Again the "slow rolling" condition caused by Hubble friction is important to trap φ in the local minimum.

The crucial point is that the local minimum where φ settles is close to where $m_h \approx 0$, far from Λ and the Planck scale, thereby offering a possible dynamical solution to the hierarchy problem.

2.5.3 UBDM from Extra Dimensions

String theory [88] provides a ubiquitous font of inspiration for new and exotic physics, and the case of UBDM scenarios is no exception. String theory dictates that physics takes place not in the usual four dimensions of spacetime, but in ten. In general relativity (GR) the geometry of the extra dimensions of spacetime should be described by new functions in the metric tensor, which themselves depend on space and time. Furthermore, the curvature of space itself gravitates and carries energy. The extra dimensions of spacetime in string theory must be small enough such that we have not noticed them. However, since the curvature of these extra dimensions can change from place to place, we might feel the gravitational influence of these changes. This is one way in which string theory realizes UBDM, giving rise to scalar *moduli* and pseudoscalar *axions*.

Let us look at a simple example, which occurs in string theory, but also in any theory with extra spacetime dimensions (such as Kaluza–Klein theory [89, 90], Randall–Sundrum theory [91, 92], and various higher dimensional supergravity theories [93]). Consider the case with spacetime being D-dimensional, given by (3+1) dimensional flat Minkowski space (the manifold $\mathcal{M}_4$), with coordinates t, $\boldsymbol{x}$, and one extra compact dimension (the manifold $\mathcal{S}^1$, topologically the circle), with coordinate θ around it. In GR, this is specified by the metric:

$$ds^2 = -dt^2 + d\boldsymbol{x}^2 + \rho(\boldsymbol{x}, \theta, t)^2 L^2 d\theta^2 \,. \tag{2.126}$$

The dimensionless scalar function ρ specifies how the radius of the "circle" varies compared to a reference length scale L (the typical size of the extra dimension, which should be small). ρ can vary along the circle's circumference as θ changes, and is also a function of space and time in "our" dimensions of Minkowski space. The field ρ is known as the *radion*. Such a situation is possible to picture if we imagine that space is a single dimension like a tightrope, and ρ describes how the cross section of the tight rope varies along its length. If we walk along the tightrope, we cannot see the change in ρ, but a small creature like an ant could, by circling the rope. We may, however, indirectly notice a change in the thickness of the rope, its texture, or some other property.

General relativity tells us that the physics of the theory described by Eq. (2.126) is determined by the Einstein–Hilbert action:

$$S = \frac{M_D^{D-2}}{2} \int dt\, d^3x\, L d\theta \sqrt{-g_D}\, \mathcal{R}_D \,, \tag{2.127}$$

where D is the total number of spacetime dimensions, M_D is the D-dimensional reduced Planck mass, g_D is the D-dimensional metric determinant, and $\mathcal{R}_D$ is the D-dimensional Ricci scalar.

Without going into the details, all we need to know is that the Ricci scalar is a function which is second order in derivatives of the metric components, in this case ρ. The θ dependence of ρ can be found by expanding in terms of the eigenfunctions of $\mathcal{S}^1$, in this case leading simply to a Fourier series:

$$\rho(\boldsymbol{x}, \theta, t) = \sum_n \rho_n(\boldsymbol{x}, t) \cos(n\theta) \,. \tag{2.128}$$

The components ρ_n are four dimensional scalars known as the Kaluza–Klein "tower." It is now possible, if a little cumbersome, to analytically perform the integral $d\theta$ in Eq. (2.127), leaving an action that is second order in derivatives of the scalar fields ρ_n. This process of doing the integral over the compact coordinates, in this case θ, goes under the fancy name of "dimensional reduction"—but it is simply an integral of a series expansion.

A little thought should convince you that derivatives with respect to θ in the Ricci scalar pull down powers of n for $n > 0$. Thus the modes in the tower with $n > 0$ have terms in the action like $(n^2/L^2)\rho_n^2$: this looks like a mass term for ρ_n, which is large if L is small. Thus, for low energy physics we typically neglect the higher modes in the Kaluza–Klein tower. The lowest order solution with $n = 0$ is simply a theory quadratic in derivatives of ρ_0, i.e., we have the action of a massless scalar field! In other words, in our four dimensional Minkowski space, we "see" the change in size of the extra dimension as we move from place to place and in time as the changing value of a massless scalar field.

Including more physics, the field ρ_0 can also pick up a small mass, like in the examples of small "explicit symmetry breaking" discussed in previous sections, giving a perfect arena for UBDM to emerge. In a more complex example, we could envisage extra dimensions with weird and wonderful topologies beyond S^1. In this case we require many fields like ρ to describe the compact space, and these fields are called *moduli*. Our metric, Eq. (2.126), made a particular symmetry assumption with no "off-diagonal" components. If we include these, as in the original Kaluza–Klein theory, we obtain new vector fields (i.e., hidden photons) in four dimensions. In string theory, there can be many hundreds of such fields. Finally, if we add supersymmetry and other string theory physics into the mix, then we end up not just with scalars but also with pseudoscalar ALPs and many other weird and wonderful fields that "come along for the ride."

2.6   Non-thermal Production of UBDM

As discussed previously, due to the very small mass of UBDM candidates, cold populations that can provide all the dark matter of the Universe must be created out-of-equilibrium. If thermally produced [94], such particles will have too high a kinetic energy to serve as cold dark matter. Cold populations of UBDM can be produced via a non-equilibrium process known as vacuum misalignment [95–98]. When inflation causes the UBDM field to be homogeneous within our horizon, vacuum misalignment is the dominant production mechanism for UBDM particles. If the UBDM candidate is the product of a phase transition which occurs after inflation, the production of the UBDM particle from cosmic strings and domain walls must also be considered (which is, in essence, another form of vacuum misalignment, but for the UBDM field as a whole).

2.6.1   Vacuum Misalignment

The essence of the vacuum misalignment mechanism is that the initial value of the field is different from the minimum of the field's potential, the vacuum expectation value. When this occurs, the field can oscillate around the minimum of the potential, and the energy density in the oscillating field is the UBDM. This process is commonly called *vacuum misalignment*, as the initial value of the field is misaligned with the potential minimum. (The process is also referred to as vacuum realignment in the literature.)

On large scales, the Universe is known to be isotropic, homogeneous, and expanding, which means it can be described by a Friedmann–Robertson–Walker (FRW) metric, i.e.,

$$-ds^2 = -dt^2 + R^2(t)d\mathbf{x} \cdot d\mathbf{x} , \tag{2.129}$$

where $(t, \mathbf{x})$ are co-moving coordinates and $R(t)$ is the scale factor. For a scalar field, ϕ, with an effective potential, $V(\phi)$, the equation of motion can be derived by writing the Lagrangian using the FRW metric instead of the metric for flat spacetime, yielding:

$$\left(\frac{\partial^2}{\partial t^2} + 3\frac{\dot{R}(t)}{R(t)}\frac{\partial}{\partial t} - \frac{1}{R^2(t)}\nabla^2 \right)\phi(t, \mathbf{x}) + \frac{\partial V}{\partial \phi} = 0 . \tag{2.130}$$

In the case of the hidden photon dark matter candidate, we will shortly discuss that the spatial parts of the vector boson field obey an equation of this form. Equation (2.130) is the equation of a harmonic oscillator in an FRW spacetime. When the field is homogeneous over the scale of interest, the spatial derivative in Eq. (2.130) can be neglected. Identifying the Hubble parameter, $H(t) = \dot{R}(t)/R(t)$ (determined from the energy density of radiation in the early universe), the resulting equation is

$$\left(\frac{\partial^2}{\partial t^2} + 3H(t)\frac{\partial}{\partial t} \right)\phi(t, \mathbf{x}) + \frac{\partial V}{\partial \phi} = 0 . \tag{2.131}$$

When the condition

$$\frac{3}{2}H(t) \gg \sqrt{\frac{1}{\phi}\frac{\partial V}{\partial \phi}} \tag{2.132}$$

is met, the field is overdamped and does not oscillate. Essentially, one wavelength of the field does not fit inside the horizon, and the field is thus "frozen in" and unable to oscillate. When the potential meets the criterion

$$\frac{3}{2}H(t) \simeq \sqrt{\frac{1}{\phi}\frac{\partial V}{\partial \phi}} \ , \tag{2.133}$$

a wavelength of the field is contained within the horizon, and it becomes free to oscillate. The energy in these oscillations is determined by the initial condition, which is the displacement, or misalignment, of the field from the potential minimum (see discussion in the tutorial at the end of Sect. 2.3, where the potential develops a periodic dependence on the phase angle θ describing the bosonic field due to explicit symmetry breaking). We denote this angle θ_i, which corresponds to a field value ϕ_i. The field can relax so that the rms value is zero, and the vacuum is effectively realigned.

2.6.2 Vector Field Misalignment

For a vector UBDM candidate arising from kinetic mixing, a phase transition does not occur and a cold population of hidden photons can be entirely produced by vacuum misalignment. While this mechanism was originally discussed in terms of the axion [95–97], we will cover the hidden photon here first, as it is a more simple case. That the spatial component of a light vector boson can also satisfy Eq. (2.131) and result in a cold population was first discussed in Ref. [98].

The hidden photon field, X_μ, will be uniform over the scale of the horizon after inflation, with an initial random value. As it is spatially uniform, $\partial_i X_\mu \sim 0$, and the resulting equation of motion is [98]

$$\left(\frac{\partial^2}{\partial t^2} + 3H(t)\frac{\partial}{\partial t}\right)X_i(\mathbf{x}) + m_{\gamma'}^2 X_i(\mathbf{x}) = 0 \tag{2.134}$$

with the mass term giving an effective potential when $m_{\gamma'} \neq 0$. When the condition of Eq. (2.133) is met and $H(t) \sim m_{\gamma'}$, the field can begin to oscillate and act as cold dark matter.

A simple bound on $m_{\gamma'}$ can be obtained by requiring that the particle's Compton wavelength permit structure formation on kiloparsec scales [98, 99]. Then the requirement that 1 kpc $< \hbar/(m_{\gamma'} v_{esc})$, where v_{esc} is the escape velocity of the structure, gives a bound $m_{\gamma'} c^2 \geq 1.7 \times 10^{-24}$ eV. More detailed bounds can be obtained from considering decays, interactions of the hidden photon with other particles, and experimental observations [98]. Further discussion is in Chap. 3.

2.6.3   Scalar Field Misalignment

For scalar (or pseudoscalar) fields that occur as the pseudo-Nambu-Goldstone boson, such as axions and ALPs, there are two temperature scales that govern the non-equilibrium production mechanisms of the particles in the early Universe. These are the temperature at which spontaneous symmetry breaking occurs, T_{SB}, and the temperature at which the boson field acquires an effective potential, $T_{\rm eff}$. For the QCD axion, T_{SB} is the temperature at which the Peccei-Quinn symmetry is spontaneously broken, T_{PQ}. In addition to vacuum misalignment, other topological effects may contribute to the cold population of axions in the Universe, depending on the relationship between T_{PQ} and the inflationary reheating temperature, T_R. For ALPs from string models, T_{SB} is the Kaluza–Klein scale,[10] generally assumed to be far above the inflationary reheating temperature. Thus, for ALPs it is commonly accepted that vacuum misalignment is the method by which a potential ALP dark matter population is produced in the early Universe.

In the scalar cases, at T_{SB}, a global chiral symmetry is spontaneously broken, and the phase can take on any value, θ_i. If $T_{SB} > T_R$, a value of the initial misalignment angle in one region of space can be inflated such that the misalignment angle has the same value everywhere within the horizon. In this case, non-equilibrium production of the scalar particles is similar to that of a cold population of hidden photons occurring due to vacuum misalignment as discussed in Sect. 2.6.2. For axions, if $T_{PQ} < T_R$, fluctuations in local temperature mean that spontaneous symmetry breaking will be seeded at different locations within the horizon, and each location will select a different value of ϕ_i. At the interface of regions with different ϕ_i, topological axion strings and domain walls will occur. These are not observed, so we surmise that they have decayed via the various available channels. In the following, we will discuss vacuum misalignment in detail, similar to Ref. [100], and touch on the other production mechanisms. A more in-depth discussion of axion cosmology is given by Ref. [101]. In the following, we will refer to the axion, but the discussion also applies to ALPs.

The second temperature scale for the axion, $T_{\rm eff}$, is when a significant mass term for the axion arises. The chiral anomaly couples the axion to the gauge field, and the gauge field instantons induce a potential and hence a mass for the axion through soft explicit symmetry breaking (following the basic ideas discussed in Sect. 2.3). This occurs at the scale when the quark-gluon plasma condenses to hadrons. We denote this time t_1 and at this temperature, $m_a t_1 \sim 1$ [95–97]. Note that $T_{QCD} \simeq 1$ GeV. For ALPs from string theory, similar non-perturbative effects create a potential for the ALP and, consequently, a mass.

When m_a becomes significant, the axion field gains an effective periodic potential, analogous to that described by Eq. (2.54),

---

[10] The Kaluza–Klein scale is the energy scale associated with the size of the compactified or "curled-up" extra dimensions in string theory [88].

$$V(\phi) = m_a^2(T) f_a^2 \left(1 - \cos\left(\frac{\phi}{f_a} \right) \right) = m_a^2(T) f_a^2 (1 - \cos\theta) , \qquad (2.135)$$

where $\theta = \phi/f_a$. At low temperatures, the axion mass is given by Eq. (2.115)

$$m_a \simeq 6 \times 10^{-6} \text{ eV} \left(\frac{10^{12} \text{ GeV}}{f_a} \right) ,$$

as discussed in the tutorial at the end of Sect. 2.5.1. At higher temperatures—while the potential is effectively "turning on"—the axion mass has a temperature dependence (which can be calculated using lattice QCD, see discussion in Ref. [75]).

Using the effective potential given by Eq. (2.135) with Eq. (2.131), the equation of motion governing the axion field dynamics is

$$\left(\frac{\partial^2}{\partial t^2} + 3H(t) \frac{\partial}{\partial t} \right) \phi(t, \mathbf{x}) + m_a^2(T(t)) f_a \sin\theta = 0 . \qquad (2.136)$$

The dependence of temperature on time in the early universe is discussed in Chap. 3. Using Eq. (2.136), the density of cold axions can be estimated as follows. For small oscillations near $\theta = 0$, $\sin\theta \approx \theta$ and

$$\left(\frac{\partial^2}{\partial t^2} + 3H(t) \frac{\partial}{\partial t} \right) \phi(t, \mathbf{x}) + m_a^2(t) \phi(t, \mathbf{x}) = 0 . \qquad (2.137)$$

At temperatures above T_{eff}, θ is approximately constant, and m_a can be neglected. When $m_a t_1 \sim 1$, the field begins to oscillate, which corresponds to the time [101]

$$t_1 \simeq 2 \times 10^{-7} \text{ s} \left(\frac{f_a}{10^{12} \text{ GeV}} \right)^{\frac{1}{3}} \qquad (2.138)$$

and

$$T_{\text{eff}} \simeq 1 \text{ GeV} \left(\frac{10^{12} \text{ GeV}}{f_a} \right)^{\frac{1}{6}} . \qquad (2.139)$$

Alignment of the field will occur on the order of the same timescale, and thus its momentum is on the order of

$$p_a(t_1) \sim \frac{1}{t_1} . \qquad (2.140)$$

If $f_a \sim 10^{12}$ GeV, then $m_a \sim 6 \ \mu$eV, and the field momentum will be $p_a \sim 10^{-9}$ eV. From this estimate, it is easily seen that the initial momentum of a population of axions from vacuum misalignment is much less than the axion mass, thus the population is nonrelativistic, or cold.

The question of whether or not a sufficient number of axions are produced to account for all the dark matter in the Universe can be addressed by estimating the energy density. Expanding around the potential minimum, this density is

$$\rho = \frac{f_a^2}{2} \left(\dot{\theta}^2 + m_a^2(t)\theta^2 \right) . \tag{2.141}$$

The virial theorem gives

$$\langle \dot{\theta}^2 \rangle = m_a^2 \langle \theta^2 \rangle = \frac{\rho}{f_a^2} . \tag{2.142}$$

The energy density of these nonrelativistic axions (for the given potential) scales with the expansion of the Universe (see Problem 3.1) as

$$\rho \propto \frac{m_a(t)}{R^3(t)}. \tag{2.143}$$

For the initial misalignment angle, θ_i, the energy density in coherent axion oscillations is

$$\rho_i = \frac{1}{2} m_a^2(t_1) f_a^2 \theta_i^2 = \frac{1}{2} m_a^2(t_1) \phi_i^2 . \tag{2.144}$$

Given matter dominated expansion of the Universe until today, the axion density scales with Eq. (2.143), to give today's average axion density,

$$\rho_0 \sim \rho_i \frac{m_a(t_0)}{m_a(t_i)} \frac{R^3(t_i)}{R^3(t_0)} , \tag{2.145}$$

or

$$\rho_0 \sim \frac{1}{2} f_a^2 \frac{m_a}{t_1} \frac{R^3(t_1)}{R^3(t_0)} \phi_i^2 , \tag{2.146}$$

using Eq. (2.144) and $m_a t_1 \sim 1$. The initial misalignment angle, θ_i, has a single value if T_{PQ} is greater than the inflationary reheat temperature, T_R. In the case when $T_{PQ} < T_R$, ϕ_i can have several different values within the horizon, and additionally, higher-order modes of Eq. (2.136) can be occupied. Under these circumstances, Eq. (2.146) gives the correct expression for the zero-momentum mode if we replace θ_i with its average within the horizon, expected to be $O(1)$. Using Eqs. (2.115), (2.138), and (2.139), and assuming $m_a(T) \propto T^{-4}$, the energy density in axions from this population today is

$$\Omega_a \sim \left(\frac{f_a}{10^{12}\,\text{GeV}}\right)^{\frac{7}{6}}.$$ (2.147)

For $T_{PQ} > T_R$ and $\theta_i \sim 1$, this gives the cold axion population today. For $T_{PQ} < T_R$, it is expected that there is an equal contribution from the sum of all higher-order modes, and possible contributions from string and wall decay. A thorough discussion of all these contributions can be found in Ref. [101].

References

1. T. Lancaster, S.J. Blundell, *Quantum Field Theory for the Gifted Amateur* (Oxford University Press, Oxford, 2014)
2. A. Zee, *Quantum Field Theory in a Nutshell*, vol. 7 (Princeton University Press, Princeton, 2010)
3. S. Weinberg, *The Quantum Theory of Fields*, vol. 1–2 (Cambridge University Press, Cambridge, 1995)
4. M. Peskin, *An Introduction to Quantum Field Theory* (CRC Press, Boca Raton, 2018)
5. M. Srednicki, *Quantum Field Theory* (Cambridge University Press, Cambridge, 2007)
6. G. Aad, T. Abajyan, B. Abbott, J. Abdallah, S.A. Khalek, A.A. Abdelalim, R. Aben, B. Abi, M. Abolins, O. AbouZeid, et al., Phys. Lett. B **716**, 1 (2012)
7. S. Chatrchyan, V. Khachatryan, A.M. Sirunyan, A. Tumasyan, W. Adam, E. Aguilo, T. Bergauer, M. Dragicevic, J. Erö, C. Fabjan, et al., Phys. Lett. B **716**, 30 (2012)
8. G. Aad, T. Abajyan, B. Abbott, J. Abdallah, S.A. Khalek, R. Aben, B. Abi, M. Abolins, O. AbouZeid, H. Abramowicz, et al., Phys. Lett. B **726**, 120 (2013)
9. T. Clifton, P.G. Ferreira, A. Padilla, C. Skordis, Phys. Rep. **513**, 1 (2012)
10. G. Sardanashvily, *Noether's Theorems* (Springer, Berlin, 2016)
11. S.R. Choudhury, S. Choubey, J. Cosmol. Astropart. Phys. **2018**, 017 (2018)
12. C.P. Burgess, Phys. Rep. **330**, 193 (2000)
13. J. Goldstone, A. Salam, S. Weinberg, Phys. Rev. **127**, 965 (1962)
14. R. Peccei, Lect. Notes Phys. **741**, 3 (2008)
15. L.D. Duffy, K. Van Bibber, New J. Phys. **11**, 105008 (2009)
16. J.E. Kim, G. Carosi, Rev. Mod. Phys. **82**, 557 (2010)
17. P. Sikivie, Comptes Rendus Physique **13**, 176 (2012)
18. P. Arias, D. Cadamuro, M. Goodsell, J. Jaeckel, J. Redondo, A. Ringwald, J. Cosm. Astropart. Phys. **2012**, 013 (2012)
19. P.W. Graham, I.G. Irastorza, S.K. Lamoreaux, A. Lindner, K.A. van Bibber, Ann. Rev. Nucl. Part. Sci. **65**, 485 (2015)
20. P.W. Graham, D.E. Kaplan, J. Mardon, S. Rajendran, W.A. Terrano, Phys. Rev. D **93**, 075029 (2016)
21. D.J. Gross, R.D. Pisarski, L.G. Yaffe, Rev. Mod. Phys. **53**, 43 (1981)
22. M. Dine, P. Draper, L. Stephenson-Haskins, D. Xu, Phys. Rev. D **96**, 095001 (2017)
23. T. Damour, G. Gibbons, C. Gundlach, Phys. Rev. Lett. **64**, 123 (1990)
24. M. Safronova, D. Budker, D. DeMille, D.F. Jackson Kimball, A. Derevianko, C.W. Clark, Rev. Mod. Phys. **90**, 025008 (2018)
25. C.S. Wu, E. Ambler, R. Hayward, D. Hoppes, R.P. Hudson, Phys. Rev. **105**, 1413 (1957)
26. J.H. Christenson, J.W. Cronin, V.L. Fitch, R. Turlay, Phys. Rev. Lett. **13**, 138 (1964)
27. L. Barkov, M. Zolotorev, JETP Lett. **27**, 357 (1978)
28. L. Barkov, M. Zolotorev, JETP Lett. **28**, 50 (1978)
29. R. Conti, P. Bucksbaum, S. Chu, E. Commins, L. Hunter, Phys. Rev. Lett. **42**, 343 (1979)

30. J.D. Jackson, *Classical Electrodynamics*, 2nd edn. (Wiley, New York, 1975)
31. H. Primakoff, Phys. Rev. **81**, 899 (1951)
32. P. Sikivie, Phys. Rev. Lett. **51**, 1415 (1983)
33. P. Sikivie, Phys. Rev. D **32**, 2988 (1985)
34. J. Ouellet, Z. Bogorad, Phys. Rev. D **99**, 055010 (2019)
35. Y. Kim, D. Kim, J. Jeong, J. Kim, Y.C. Shin, Y.K. Semertzidis, Phys. Dark Universe **26**, 100362 (2019)
36. G.B. Arfken, H.J. Weber, *Mathematical Methods for Physicists* (Elsevier, Amsterdam, 1999)
37. M.L. Boas, *Mathematical Methods in the Physical Sciences* (Wiley, Hoboken, 2006)
38. D. Budker, P.W. Graham, M. Ledbetter, S. Rajendran, A.O. Sushkov, Phys. Rev. X **4**, 021030 (2014)
39. N. Crescini, D. Alesini, C. Braggio, G. Carugno, D. D'Agostino, D. Di Gioacchino, P. Falferi, U. Gambardella, C. Gatti, G. Iannone, et al., Phys. Rev. Lett. **124**, 171801 (2020)
40. M. Pospelov, S. Pustelny, M.P. Ledbetter, D.F. Jackson Kimball, W. Gawlik, D. Budker, Phys. Rev. Lett. **110**, 021803 (2013)
41. S. Pustelny, D.F. Jackson Kimball, C. Pankow, M.P. Ledbetter, P. Wlodarczyk, P. Wcislo, M. Pospelov, J.R. Smith, J. Read, W. Gawlik, D. Budker, Ann. Phys. **525**, 659 (2013)
42. A. Arvanitaki, A.A. Geraci, Phys. Rev. Lett. **113**, 161801 (2014)
43. D. Budker, P.W. Graham, M. Ledbetter, S. Rajendran, A.O. Sushkov, Phys. Rev. X **4**, 021030 (2014)
44. T. Wu, J.W. Blanchard, G.P. Centers, N.L. Figueroa, A. Garcon, P.W. Graham, D.F. Jackson Kimball, S. Rajendran, Y.V. Stadnik, A.O. Sushkov, et al., Phys. Rev. Lett. **122**, 191302 (2019)
45. A. Garcon, J.W. Blanchard, G.P. Centers, N.L. Figueroa, P.W. Graham, D.F. Jackson Kimball, S. Rajendran, A.O. Sushkov, Y.V. Stadnik, A. Wickenbrock, et al., Sci. Adv. **5**, eaax4539 (2019)
46. S. Afach, D. Budker, G. DeCamp, V. Dumont, Z.D. Grujić, H. Guo, D.F. Jackson Kimball, T. Kornack, V. Lebedev, W. Li, et al., Phys. Dark Universe **22**, 162 (2018)
47. R. Peccei, H.R. Quinn, Phys. Rev. Lett. **38**, 1440 (1977)
48. R. Peccei, H.R. Quinn, Phys. Rev. D **16**, 1791 (1977)
49. P.W. Graham, D.E. Kaplan, S. Rajendran, Phys. Rev. Lett. **115**, 221801 (2015)
50. K. Harigaya, et al., Phys. Rev. Lett. **124**, 111602 (2020)
51. R.T. Co, L.J. Hall, K. Harigaya, J. High Energ. Phys. **2021**, 172 (2021)
52. P. Svrcek, E. Witten, J. High Energy Phys. **06**, 051 (2006)
53. A. Arvanitaki, S. Dimopoulos, S. Dubovsky, N. Kaloper, J. March-Russell, Phys. Rev. D **81**, 123530 (2010)
54. B. Holdom, Phys. Lett. B **166**, 196 (1986)
55. M. Cvetic, P. Langacker, Phys. Rev. D **54**, 3570 (1996)
56. S. Weinberg, Phys. Rev. Lett. **40**, 223 (1978)
57. F. Wilczek, Phys. Rev. Lett. **40**, 279 (1978)
58. C. Abel, S. Afach, N.J. Ayres, C.A. Baker, G. Ban, G. Bison, K. Bodek, V. Bondar, M. Burghoff, E. Chanel, et al., Phys. Rev. Lett. **124**, 081803 (2020)
59. T. Chupp, P. Fierlinger, M. Ramsey-Musolf, J. Singh, Rev. Mod. Phys. **91**, 015001 (2019)
60. B. Graner, Y. Chen, E. Lindahl, B. Heckel, et al., Phys. Rev. Lett. **116**, 161601 (2016)
61. J.E. Kim, Phys. Rep. **150**, 1 (1987)
62. H.Y. Cheng, Phys. Rep. **158**, 1 (1988)
63. M.S. Turner, Phys. Rep. **197**, 67 (1990)
64. G.G. Raffelt, Phys. Rep. **198**, 1 (1990)
65. J. de Vries, E. Mereghetti, A. Walker-Loud, Phys. Rev. C **92**, 045201 (2015)
66. J. Bsaisou, J. de Vries, C. Hanhart, S. Liebig, U.G. Meißner, D. Minossi, A. Nogga, A. Wirzba, J. High Energy Phys. **2015**, 104 (2015)
67. S.L. Adler, Phys. Rev. **177**, 2426 (1969)
68. J. Bell, R. Jackiw, Nuovo Cim. A **60**, 47 (1969)
69. S. Kanemaki, I. Furuoya, Prog. Theor. Phys. **89**, 1235 (1993)

70. W.A. Bardeen, S.H. Tye, Phys. Lett. B **74**, 229 (1978)
71. L. Di Luzio, M. Giannotti, E. Nardi, L. Visinelli, Phys. Rep. **870**, 1 (2020)
72. G.G. di Cortona, E. Hardy, J.P. Vega, G. Villadoro, J. High Energy Phys. **2016**, 34 (2016)
73. D.J. Gross, R.D. Pisarski, L.G. Yaffe, Rev. Mod. Phys. **53**, 43 (1981)
74. O. Wantz, E.P.S. Shellard, Phys. Rev. D **82**, 123508 (2010)
75. S. Borsanyi, et al., Nature **539**, 69 (2016)
76. H. Davoudiasl, C.W. Murphy, Phys. Rev. Lett. **118**, 141801 (2017)
77. G. Degrassi, S. Di Vita, J. Elias-Miro, J.R. Espinosa, G.F. Giudice, G. Isidori, A. Strumia, J. High Energy Phys. **2012**, 98 (2012)
78. Y. Hamada, H. Kawai, K.Y. Oda, Phys. Rev. D **87**, 053009 (2013)
79. J. Elias-Miro, J.R. Espinosa, G.F. Giudice, G. Isidori, A. Riotto, A. Strumia, Phys. Lett. B **709**, 222 (2012)
80. S. Dimopoulos, H. Georgi, Nucl. Phys. B **193**, 150 (1981)
81. N. Arkani-Hamed, S. Dimopoulos, G.R. Dvali, Phys. Lett. B **429**, 263 (1998)
82. L. Randall, R. Sundrum, Phys. Rev. Lett. **83**, 3370 (1999)
83. N. Fonseca, E. Morgante, Phys. Rev. D **100**, 055010 (2019)
84. A. Banerjee, H. Kim, G. Perez, Phys. Rev. D **100**, 115026 (2019)
85. R. Gupta, J. Reiness, M. Spannowsky, Phys. Rev. D **100**, 055003 (2019)
86. R.S. Gupta, Z. Komargodski, G. Perez, L. Ubaldi, J. High Energy Phys. **2016**, 166 (2016)
87. J. Espinosa, C. Grojean, G. Panico, A. Pomarol, O. Pujolas, G. Servant, Phys. Rev. Lett. **115**, 251803 (2015)
88. B. Zwiebach, *A First Course in String Theory* (Cambridge University Press, Cambridge, 2004)
89. E. Witten, Nucl. Phys. B **186**, 412 (1981)
90. J.M. Overduin, P.S. Wesson, Phys. Rep. **283**, 303 (1997)
91. L. Randall, R. Sundrum, Phys. Rev. Lett. **83**, 3370 (1999)
92. L. Randall, R. Sundrum, Phys. Rev. Lett. **83**, 4690 (1999)
93. D.Z. Freedman, A. Van Proeyen, *Supergravity* (Cambridge University Press, Cambridge, 2012)
94. E.W. Kolb, M.S. Turner, Nature **294**, 521 (1981)
95. L. Abbott, P. Sikivie, Phys. Lett. B **120**, 133 (1983)
96. J. Preskill, M.B. Wise, F. Wilczek, Phys. Lett. B **120**, 127 (1983)
97. M. Dine, W. Fischler, Phys. Lett. B **120**, 137 (1983)
98. A.E. Nelson, J. Scholtz, Phys. Rev. D **84**, 103501 (2011)
99. P. van Dokkum, et al., Astrophys. J. Lett. **677**, L5 (2008)
100. L.D. Duffy, K. van Bibber, New J. Phys. **11**, 105008 (2009)
101. P. Sikivie, Lect. Notes Phys. **741**, 19 (2008)

Chapter 3
Astrophysical Searches and Constraints

David J. E. Marsh and Sebastian Hoof

Abstract Starting from the evidence that dark matter (DM) indeed exists and permeates the entire cosmos, various bounds on its properties can be estimated. Beginning with the cosmic microwave background and large-scale structure, we summarize bounds on the ultralight bosonic dark matter (UBDM) mass and cosmic density. These bounds are extended to larger masses by considering galaxy formation and evolution and the phenomenon of black hole superradiance. We then discuss the formation of different classes of UBDM compact objects including solitons/axion stars and miniclusters. Next, we consider astrophysical constraints on the couplings of UBDM to Standard Model particles, from stellar cooling (production of UBDM) and indirect searches (decays or conversion of UBDM). Throughout, there are short discussions of "hints and opportunities" in searching for UBDM in each area.

3.1 Astrophysical Search Channels

Astrophysics and cosmology, as outlined in Chap. 1, give convincing evidence that dark matter (DM) exists in the form of new particles beyond the Standard Model of particle physics. The space of possible theories in Chap. 2, even for the subclass of ultralight bosonic DM (UBDM) models considered in this book, is vast. Beyond the basic fact of the existence of DM, astrophysics can be used to reign in this vast theoretical parameter space, with a view to direct detection and measurement of model parameters.

The most basic astrophysical route to constrain UBDM is via the relic density. There are three channels for UBDM production:

D. J. E. Marsh (✉)
Department of Physics, King's College London, London, UK
e-mail: david.j.marsh@kcl.ac.uk

S. Hoof
Institut für Astrophysik, Georg-August-Universität Göttingen, Göttingen, Germany
e-mail: hoof@uni-goettingen.de

© The Author(s) 2023
D. F. Jackson Kimball, K. van Bibber (eds.), *The Search for Ultralight Bosonic Dark Matter*, https://doi.org/10.1007/978-3-030-95852-7_3

1. Coherent field oscillations

    (a) Vacuum realignment
    (b) Topological defect decay

2. Thermal production
3. Non-thermal production by direct decay

Without going into the specifics (see Ref. [1]), it suffices to say that only channel 1 produces UBDM with the required properties as outlined in Chap. 1. Production channels 2 and 3 produce hot DM, or indeed dark radiation, each of which is strongly constrained by the CMB anisotropies [2, 3].

In channel 1a (vacuum realignment), the UBDM relic density is a function of two parameters, (m, ϕ_i), where ϕ_i is the initial field displacement, i.e., the location of the field in its potential relative to the minimum at "the initial time" (in practice, at the end of inflation). In this scenario, the initial field displacement is taken to be completely uniform throughout space, this state of affairs having been arranged by the same mechanism that causes the large-scale observed homogeneity of the cosmic microwave background (CMB), inflation, or otherwise. The correct relic abundance can be achieved across many orders of magnitude, covering all the masses of interest $(10^{-33}\,\text{eV}, 10^{-1}\,\text{eV})$ for $\phi_i \leq M_{pl}$.[1] For an axion-like particle (ALP), the fundamental parameter from theory is f_a: the scale of spontaneous symmetry breaking, also called the axion decay constant. The parameter θ_i, defined via $\phi_i \equiv \theta_i f_a$, is the initial angle that the axion field takes (recall that the axion is the phase of a complex field). At early times, the axion possesses a shift symmetry, $\phi \rightarrow \phi + \text{constant}$, and thus θ_i has no preferred value and can be considered a free random variable (although very small values or values very close to π are considered fine-tuned). Because θ_i is undetermined, there is a wide range of allowed values for the fundamental parameters (m, f_a) consistent with the required relic density. In particular, in this channel, large values of the decay constant at the grand unified scale ($\sim 10^{16}\,\text{GeV}$) or the reduced Planck scale ($\sim 10^{18}\,\text{GeV}$) are allowed.

Production via channel 1b (topological defect decay) is possible only for UBDM that is a Goldstone boson of a spontaneously broken global symmetry (the "Kibble-Zurek mechanism" [4, 5] described in Sect. 3.3.2). In particular, it applies to the QCD axion and other ALPs, where topological strings and domain walls are formed when the global $\mathbb{U}(1)$ symmetry is spontaneously broken. If symmetry breaking occurs after inflation, then the defects cannot be smoothed out and inflated away, and the axion field takes on a very inhomogeneous distribution (in contrast to the case of vacuum realignment). The defects later decay when non-perturbative effects give the ALP a mass. This process must be simulated using classical lattice field theory and has only been studied in detail for the QCD axion [6–8]. Large numerical uncertainties related to extrapolation to physical couplings

---

[1] $M_{pl} = 1/\sqrt{8\pi G_N}$ is the *reduced* Planck mass, related to the mass scale given in Table 7 of the "Units and Conversions" section by the factor of $\sqrt{8\pi}$ coming from Einstein's equation in general relativity.

prevent an agreed estimation of the relic density. The correct relic abundance can be achieved within numerical and model uncertainty (extrapolation, domain wall number, explicit symmetry breaking) for all values of $f_a \lesssim 10^{12}$ GeV [9].

The production mechanism channel 1b works for $f_a < T_{\max}$, where $T_{\max}$ is the maximum thermalization temperature of the Universe, and the bound arises since defects only form if symmetry breaking occurs during the ordinary thermal history of the Universe. $T_{\max}$ is bounded from above due to observational constraints on the theory of inflation. In particular, H_I, the inflationary Hubble rate, is bounded from above by the fact that tensor-type CMB anisotropies have relative amplitude $r \lesssim 0.1$ compared to scalar-type perturbations leading to the constraint $H_I \lesssim 10^{14}$ GeV. H_I sets the temperature of the Universe during inflation to be the Gibbons–Hawking temperature, $T_{\mathrm{GH}} = H_I / 2\pi$. The maximum thermalization temperature could actually be larger than this, which can easily be seen from the Friedmann equation during radiation domination, $3H^2 M_{pl}^2 = \pi^2 g_\star T^4/30$, where the quantity $g_\star$ counts the effective number of relativistic degrees of freedom [10]:

$$ g_\star = \frac{7}{8} \sum_{i \in \text{fermions}} g_i \left(\frac{T_i}{T} \right)^4 + \sum_{i \in \text{bosons}} g_i \left(\frac{T_i}{T} \right)^4 , \qquad (3.1) $$

where g_i is the degrees of freedom of species i (e.g., two polarizations for the photon) and T_i is the temperature of species i, and T is the photon bath temperature. The value of $g_\star$ at very high temperatures is bounded from below by the Standard Model contribution, $g_{\star,\mathrm{SM}} = 106.75$. H monotonically decreases, and so $H_{\max} = H_I$. If reheating after inflation is instantaneous and 100% efficient, we find an upper bound for $T_{\max} \lesssim 8 \times 10^{15}$ GeV. ALPs with values of f_a larger than this upper bound on $T_{\max}$ cannot be produced by mechanism 1b and must be produced by mechanism 1a. The observational lower bound on $T_{\max}$ arises from demanding successful Big Bang nucleosynthesis, $T_{\max} \gtrsim 1$ MeV. For values of f_a in this very large range of allowed $T_{\max}$ values, it is not determined whether ALPs are produced by mechanism 1a or 1b, either being possible depending on the model of inflation and reheating.

There are various astrophysical search channels we can use to constrain UBDM:

1. Gravitational probes
2. "Indirect detection"

   (a) Production of UBDM (e.g., in stars or from background radiation)
   (b) Decay/conversion of existing UBDM

Gravitational probes are the most general form of constraints on UBDM and give us powerful bounds on the key parameters of mass and density (both cosmic and local), which are important for the design of direct DM searches. Indirect detection depends on the UBDM interactions with ordinary matter: null results provide baseline constraints on couplings to which laboratory searches are compared, and anomalous results give hints for promising regions of parameter space to search.

In this chapter, unless stated otherwise, we use natural units where $\hbar = c = k_B = 1$ and express all quantities in electronvolts (eV). We use the Einstein summation convention for repeated indices. Roman indices i, j, etc. run from 1 to 3, while Greek indices μ, ν, etc. run from 0 to 3, with zero labelling the time-like direction. In relativity, we distinguish covariant (lower) and contravariant (upper) indices, with the metric being responsible for raising and lowering: $x_\mu = g_{\mu\nu} x^\nu$.

3.2   Gravitational Probes of UBDM

The goal of this section is to assess the validity of UBDM as a model of DM. Since all current observations are consistent with cold dark matter (CDM, defined as a pressureless fluid), the bounds we estimate on the UBDM mass m can be thought of as answering the question: "is UBDM observationally equivalent to CDM?" The answer to this question depends on the observable and leads to lower bounds on m (and upper bounds on the UBDM density if we allow for multi-component DM). In order to derive our bounds, we must specify the ways in which UBDM is not equivalent to CDM. These differences further suggest astrophysical phenomena that could distinguish between UBDM and CDM in the future, possibly providing evidence for one model over the other.

3.2.1   The CMB and Linear Structure Formation

Considering how the gravitational effects of DM dominate the formation of structure in the Universe, one can derive bounds on the UBDM properties from the theory of cosmological structure formation in general relativity [11]. Consider a flat, homogeneous, and isotropic spacetime described by the Friedmann–Robertson–Walker metric:

$$g = \text{diag}[-1, a(t)^2, a(t)^2, a(t)^2]. \tag{3.2}$$

The scale factor is $a(t)$, which obeys Friedmann's equation for the Hubble rate $H(t) = \dot{a}/a$:

$$H(t)^2 = \frac{8\pi G_N}{3} \bar{\rho}, \tag{3.3}$$

where $\bar{\rho}$ is the total, spatially averaged, energy density. ρ is composed of photons, "baryons" (by convention in cosmology, we do not separately consider the small mass density of electrons), neutrinos, DM, and the cosmological constant or dark energy. Objects "on the Hubble flow," i.e., feeling negligible local gravitational potentials, appear to recede from an observer at the origin with a velocity $v_H =$

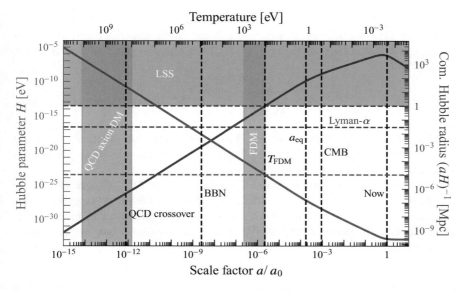

Fig. 3.1 The evolution of cosmic quantities as a function of scale factor or temperature. We show the evolution of the Hubble parameter (red line, left axis) and the comoving Hubble radius (blue line, right axis) together with various relevant cosmological events. The blue shaded area approximately encompasses the large-scale structure (LSS) of the Universe, while grey shaded areas indicate where QCD axion (with $f_a \in [10^6 \text{ GeV}, 10^{18} \text{ GeV}]$) and fuzzy dark matter (FDM) start to become dynamical. Note that the temperature scale on the top is not exactly regular due to the scaling with the number of relativistic degrees of freedom for entropy, $g_{\star,S}$. The quantity $g_{\star,S}$ gives the number of effective relativistic degrees of freedom contributing to the entropy density; $g_{\star,S}$ takes the same form as Eq. (3.1) with the fourth powers replaced by cubes (see, e.g., Ref. [10], Chap. 3)

$Hr\hat{r}$, where r and $\hat{r}$ are the distance and direction from the observer to the object, respectively. We begin with a Newtonian approximation to cosmology (see, e.g., Ref. [12]). Consider an observer at the origin and a single particle of UBDM on the Hubble flow. The UBDM de Broglie wavelength is $\lambda_H = 1/(mv) = 1/(mHr)$, which gives the radial position uncertainty, Δr. A net gravitational force in the positive direction along the line of centres between the observer and the UBDM requires $\Delta r \lesssim r \Rightarrow r \gtrsim (mH)^{-1/2}$, which defines a critical separation $r_{\text{crit}} = (mH)^{-1/2}$. On average, UBDM separations larger than r_{crit} undergo gravitational clustering, and those smaller than it do not.

The cosmological horizon size is approximated by the Hubble length scale $R_H = H^{-1}$. In order for UBDM to have any inhomogeneous gravitational effect within this radius requires $r_{\text{crit}} < R_H$. We show the cosmological evolution of $H = R_H^{-1}$, and the related comoving Hubble radius $(aH)^{-1}$, as functions of temperature and scale factor in Fig. 3.1. The bounds and other cosmological events mentioned in what follows can often be read off directly from that figure, and we will occasionally highlight this fact going forward.

Evaluating the Hubble length scale today, and using that $H_0 = 100\,h\,\mathrm{km\,s}^{-1}$ $\mathrm{Mpc}^{-1} = 2.13 \times 10^{-33}\,\mathrm{eV} \times h$ (where h is the dimensionless Hubble parameter with approximate observed value $h \sim 0.7$), we arrive at our first bound on the UBDM mass:

$$m > 1.5 \times 10^{-33}\,\mathrm{eV} \times \left(\frac{h}{0.7}\right) \quad \text{(size of the observable universe).} \quad (3.4)$$

UBDM violating this bound does not cluster within our cosmological horizon, is thus indistinguishable from the cosmological constant, and will not concern us in this book.[2]

Assuming that UBDM constitutes the entirety of the DM, we can extend the bound to any redshift of interest where we know that DM exerted a discernible gravitational effect by simply substituting the Hubble parameter at that redshift. For temperatures below about 1 MeV, we can use the expression for the Hubble parameter [11]:

$$H(z) = H_0 E(z) = H_0 \sqrt{\Omega_\Lambda + \Omega_m \left[(1+z)^3 + \frac{(1+z)^4}{1+z_{\mathrm{eq}}}\right]}, \quad (3.5)$$

where the second equality defines the energy function $E(z)$. The quantities Ω_m and Ω_Λ are the density parameters of matter and the cosmological constant, defined as the density divided by the critical density, i.e., $\Omega_i = \bar{\rho}_i/\rho_{\mathrm{crit}}$ and $\rho_{\mathrm{crit}} = 3M_{pl}^2 H_0^2$. The last term in the brackets arises from the radiation energy density, which is defined relative to the matter density via the redshift of matter–radiation equality, z_{eq}. The epoch of matter-radiation equality can be found via the relative redshifting of matter and radiation components: $\rho_m(1+z_{\mathrm{eq}})^3 = \rho_r(1+z_{\mathrm{eq}})^4$, with the density parameters defined today. CMB observations fix $z_{\mathrm{eq}} \approx 3390$, and it is thus slightly earlier in cosmic history than decoupling, $z_{\mathrm{dec}} \approx 1100$.

The *baryon acoustic oscillations* (BAO, see Sect. 1.1) observed in the CMB and galaxy surveys like the *Sloan Digital Sky Survey* [15] require that DM was gravitationally relevant at and before matter–radiation equality: if it were not, because baryons are coupled to the photons at early times and perturbations in them cannot grow in the radiation era, the amplitude of galactic fluctuations on scales of order 1 Mpc would not be consistent with the amplitude and scale dependence of the CMB anisotropies. Again assuming that UBDM is all the DM and substituting

---

[2] Very light scalar fields that are homogeneous on the scale of the cosmological horizon provide models for dark energy. The simplest such models are described by a canonical kinetic term in the Lagrangian, and a scalar potential $V(\phi)$, and are known as "quintessence" [13]. An ultralight bosonic field with a mass less than the bound from Eq. (3.4) is one such very simple model, with $V(\phi) = m^2\phi^2/2$. More complex models invoke different potentials, or more fields, or even generalize the kinetic term, at which point they cross over into theories of "modified gravity" and "beyond Horndeski" scalar–tensor theory [14].

$H(z_{eq})$, we arrive at the tighter bound (cf. Fig. 3.1):

$$m > 1.6 \times 10^{-28} \text{ eV} \times \left(\frac{h}{0.676}\right) \left(\frac{\Omega_m}{0.311}\right)^{1/2} \left(\frac{1 + z_{eq}}{3390}\right)^{3/2}$$

$$\text{(matter–radiation equality)}, \quad (3.6)$$

where we have neglected the small contribution of Ω_Λ at equality and taken reference parameters from the CMB+BAO combination in Ref. [3].[3]

The matter–radiation equality bound, Eq. (3.6), is the UBDM equivalent of saying that DM is not "hot" [16]: gravitational clustering is required before matter–radiation equality in order for bottom-up hierarchical structure formation (rather than top-down fragmentation) of galaxies, consistent with observations of extremely high redshift galaxies. We could progress further with such estimates (and we will in due course), but now we must make our model more precise.

Tutorial: The Growth of Cosmic Structure

The challenge in cosmological perturbation theory [17] is to compute the *transfer function*, $T_X(t, k)$ for the mode evolution of each cosmological species X (baryons, photons, neutrinos, dark matter) with Fourier wavenumber k, which fully specifies linear evolution of cosmological fields from Gaussian initial conditions. That is,

$$\zeta_X(k, t) = \zeta_{X,i}(k) T_X(t, k) \xi_X, \quad (3.7)$$

where $\zeta_{X,i}(k)$ is the initial condition of the field and ξ_X is a Gaussian random field defining the initial correlation functions of the field ζ_X.

The codes CAMB [18] and CLASS [19] are the standards for numerical computation for CDM (and many other things), while AXIONCAMB [20][4] can be used for UBDM that is a real scalar field with the self-interaction potential approximated by $V(\phi) = m^2 \phi^2 / 2$. This tutorial gives a brief overview of the most relevant aspects of cosmological perturbation theory for UBDM constraints.

Cosmological perturbation theory deals with the evolution of fluctuations relative to a homogeneous and isotropic background. Background quantities are labeled with an overbar, since they represent the spatial average, and thus depend only on cosmic time t. The perturbation modes have spatial dependence captured by their wavenumber, and perturbations at the initial time all have relative amplitude much less than one with respect to the background quantities. The fields ζ of interest are the components of the energy momentum tensor, written as $T^0_0 = -(\bar{\rho} + \delta\rho)$, $T^i_j = (\bar{P} + \delta P)\delta^i_j + \Sigma^i_j$, $ik^i T^0_i = (\bar{\rho} + \bar{P})\theta$, which defines the energy density, ρ,

---

[3] Using these reference parameters further assumes that UBDM is sufficiently CDM-like that we can use the standard CMB parameters (which are derived under the assumption of ΛCDM).

[4] Available at https://github.com/dgrin1/axionCAMB.

pressure, P, and heat flux, $\theta = \nabla \cdot \boldsymbol{v}$, and we assume anisotropic stresses Σ^i_j vanish. This gives the fields $\delta_X = \delta\rho_X/\bar{\rho}_X$ and θ_X, while pressure is typically described in terms of a sound speed, $c^2_s = \delta P/\delta\rho$.

Next, perturb the metric from Eq. (3.2), and switch to conformal time, τ, via $dt = ad\tau$. The Newtonian gauge considers only scalar metric perturbations:

$$g = a^2 \text{diag}[-(1+2\Psi),\ 1-2\Phi,\ 1-2\Phi,\ 1-2\Phi]. \tag{3.8}$$

The potential Φ is the usual Newtonian potential, and Ψ is the curvature perturbation: they are equal in the non-relativistic limit. The energy momentum tensor is coupled to the metric degrees of freedom by the Einstein equation:

$$G_{\mu\nu} = 8\pi G_N T_{\mu\nu}, \tag{3.9}$$

where $G_{\mu\nu}$ is the Einstein tensor, and it depends on the metric potentials and their derivatives. This is the dynamical equation determining the evolution of the metric.

The equation of motion for the UBDM field with self-interaction potential $V(\phi)$ is

$$\Box\phi - \partial_\phi V = 0, \tag{3.10}$$

where the d'Alembertian ($\Box$) is

$$\Box = \frac{1}{\sqrt{-g}}\partial_\mu\sqrt{-g}g^{\mu\nu}\partial_\nu, \tag{3.11}$$

where g and $g^{\mu\nu}$ are the metric determinant and the inverse of the metric, respectively. Setting $V = \frac{1}{2}m^2\phi^2$ for simplicity, this leads to the equations of motion for the UBDM background field, $\bar{\phi}$, and fluctuation mode $\delta\phi_k$:

$$\bar{\phi}'' + 2\mathcal{H}\bar{\phi}' + a^2m^2\bar{\phi} = 0, \tag{3.12}$$

$$\delta\phi_k'' + 2\mathcal{H}\delta\phi_k' + (m^2a^2\delta\phi_k + k^2)\delta\phi_k = (\Psi' + 3\Phi')\bar{\phi}' - 2m^2a^2\Psi\bar{\phi}, \tag{3.13}$$

where primes denote derivatives with respect to conformal time, and $\mathcal{H} = a'/a = aH$. For the UBDM field, we find $T^{\mu\nu} = \delta S/(\delta g_{\mu\nu})$ by variation of the action with respect to the metric tensor, giving

$$T^{\mu\nu} = g^{\mu\alpha}\partial_\alpha\phi\partial^\nu\phi - g^{\mu\nu}\left[\frac{1}{2}g^{\alpha\beta}\partial_\alpha\phi\partial_\beta\phi + V(\phi)\right]. \tag{3.14}$$

Working to first order in the metric perturbations and $\delta\phi$, and with potential $V = m^2\phi^2/2$, the components are

$$\bar{\rho} = \frac{1}{2}a^{-2}(\bar{\phi}')^2 + \frac{1}{2}m^2\bar{\phi}^2 \,, \tag{3.15}$$

$$\bar{P} = \frac{1}{2}a^{-2}(\bar{\phi}')^2 - \frac{1}{2}m^2\bar{\phi}^2 \,, \tag{3.16}$$

$$\delta\rho = a^{-2}[\bar{\phi}'\delta\phi'_k - \Psi(\bar{\phi}')^2] + m^2\bar{\phi}\delta\phi_k \,, \tag{3.17}$$

$$\delta P = a^{-2}[\bar{\phi}'\delta\phi'_k - \Psi(\bar{\phi}')^2] - m^2\bar{\phi}\delta\phi_k \,, \tag{3.18}$$

$$(\bar{\rho} + \bar{P})\theta = a^{-2}ik^2\bar{\phi}'\delta\phi_k \,. \tag{3.19}$$

? Problem 3.1 Background Evolution of UBDM

Assuming a single-fluid Universe with constant equation of state w satisfying $\dot{\rho} = -3H(1 + w)\rho$, first solve Friedmann's equation, Eq. (3.3), for $a(t)$ and thus $H(t)$. Then, change variables in Eq. (3.12) to physical time $dt = ad\tau$. Substituting your solution for $H(t)$, derive the solution for $\bar{\phi}(t)$ (you may use exact functions or asymptotic methods). Given that the energy density and pressure of UBDM are $\bar{\rho} = \frac{1}{2}\dot{\phi}^2 + V(\phi)$ and $\bar{P} = \frac{1}{2}\dot{\phi}^2 - V(\phi)$, derive the behaviour of the equation of state for UBDM, $w_{\mathrm{UBDM}} = \bar{P}/\bar{\rho}$. What is the asymptotic value of w_{UBDM} for $m \ll H$ and $\langle w \rangle$ for $m \gg H$ (brackets denote period average)? Repeat this exercise for a $\lambda\phi^4$ potential. Comment on the results for w_{UBDM} in relation to the approximate UBDM mass bounds above.

Solution on page 315.

CDM is defined as a collisionless and uncoupled fluid, $w_c = c_c^2 = 0$. Baryons have a sound speed $c_b^2 \neq 0$ (computed from the evolution of the baryon temperature) and an equation of state $w_b = 0$ (on average the baryons have negligible pressure) and are coupled to photons via Thomson scattering. The photon equation of motion is derived from the Boltzmann equation, which is expanded in Legendre polynomials to capture the dependence on the angle between the momentum coordinate on phase space and the wavevector. The hierarchy of moment equations is labeled by the order (l) of the Legendre polynomial: the zeroth moment gives the equation of motion for the density, the first, for the velocity, the second, for the anisotropic stress, and so on (a recursion relation can be used to approximately close the hierarchy above some $l_{\max}$). Truncating this Boltzmann hierarchy at the velocity moment, the photons resemble a fluid with $w = c_s^2 = 1/3$, collisionally coupled to the baryons. We consider perturbations to the energy density $\delta_X = \delta\rho_X/\bar{\rho}_X$ and heat flux θ_X, defined via $\bar{\rho}_X(1 + w_X)\theta_X = ik^j\delta(T^0{}_j)_X$, where $(T^\mu{}_\nu)_X$ is the X energy momentum tensor.

Let us now consider a number of limits of the full equations of motion, which can be found in Ref. [17]. At early times, photons have enough energy to keep hydrogen and other atoms ionized, giving rise to a large free electron density. Thus, the photons and baryons are tightly coupled by Thomson scattering and can be treated as a single fluid with $\theta_\gamma = \theta_b$. Considering only sub-horizon modes ($k \gg aH$), and using the Poisson equation and the ii pressure component of the Einstein equation, Eq. (3.9), the photon fluid at early times obeys the equation of motion:

$$\delta_\gamma'' + \left(c_{s,\gamma}^2 k^2 - \frac{16\pi}{3} G_N a^2 \rho_\gamma \right) \delta_\gamma = 4\pi G_N a^2 \sum_i (1 + c_i^2) \rho_i \delta_i \,, \qquad (3.20)$$

where the photon sound speed is $c_{s,\gamma} = 1/\sqrt{3}$ (speed of pressure perturbations in a gas of photons in thermodynamic equilibrium). At very early times, all ρ_i in the driving term on the right-hand side can be neglected. Then, this equation has sound wave solutions for $k > (16\pi G_N a^2 \rho_\gamma)^{1/2} = \sqrt{6} aH$. This defines the Jeans scale of the photon–baryon fluid, which is of order the comoving horizon size. Perturbations with wavelength shorter than the Jeans scale undergo coherent, pressure supported oscillations. Perturbations with wavelength longer than the Jeans scale grow due to gravitational instability. The sound waves prevent the formation of gravitationally bound structures in the photon–baryon fluid and lead to BAO. At *recombination* temperatures of around 0.2 eV (redshift $z \approx 1100$) [10], the energy of the ambient photon fluid is no longer sufficient to keep neutral hydrogen from forming. At this time, the free electron density drops to zero, the photon–baryon fluid decouples, and the sound wave stalls. This *sound horizon* for the BAO is given by

$$r_s = \int_0^{t_0} \frac{\mathrm{d}t}{a} c_{s,b} \approx \frac{1}{\sqrt{3}} \int_0^{t_{\text{dec}}} \frac{\mathrm{d}t}{a} \,, \qquad (3.21)$$

where $c_{s,b}$ is the baryon sound speed in the plasma, t_0 is the time today, and t_{rec} is the time at recombination when $c_{s,b}$ drops rapidly from $c_{s,\gamma}$ to zero. The BAO scale leads to oscillations in the CMB angular power spectrum, which we have seen already in Chap. 1. The gauge invariant temperature anisotropy of the CMB is given by[5]

$$\frac{\delta T}{T} = \int_0^{\tau_0} \left[\dot{\mu} \left(\Phi + \frac{\delta_\gamma}{4} + \hat{n} \cdot v_b + 2\dot{\Phi} \right) \right] e^{-\mu} \mathrm{d}\tau \,, \qquad (3.22)$$

where μ is the Thomson scattering opacity, v_b is the baryon velocity, $\hat{n}$ is a unit vector giving the sky position, and the integral is along the line of sight. The four terms in Eq. (3.22) correspond, respectively, to the gravitational redshift, the photon anisotropy, and the Doppler effect, and the final term gives rise to the integrated Sachs–Wolfe effect, which is an additional form of gravitational redshift.

---

[5] This equation ignores the effect of gravitational lensing along the line of sight. This second-order effect is important at high multipoles and is sensitive to the UBDM sound speed and structure growth; see Refs. [21, 22].

Decoupling occurs at a redshift $z_{dec} \approx 1100$, which gives the angular scale of the first CMB acoustic peak. The driving term on the right-hand side of Eq. (3.20) is dominant for $z < z_{eq} \approx 3400$, corresponding to angular scales slightly smaller than the first peak, including the second and the third peak. Thus, the relative heights of these peaks can be used to measure the matter content and its behaviour near matter–radiation equality.

How do UBDM perturbations evolve? The first transition in behaviour is in the equation of state, which becomes zero (i.e., pressureless) shortly after $H(a_{osc}) = m$ (this defines the value of the scale factor a_{osc} when the background UBDM field, $\bar{\phi}$, begins to undergo coherent oscillations, see Problem 3.1). Prior to this time, the UBDM is relativistic and perturbations cannot grow.[6] For $H \ll m$, the UBDM perturbations, Eq. (3.13), can be approximated as a fluid with sound speed [24]:[7]

$$c_{UBDM}^2 = \frac{k^2/4m^2a^2}{1 + k^2/4m^2a^2} \, . \qquad (3.23)$$

The non-relativistic limit of this expression is derived later on in this chapter from the Schrödinger–Poisson equation, see Sect. 3.2.2 and Problem 3.2.

Now compare the behaviour of UBDM and CDM+baryons for sub-horizon modes in the matter-dominated era. The baryon sound speed can be neglected after decoupling, so CDM and baryons can be combined into a single pressureless fluid. In the sub-horizon $k \gg aH$, super-Compton $k \ll m$ limit, the CDM+baryon and UBDM fluids obey the coupled equations of motion:

$$\ddot{\delta}_{c+b} + 2H\dot{\delta}_{c+b} = -\frac{k^2}{a^2}\Phi \, , \qquad (3.24)$$

$$\ddot{\delta}_{UBDM} + 2H\dot{\delta}_{UBDM} + \frac{k^4}{4m^2a^4}\delta_{UBDM} = -\frac{k^2}{a^2}\Phi \, , \qquad (3.25)$$

$$\frac{k^2}{a^2}\Phi = 4\pi G_N (\bar{\rho}_{c+b}\delta_{c+b} + \bar{\rho}_{UBDM}\delta_{UBDM}) \, . \qquad (3.26)$$

Setting $\bar{\rho}_{UBDM} = 0$ and substituting the Poisson equation, Eq. (3.26), into Eq. (3.24) give the solution $\delta_{c+b} = A_+(k)a + A_-(k)a^{-3/2}$. The growing mode initial conditions set $A_-(k) = 0$, and the inflationary initial conditions and matter transfer function fix $A_+(k)$. Due to the zero pressure and sound speed of CDM, all the k-dependence in the solution is fixed by the initial conditions, and the dynamics are scale invariant.

---

[6] For an axion-like potential, the equation of state is $w = -1$ prior to a_{osc}. For a scalar field with potential $V = m^2\phi^2/2 + \lambda\phi^4$, the equation of state is $w = 1/3$ at early times for large ϕ initial conditions. For a complex scalar, the early time equation of state is $w = 1$ due to the conserved charge and Goldstone mode [23]. In each case, perturbations are suppressed relative to pressureless CDM.

[7] This expression is exact in the UBDM comoving gauge. Additional terms due to the gauge transformation to a standard gauge, e.g., Newtonian or synchronous, decay on sub-horizon scales as all gauge artifacts do in cosmological perturbation theory [20].

Now, consider a UBDM-dominated Universe by taking $\bar{\rho}_{c+b} = \bar{\rho}_b \ll \bar{\rho}_{UBDM}$ (i.e., no CDM and treating the baryons as sub-dominant) in Eq. (3.25) and again substituting the Poisson equation. The substitution of the Poisson equation gives rise to a negative contribution on the left-hand side proportional to δ_{UBDM}, which drives growth of δ_{UBDM}, while the positive contribution from the sound speed term leads to acoustic oscillations. The sign of the term proportional to δ_{UBDM} depends on k and as such different modes evolve differently. That is, we find Eq. (3.25) exhibits a *Jeans scale*, k_J, separating growing/decaying and oscillating modes. The exact solution for pure UBDM is $\delta_{UBDM} = A_+(k)D_+(k, a) + A_-(k)D_-(k, a)$, where the growth functions are

$$D_+(k, a) = \frac{3\sqrt{a}}{\tilde{k}^2} \sin\left(\frac{\tilde{k}^2}{\sqrt{a}}\right) + \left[\frac{3a}{\tilde{k}^4} - 1\right] \cos\left(\frac{\tilde{k}^2}{\sqrt{a}}\right), \quad (3.27)$$

$$D_-(k, a) = \left[\frac{3a}{\tilde{k}^4} - 1\right] \sin\left(\frac{\tilde{k}^2}{\sqrt{a}}\right) - \frac{3\sqrt{a}}{\tilde{k}^2} \cos\left(\frac{\tilde{k}^2}{\sqrt{a}}\right). \quad (3.28)$$

$$\tilde{k} = k/\sqrt{mH_0}. \quad (3.29)$$

Consider the evolution of three wavenumbers in the pure UBDM case: the horizon size, $k_\star = aH$, the Jeans scale, $k_J = a\sqrt{Hm}$, and the Compton scale, $k_c = ma$. The Compton scale defines relativistic modes where $c_{UBDM}^2 = 1$; k_c increases with time, and more modes become non-relativistic. If $k_\star < k_c$, then a mode is non-relativistic when it enters the horizon and behaves as CDM ("long modes"). If a mode is relativistic when it enters the horizon ("short modes"), then the sound speed cannot be neglected, and modes will not grow until the later time when the Jeans wavenumber enters the horizon. The evolution of these three modes is illustrated in Fig. 3.2. All modes intersect at the time a_{osc}, which defines the special mode k_m, the horizon size when the UBDM background becomes non-relativistic. All $k < k_m$ evolve similarly to CDM. All $k > k_m$ have suppressed growth.

The scale that determines suppression of growth compared to CDM is the Jeans scale at matter–radiation equality. Using Eq. (3.25) in the pure UBDM limit with $c_{UBDM} \approx k^2/4m^2a^2$, substituting the Poisson equation, and solving for k_J where the effective mass term in the oscillator equation for the overdensity vanishes, we find

$$k_{J,eq} = 9.0 \left(\frac{3390}{1 + z_{eq}}\right)^{1/4} \left(\frac{\Omega_{UBDM}}{0.12}\right)^{1/4} \left(\frac{m}{10^{-22} \text{ eV}}\right)^{1/2} \text{Mpc}^{-1}. \quad (3.30)$$

Recall that by definition CDM has zero sound speed. Thus, CDM possesses no Jeans scale (the growing mode solution above is scale invariant), and we see that UBDM is only equivalent to CDM exactly in the limit $m \to \infty$. In practice, they are equivalent as long as k_J does not play a role in any observation.

An observable related to the matter clustering is the *matter power spectrum* defined by $\langle \delta_m(\mathbf{k}_1)\delta_m(\mathbf{k}_2)\rangle = (2\pi)^3 \delta_D(\mathbf{k}_1 - \mathbf{k}_2)P(k)$, where δ_m is the total matter (baryon+CDM+UBDM+neutrino) overdensity, and δ_D is the Dirac delta

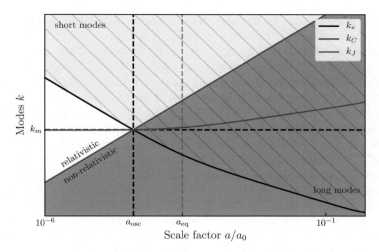

Fig. 3.2 Evolution of scales for linear perturbations with $m = 10^{-26}$ eV. The Jeans scale, Compton scale, and horizon scale all intersect at a_{osc} when the field begins to oscillate. This defines the scale of power suppression as the comoving horizon size at this time, $k_m = a_{\text{osc}} H_{\text{osc}} = R_H(a_{\text{osc.}})^{-1}$. Due to the slow evolution of k_J with a and the logarithmic growth of density perturbations during the radiation epoch, the suppression scale is also approximated by the Jeans scale at matter–radiation equality. Adapted from Ref. [25]

distribution. The presence of the sound speed and consequent Jeans scale for UBDM leads to a suppression of $P(k)$ relative to CDM at large wavenumbers. A fit for the relative suppression in $P(k)$ for UBDM with $V(\phi) = m^2\phi^2/2$ versus CDM is [26]

$$P_{\text{UBDM}}(k) = T_{\text{UBDM}}(k)^2 P_{\text{CDM}}, \tag{3.31}$$

$$T_{\text{UBDM}}(k) = \frac{\cos x_J^3(k)}{1 + x_J^8(k)}, \tag{3.32}$$

$$x_J(k) = 1.61 \left(\frac{m_a}{10^{-22} \text{ eV}}\right)^{1/18} \frac{k}{k_{J,\text{eq}}}. \tag{3.33}$$

For the mixed CDM-UBDM system, the behaviour of $P(k)$ can also be derived [27, 28]: perturbations with $k > k_m$ experience a finite amplitude suppression which increases with the ratio $\Omega_{\text{UBDM}}/\Omega_m$.

End of Tutorial

As we have just seen in the above tutorial, two effects distinguish UBDM from other ingredients in the ΛCDM model: (1) the background expansion rate, $H(z)$, driven by the transition in the equation of state w_{UBDM} at the epoch a_{osc}, and (2) the growth of perturbations, driven by the gradient energy in the Klein–Gordon equation and manifested as an effective sound speed, c_{UBDM}^2.

Depending on the value of a_{osc}, the change in $H(z)$ affects different CMB multipoles. This can be understood by considering Eqs. (3.20) and (3.21) in the tutorial. First, consider UBDM *violating* the bound from Eq. (3.6). We know such UBDM must be a sub-dominant component of the DM. How does the CMB tell us this? Such UBDM changes the expansion rate *after* matter–radiation equality. This changes the distance to the surface of last scattering and the angular size of the BAO in the CMB: it moves the first acoustic peak from its observed position $\ell \approx 200$. This can be compensated by a change in the value of the Hubble constant, H_0. After such a compensation, there is a residual *integrated Sachs–Wolfe effect*, which differs from ΛCDM. If $w \neq 0$ in the post-recombination Universe, then the gravitational potential $\dot{\Phi} \neq 0$ into Eq. (3.22). Due to the fact that the equation of state $w_{\mathrm{UBDM}} \neq 0, -1$ (the two available equations of state in ΛCDM), the evolution of Φ is different in the presence of a small contribution of UBDM, and the shape of the $\ell < 200$ CMB multipoles is very sensitive to the value of Ω_{UBDM}.[8]

Now, consider UBDM satisfying the bound given by Eq. (3.6). The change in the expansion rate compared to ΛCDM now occurs during the radiation dominated epoch. The horizon size at the time a_{osc} was smaller than one degree on the sky, corresponding to multipoles $\ell > 200$, i.e., the higher acoustic peaks. UBDM changes the distance scales for sound waves in the photon–baryon plasma and alters the radiation driving term by changing the relative densities of matter (including UBDM) and radiation. These effects change the relative heights of the CMB acoustic peaks. An additional effect occurs in the *diffusion damping* (Silk damping) at larger multipoles, since the diffusion scale depends on the expansion rate during the radiation era.

Due to the abovementioned effects, the CMB is sensitive to the relative contribution of $\Omega_{\mathrm{UBDM}}(a_{\mathrm{osc}})$. However, any fluid component with $w < 1/3$ becomes increasingly sub-dominant to the radiation at early times (as is the case for axion-like UBDM) and so Ω_{UBDM} decreases moving deeper into the radiation era.[9] Because of this decrease in $\rho_{\mathrm{UBDM}}/\rho_{\gamma}$, the CMB is unable to distinguish between axion-like UBDM and CDM for $a_{\mathrm{osc}} \lesssim 10^{-5}$ [29]. Plugging $z = 10^5$ in Eq. (3.5) and requiring $m > H(z_{\mathrm{osc}})$ give the bound (see Fig. 3.1)

$$m > 2.6 \times 10^{-25}\,\mathrm{eV} \quad \text{(primary CMB anisotropies)}, \qquad (3.34)$$

using the same reference parameters as Eq. (3.6). UBDM effects on the CMB are illustrated in Fig. 3.3. A detailed study of these effects on the *Planck* CMB anisotropies constrains axion-like UBDM violating Eq. (3.34) [but satisfying Eq. (3.4)] to be at most a few percent of the total DM density [20, 22, 29]. We

---

[8] This is one of the ways the CMB is used to constrain the equation of state of dark energy.

[9] A complex scalar with $w = 1, 1/3$ prior to a_{osc} *increases* its energy density relative to radiation at early times. The effect in the expansion rate is similar to adding additional neutrino species, which are also strongly constrained by the CMB [23].

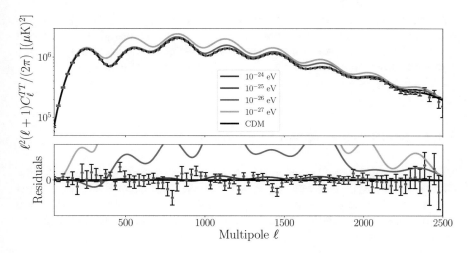

Fig. 3.3 UBDM effects on the CMB temperature power spectrum. UBDM changes the expansion rate compared to CDM in the early radiation dominated epoch, $z \gtrsim 3000$, which affects the damping of the BAO, visible through the heights of the power spectrum peaks at large multipoles. By eye, it is clear that the *Planck* data strongly exclude UBDM with $m \leq 10^{-26}$ eV. Combining the temperature data with polarization, lensing, and cross-correlations [22] tightens the bound to be roughly consistent with our estimate, Eq. (3.34). On the other hand, UBDM with $m \geq 10^{-24}$ eV is indistinguishable from the black best-fit CDM curve. Note that this plot rescales the y-axis in the upper panel by one power of ℓ compared to the usual convention, to enhance the visibility of high-ℓ features, and that the x-axis begins at $\ell = 50$, since the large scales are not sensitive to this particular physics

have spent a considerable time deriving what will turn out to be a rather weak lower bound on m. However, this bound is extremely rigorous in practice, in a way that our later bounds are not. The bound expressed in Eq. (3.34) relies only on linear physics and on the extremely well understood statistics of the CMB that give us our most rigorous evidence for the existence of DM in the first place.

UBDM Hints: Precision Cosmology and ALPs from the GUT Scale

The realignment production mechanism of ALPs gives the relic density Ω_a as a function of mass and initial field value, ϕ_i. Taking ϕ_i to be near the GUT scale, $\phi_i \in [10^{15}, 10^{17}]$ GeV gives a DM relic density compatible with the observed value $\Omega_d h^2 \leq 0.12$ for all masses $m \lesssim 10^{-18}$ eV. At lower masses, a sub-dominant population is predicted, with the fraction of ALP DM saturating at around 0.1%. Upcoming cosmological surveys, including lensing tomography and intensity mapping, will greatly increase the sensitivity to sub-dominant components of the DM. The CMB is a 2D probe, and the number of modes measured with a cosmic variance precision is ℓ_{max}^2. An intensity mapping survey is 3D, measuring in the line-of-sight redshift direction, and thus has many more modes. The combination of a next generation CMB survey like the Simons Observatory or CMB-S4 with an intensity mapping survey by the *Square Kilometre Array* [30] or HIRAX [31] could make significant inroads into the GUT scale predictions [25], as will next generation Lyman-α forest surveys (see below) and "pulsar timing arrays" [32, 33]. These forecasted opportunities are shown as open regions in Fig. 3.5.

3.2.2 Schrödinger–Poisson Equations

The UBDM condensate[10] coupled to general relativity obeys the Einstein–Klein–Gordon equations, derived from variation of the relevant fundamental action. In the non-relativistic limit (the Newtonian approximation), for all forms of UBDM (be they ALPs, real, or complex scalars), these equations reduce to the Schrödinger–Poisson equations (SPEs):

$$i\dot\psi + \frac{\nabla^2}{2m}\psi - m\Phi\psi + \frac{\lambda_{\mathrm{GP}}}{m}|\psi|^2\psi = 0\,, \tag{3.35}$$

$$\nabla^2\Phi = 4\pi G_N m^2 \left(|\psi|^2 - \int \mathrm{d}^3 x|\psi|^2\right), \tag{3.36}$$

where we are using the convention that the Newtonian potential is dimensionless, and the field ψ has canonical mass dimension one such that the average number density is

$$\bar{n} = m \int \mathrm{d}^3 x|\psi|^2\,. \tag{3.37}$$

The subtraction of the background density in the Poisson equation follows from the background-perturbation split of the Einstein equations on the Friedmann background.

Equations (3.35) and (3.36) are a nonlinear Schrödinger equation for the UBDM condensate, with Gross–Pitaevski self-coupling λ_{GP}, which can be computed from the relativistic self-interaction potential, V. The SPEs fully describe the nonlinear, non-relativistic, structure formation in most astrophysical environments at low redshifts ($a \gg a_{\mathrm{osc}}$, $L \gg 1/m$, $v \ll 1$, $\Phi \ll 1$), i.e., the gravitational structure of UBDM at the coherence scale. One should avoid letting the name "Schrödinger" cause confusion; these equations have nothing quantum about them: ψ is not a probability density, and there is no measurement problem or wavefunction collapse. The SPEs are simply the non-relativistic limit of the classical field equations, valid whenever the particle number is large: they are the UBDM equivalent of Maxwell's equations.

---

[10] In the sense that all classical fields can be thought of as condensates.

? Problem 3.2 Derivation of the Schrödinger–Poisson Equations for UBDM

Take the metric from Eq. (3.8) in the non-relativistic limit ($\Phi = \Psi$) on a non-expanding background ($a = 1$). Evaluate the d'Alembertian, Eq. (3.11), to first order in Ψ. Substitute the ansatz:

$$\phi = \frac{1}{m\sqrt{2}} \left(\psi e^{imt} + \psi^* e^{-imt} \right), \qquad (3.38)$$

into the Klein–Gordon equation with potential $V(\phi) = m^2\phi^2/2 + \lambda\phi^4$. In the Wentzel–Kramers–Brillouin (WKB) limit, $\dot{\psi}/(m\psi) \ll 1$, and making the non-relativistic approximations $k/m \ll 1$ and $\dot{\Psi}/m \ll 1$, show that the complex field amplitude ψ obeys the Schrödinger equation, Eq. (3.35). Now, take the general form of the stress energy tensor, Eq. (3.14), and show that in the same limits $\rho = |\psi|^2$ at leading order and hence that the Poisson equation, Eq. (3.36), is obeyed for the overdensity $\delta\rho$.

Solution on page 318.

An instructive change of variables on the SPEs makes use of the *Madelung transformation*, $\psi = \sqrt{\rho}e^{i\theta}/m$ to write the wave function as a fluid with density ρ and velocity $\boldsymbol{v} = \nabla\theta$. Substitution into the SPEs yields the continuity and Euler equations:

$$\dot{\delta}_{\text{UBDM}} + a^{-1}\boldsymbol{v}_{\text{UBDM}} \cdot \nabla\delta_{\text{UBDM}} = -a^{-1}(1 + \delta_{\text{UBDM}})\nabla \cdot \boldsymbol{v}_{\text{UBDM}}, \quad (3.39)$$

$$\dot{\boldsymbol{v}}_{\text{UBDM}} + a^{-1}\left(\boldsymbol{v}_{\text{UBDM}} \cdot \nabla\right)\boldsymbol{v}_{\text{UBDM}} = -a^{-1}\nabla(\Phi + Q) - H\boldsymbol{v}_{\text{UBDM}}, \quad (3.40)$$

$$\text{where } Q \equiv -\frac{1}{2m^2 a^2}\frac{\nabla^2\sqrt{1 + \delta_{\text{UBDM}}}}{\sqrt{1 + \delta_{\text{UBDM}}}}. \quad (3.41)$$

The continuity and Euler equations differ from those of CDM by the presence of the so-called "quantum pressure" Q—a misleading term, as it is neither quantum nor a pressure. Expanding these equations to first order in δ_{UBDM} and going to Fourier space, one can verify that they are equivalent to the fluid equation, Eq. (3.25), for pure UBDM: in the non-relativistic and linearized limit, the quantum pressure and sound speed are equivalent.

For UBDM, the SPEs replace the normal Newtonian dynamics of particle DM. Solving gravitational collapse and dynamics in generality requires methods of solution of nonlinear partial differential equations. The challenge in this system is the non-local interaction from the Newtonian potential, the wide range of scales in gravitational collapse, and the need to accurately resolve the phase of the field ψ in low-density and large cosmic voids. Common numerical methods include lattice field theory (discretizing derivatives in real space), spectral methods (numerical Fourier analysis), or finite elements (alternative real space discretizations). A public code is PYULTRALIGHT [34]. Particle-based hydrodynamics using Eqs. (3.39)–(3.41) is also useful on some scales, but it fails to resolve interference fringes (as can be seen from the coordinate singularity in Q when $\rho \to 0$) and vortex lines, which appear generically in complex fields (the fluid has $\nabla \times v = 0$). On scales larger than the UBDM de Broglie wavelength, standard Newtonian particle mechanics is accurate, e.g., the public code GADGET [35]. The convergence of the SPEs to the ordinary collisionless limit of CDM on super-de Broglie scales can be shown rigorously via the *Schrödinger–Vlasov* correspondence [36–38] and is well known in the field of quantum hydrodynamics [39].

A kinetic description of the SPEs begins by writing the field ψ using the Wigner distribution (see, e.g., Ref. [40]), which describes the occupation probability of modes k. This distribution function obeys a collisional Boltzmann equation, with scattering timescale [41]:

$$\tau_{\text{gr}} \approx \frac{\sqrt{2}}{12\pi^3} \frac{mv^6}{G_N^2 \bar{n}^2 \log \Lambda} , \qquad (3.42)$$

where v is the typical speed in the system (i.e., the virial velocity) and $\log \Lambda = \log(r_{\text{max}}/r_{\text{min}})$ is the Coulomb scattering logarithm for r_{min} and r_{max} the minimum and maximum length scales in the problem, respectively. This gravitational scattering timescale governs the time over which wave-like effects cause UBDM to depart dynamically from CDM.

In addition to the scattering timescale, solution of the SPEs leads to UBDM having distinctive effects on scales of order the de Broglie wavelength. There are three important consequences:

1. Transient "quasi-particle" fluctuations
2. Formation of long-lived self-bound objects
3. Interference fringes

We discuss the first in Sect. 3.2.3 and the second in Sect. 3.3.1. Interference fringes are observed prominently in numerical simulations of galactic filaments composed of UBDM with $m \approx 10^{-22}$ eV [42, 43], though the observational consequences are at present unclear.

3.2.3   Galaxies and Nonlinear Structure

The scale of suppression described by Eq. (3.30) can be converted into a DM halo mass by considering the average DM density in a sphere with radius of one-half wavelength, $R_J = \pi/k_{J,\text{eq}}$:

$$\frac{M_0}{M_\odot} = 5.9 \times 10^9 \left(\frac{\Omega_{\mathrm{UBDM}} h^2}{0.12}\right)^{1/4} \left(\frac{h}{0.676}\right) \left(\frac{1 + z_{\mathrm{eq}}}{3390}\right)^{3/4} \left(\frac{m}{10^{-22}\,\mathrm{eV}}\right)^{-3/2}.$$

$$(3.43)$$

Halos that are significantly more massive than M_0 will have the same abundance as in a CDM universe, while halos much lighter than M_0 are largely absent. Our estimate for M_0 from inspection of the linear equations of motion is within a factor of two of the suppression scales found in N-body simulations of nonlinear cosmological structure formation: Ref. [44] finds $M_0 = 1.9 \times 10^{10}\, M_\odot\, (m/10^{-22}\,\mathrm{eV})^{-4/3}$, where the different scaling with m results from using the half-mode of the transfer function Eq. (3.32), $T(k_{1/2}) = 1/2$, instead of the Jeans scale. The half-mode is always at $k < k_J$ since $T(k)$ decreases below $k_{1/2}$, and $T(k_J) = 0$. It is possible for structures to form at the half-mode, though they will have suppressed number with respect to CDM. The Jeans scale represents the absolute limit below which no structures form and corresponds to lower mass halos. Thus, using the Jeans scale gives more conservative limits on m.

How can we constrain UBDM using our estimate for M_0? In hierarchical structure formation, low-mass halos form first, i.e., at high redshift. Halos with low masses can be identified at high redshift from the light emitted by the galaxies that they host, which is in the form of UV flux from stars, which in turn ionizes hot gas. An approximate relationship between UV flux and halo mass can be derived by so-called abundance matching. One assumes that there is a one-to-one mapping between UV magnitude, M_{AB} (the "AB system" for defining magnitude), and halo mass. This can be found assuming that the number of UV sources at some redshift z, $n_{\mathrm{UV}}(z)$, statistically matches the number of DM halos, $n_h(z)$. The matching depends on the observations used to calibrate it and monotonicity of each function. Current observations (e.g., Ref. [46]) are largely consistent with monotonicity (however, see Ref. [48]), which is consistent with all sources being in halos with mass above M_0. In this case, all the halos we observe are formed on scales far from the Jeans scale, and so the relationship between UV magnitude, M_{AB}, and halo mass, $M_h(M_{AB})$, is as in Fig. 3.4 (computed from simulations of CDM with no Jeans scale). The limiting magnitude of the *Hubble Ultra Deep Field* UV luminosity function at $z = 8$ is $M_{\mathrm{AB,lim}} \approx -18$, which we read off from Fig. 3.4 as giving a limiting halo mass of $M_h \approx 10^{10} M_\odot$. Demanding $M_0 > M_h(M_{\mathrm{AB,lim}})$, we find the bound:

$$m > 0.7 \times 10^{-22} \left(\frac{\Omega_{\mathrm{UBDM}} h^2}{0.12}\right)^{1/6} \left(\frac{h}{0.676}\right)^{2/3} \left(\frac{1 + z_{\mathrm{eq}}}{3390}\right)^{1/2} \mathrm{eV}$$

$$\text{(high redshift galaxies)}. \qquad (3.44)$$

The estimate given in Eq. (3.44) agrees favourably with complete analyses of similar data [44, 49, 50].

Another important bound to consider is from the Lyman-α forest flux power spectrum. This observable traces the matter power spectrum, $P(k)$, on quasi-linear scales at high redshifts. It can be used to infer the existence of a UBDM Jeans scale.

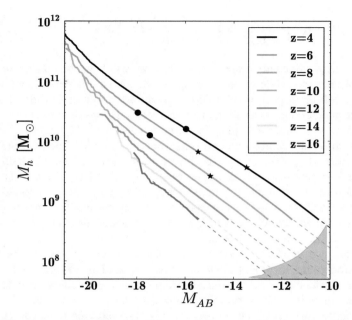

Fig. 3.4 "Abundance matching" between halo mass, M_h, measured in solar masses ($M_\odot$), and UV magnitude, M_{AB}, assuming CDM, evaluated at different redshifts, z. Taken from Ref. [45]. Filled circles show the limiting magnitudes for the *Hubble Ultra Deep Field* observation [46], while stars are for the future *James Webb Space Telescope* [47]. The dotted lines represent power law extrapolation from the simulations, while the shaded region denotes the cooling limit below which galaxies cannot form efficiently

Current observations see no evidence for such a Jeans scale and thus show that the UBDM de Broglie wavelength must be correspondingly small.

The light from distant quasars is absorbed by neutral hydrogen (HI) along the line of sight. The differing optical thickness of dense clouds of HI leads to a "forest" of absorption features: the optical depth for the absorption traces the HI density and (since HI clouds lie in gravitational potential wells) the total matter density including DM. A survey of cosmological quasars can then be used to estimate the matter power spectrum by correlation of the absorption feature. For example, *HIRES/MIKE* covers k as large as $k_{max} \approx 50\,h\,\mathrm{Mpc}^{-1}$ [51].[11] The data are well described by CDM with no evidence for a suppression of power, and so we can derive an approximate bound on the UBDM mass. Using Eq. (3.30) with the quoted k_{max} gives the bound (cf. Fig. 3.1):

---

[11] We convert from Lyman-α units for k in s km^{-1} to the more standard Mpc^{-1} by multiplication with $H_0(1+z)$. For reviews and discussions of the Lyman-α forest as a probe of the matter power spectrum, see Refs. [52–55].

$$m > 1.5 \times 10^{-21} \, \text{eV} \quad \text{(Lyman-}\alpha \text{ forest)}, \tag{3.45}$$

which again agrees well with the result derived from more careful analysis [56, 57].

Caution is advised with all our estimates on UBDM mass bounds in this section, since they assume that the observations agree perfectly with CDM and thus that on scales observed UBDM can be treated as such. Strong self-interactions of UBDM also change these bounds and any other bound based on the suppression of structure formation relative to CDM. A particular example of this is an ALP with large initial field displacement. The ALP potential is $V(\phi) = f_a^2 m^2 [1 - \cos(\phi/f_a)]$. An initial displacement $\theta = \phi/f_a = \pi - \delta\theta$ with small $\delta\theta$ leads to large self-interactions at early times, and the field is near an unstable local maximum of the potential. This *tachyonic instability*[12] in the evolution of δ leads to an *increase* in the UBDM power spectrum relative to CDM on scales close to the Jeans scale [58]. A displacement $\delta\theta \approx 0.02$ is sufficient to evade the bound described by Eq. (3.45) and allow $m \approx 10^{-22}$ eV to fit the Lyman-α power spectrum as well as CDM, while a value $\delta\theta \approx 0.003$ leads to a *better* fit than CDM [59]. The tuned values of $\delta\theta$ require smaller f_a to get the correct relic abundance than in the harmonic approximation, which could make direct detection of this type of tuned UBDM easier by increasing the matter couplings.

UBDM displays dynamics distinct from CDM on scales of order the de Broglie wavelength. A complete description of the effects of sub-de Broglie physics requires numerical simulation. However, analytical understanding is possible in varying degrees of complexity, which has largely been developed in recent years (see, e.g., Refs. [54, 60–64]). We will give only the simplest description useful for estimates.

UBDM in a gravitational potential well has a coherence length, $L \sim \lambda_{\text{dB}} = 1/mv$ ($\hbar/mv$ in physical units), and coherence time, $\tau \sim 1/mv^2$, where v is the characteristic velocity. The heuristic picture of a wave distribution with these properties is one of the quasi-particles of size L and lifetime τ. The quasi-particle mass is

$$M_{\text{qp}} \sim \lambda_{\text{dB}}^3 \bar{\rho}, \tag{3.46}$$

where $\bar{\rho}$ is the average local density in a volume encompassing a large number of quasi-particles (i.e., in the solar neighbourhood, $0.4 \, \text{GeV cm}^{-3}$). Two-body relaxation between quasi-particles leads to the relaxation time (see Problem 3.3) [54]:

$$t_{\text{relax}} \sim \frac{10^{10}}{\log \Lambda} \left(\frac{m}{10^{-22} \, \text{eV}} \right)^3 \left(\frac{v}{100 \, \text{km s}^{-1}} \right)^2 \left(\frac{R}{5 \, \text{kpc}} \right)^4 \, \text{yr}, \tag{3.47}$$

where the Coulomb logarithm in the quasi-particle picture is $\log \Lambda = \log(R/\lambda_{\text{dB}})$. On timescales longer than t_{relax}, UBDM departs from the SHM (in the sense that the

---

[12] A potential is said to have a tachyonic region if $V''(\phi) < 0$, i.e., a local maximum, and negative effective mass squared.

density distribution is not time-independent) due to heating and cooling. Note the similarity of the relaxation time, Eq. (3.47), to the gravitational scattering timescale, Eq. (3.42), in the kinetic picture if we substitute $v^2 = G_N m \bar{n} R^2$.

Heating and cooling on the timescale t_{relax} can be observed if a tracer population of stars with mass m_t is present in the UBDM halo (when the gravitational potential due to DM is dominant, stars are tracer particles). For $m_t \ll M_{\text{qp}}$, heating dominates, while for $m_t \gg M_{\text{qp}}$, cooling dominates. Let us estimate M_{qp} for some systems of interest. In the solar neighbourhood, $\bar{\rho} \approx 0.4$ GeV cm^{-3} = $10^7 M_\odot$ kpc^{-3} and $v \approx 100$ km s^{-1} $\Rightarrow \lambda = 0.2(10^{-22}$ eV/m) kpc, which gives $M_{\text{qp}} \approx 7 \times 10^4 (10^{-22}$ eV/m)$^3 M_\odot$. In the solar neighbourhood, tracers are stars with $m_t \sim 1 M_\odot$, and the transition from heating to cooling occurs for UBDM mass $m \approx 4 \times 10^{-21}$ eV, with lighter masses giving rise to heating. The Milky Way in fact possesses a "thick disk" of old stars [65], and this has been argued to provide evidence that in fact DM is composed of UBDM in this so-called *fuzzy DM* regime [54, 64] (for more information, see the "Fuzzy Dark Matter Hints" box below). On the other hand, if heating is too efficient, then the disk will be destroyed completely. Demanding that the relaxation time is shorter than the age of the Universe, i.e., 10^{10} years, and applying Eq. (3.47), we find

$$m \gtrsim 10^{-22} \text{ eV} \quad \text{(Milky Way disk heating)}, \qquad (3.48)$$

which agrees with more accurate modelling [64].

A very strong bound from UBDM heating can be derived by considering the existence of the old, centrally located star cluster in the ultrafaint dwarf galaxy Eridanus II. Observations [66, 67] indicate that the DM density is $\bar{\rho} = 0.15 M_\odot$ pc^{-3}, and the velocity dispersion is $\sigma_v = 6.9^{+1.2}_{-0.9}$ km s^{-1}. For UBDM, this gives $M_{\text{qp}} = 3(10^{-19}$ eV/m)$^3 M_\odot$, implying that heating dominates for masses $m \lesssim 10^{-19}$ eV. The star cluster has a half-light radius of $r_h = 13$ pc, an estimated age $t \sim 10^{10}$ years, and is close to the centre of Eridanus II. Using Eq. (3.47), replacing R with the half-light radius (since the star cluster is approximately centrally located), substituting for the characteristic velocity v the velocity dispersion σ_v, taking log $\Lambda \sim O(1)$, and demanding that the star cluster is stable on the timescale of its age, we obtain the bound:

$$m \gtrsim 10^{-19} \text{ eV} \quad \text{(Eridanus II)}, \qquad (3.49)$$

which again agrees very favourably with a more rigorous treatment [63].

Based on the present analysis, the bound from Eridanus II does *not*, however, apply for $m \lesssim 10^{-21}$ eV, where the fluctuation timescale becomes longer than the star cluster orbital period, and potential fluctuations become adiabatic. Another time-dependent feature of UBDM halos becomes important at $m \lesssim 10^{-21}$ eV: the central soliton (see Sect. 3.3.1) undergoes a random walk on scales of order its own radius (which is much larger than the star cluster radius in this case) due to collisions with the quasi-particles in the halo. This again leads to star cluster disruption and could exclude $m \approx 10^{-22}$ eV from the Eridanus II star cluster stability. However, the Milky Way tidal potential may lead to sufficient tidal stripping of the quasi-particle atmosphere to quell this random walk and leave $m \approx 10^{-22}$ eV safe from this bound [68].

? Problem 3.3 Relaxation of UBDM

The timescale for gravitational two-body relaxation (diffusion of a body's velocity caused by gravitational interaction in two-body close encounters) of particles with mass m moving with velocity v in a host of mass M with radius R can be written as [65]

$$t_{\text{relax}} = 0.1 \frac{R}{v} \frac{M}{m \log \Lambda}. \tag{3.50}$$

Use this to derive the relaxation timescale, t_{relax}, in Eq. (3.47).

Solution on page 321.

UBDM Hints: Fuzzy Dark Matter

We have seen a large variety of constraints on UBDM with mass $m \lesssim 10^{-22}$ eV from cosmic large-scale structure. We have also seen how heating in Eridanus II excludes the range 10^{-21} eV $\lesssim m \lesssim 10^{-19}$ eV, and we will see shortly that black hole superradiance excludes 10^{-19} eV $\lesssim m \lesssim 10^{-16}$ eV. There is only one strong bound in the range just above 10^{-22} eV coming from the Lyman-α forest flux power spectrum. This bound is sensitive to aspects of astrophysical modelling and, in particular, can be relaxed if the baryon temperature evolves non-monotonically or if significant ionizing photons are produced outside of galactic halos, e.g., in filaments (however, see the recent Ref. [69]). Another possible window is afforded by the Eridanus II bounds around 10^{-21} eV, where the statistical modelling is uncertain, and Eridanus II can survive sandwiched between orbital resonances. If either of these bounds (Ly-α or Eridanus II) can be relaxed, then there are some hints that DM may in fact be UBDM with masses between about 10^{-22} eV and 10^{-21} eV, the so-called *fuzzy dark matter* (FDM) model (cf. Fig. 3.1). These hints include:

- **The Milky Way "thick disk"**: FDM just outside the bound given in Eq. (3.48) can help explain the old thick disk in our galaxy [64].
- **Suppressed high-z galaxy formation**: The redshift of reionization is known to be around $z_{\text{reion}} \approx 8$. This relatively low value is naturally explained by FDM, which suppresses formation of galaxies at $z \gtrsim 8$.
- **Solitons and galactic cores**: Solitons in FDM halos (see Sect. 3.3.1) may help explain cored density profiles in dwarf galaxies without baryonic feedback [42, 70].

- **Relic density**: The relic density is naturally explained by an FDM ALP with f_a close to the GUT scale, as expected in certain string compactifications.

Each hint provides a method to search for FDM. Furthermore, the FDM mass range corresponds to field oscillation frequencies of order one inverse month, making it challenging, but not impossible, to search for via direct detection.

3.2.4 Black Hole Superradiance

In the following, we adopt different units: so-called geometric units where $G_N = c = 1$.

Spinning black holes (BHs) are described by the Kerr metric, which has two parameters: mass, M, and dimensionless spin, $a_J = J/M \in [0, 1]$. In "Boyer–Linquist" coordinates, the line element is[13]

$$ds_{\text{Kerr}}^2 = -\left(1 - \frac{2Mr}{\Sigma}\right) dt^2 - \frac{4Ma_Jr\sin^2\theta}{\Sigma} dt\,d\phi + \frac{\Sigma}{\Delta} dr^2 + \Sigma\,d\theta^2$$

$$+ \frac{(r^2 + a_J^2)^2 - a_J^2\Delta\sin^2\theta}{\Sigma}\sin^2\theta\,d\phi^2\,, \tag{3.51}$$

$$\Sigma \equiv r^2 + a_J^2\cos^2\theta\,, \tag{3.52}$$

$$\Delta \equiv r^2 + a_J^2 - 2Mr\,, \tag{3.53}$$

$$r_\pm \equiv M \pm \sqrt{M^2 - a_J^2}\,. \tag{3.54}$$

$$r_{\text{ergo}} \equiv M + \sqrt{M^2 - a_J^2\cos^2\theta}\,, \tag{3.55}$$

where we use spherical polar coordinates. The zero solutions of Eq. (3.53) define the two horizons $r_\pm$: an inner Cauchy (causal) horizon at r_- and the outer physical event horizon at r_+. The "ergoregion" is defined as radii smaller than r_{ergo}, where $g_{00} = 0$ (the coefficient of dt^2 in the line element). If an object enters the ergoregion between $r_+ < r < r_{\text{ergo}}$ and ejects some mass which falls into the event horizon, then the object will emerge from the ergoregion with a larger energy than it went in with, and the BH will lose a small amount of energy in the form of mass and spin. This is known as the Penrose process.

A wavepacket has a finite extent and can "eject" part of itself into the BH if it passes through the ergoregion and overlaps with the event horizon. If the wave is trapped near the BH, then this process continually extracts energy from

---

[13] An accessible introduction to general relativity can be found in Ref. [71].

the BH, growing the wavepacket amplitude and becoming "superradiant." The process only ends when the ergoregion has shrunk small enough to remove the overlap (ultimately, the process must stop if $a_J = 0$, i.e., a Schwarzschild BH with no ergoregion). Such a situation is in fact realized naturally for a massive bosonic field. Gravitational bound states trap the field near the BH, and the hydrogen-like wavefunctions overlap with the superradiant region between the ergosphere and the event horizon. The field in question must be bosonic in order that the wavepacket energy levels can continue to be filled as energy is extracted. "Black hole superradiance" (BHSR) for bosonic fields is discussed in detail in Refs. [72, 73].

Consider a scalar field near a Kerr BH. Just like in the tutorial on cosmic structure above, the field obeys the Klein–Gordon equation, Eq. (3.10), except that now the d'Alembertian ($\Box$) should be evaluated with the metric from Eq. (3.51). Let us write the field as

$$\phi = \sum_{\ell,\alpha} e^{-i\omega t + i\mu\varphi} S_{\ell\mu}(\theta)\psi_{\ell\mu}(r) + \text{h.c.}, \qquad (3.56)$$

where $S_{\ell\mu}(\theta)$ are the spheroidal harmonics (eigenfunctions of the Laplacian on the surface of a spheroid, respecting the axial symmetry of the Kerr spacetime). To avoid confusion, we have labeled the magnetic quantum number μ and the azimuthal angle φ. The Klein–Gordon equation can then be reduced to a time-independent Schrödinger equation for the radial eigenfunctions $\psi_{\ell\mu}$, with eigenvalue ω. The BH provides a background potential $V(r, \omega)$, which possesses a barrier separating the bound states from the horizon, and a potential well with size of order the boson Compton wavelength, $1/m$. The system resembles a hydrogen atom with effective fine structure constant $\alpha_{\text{eff}} \equiv G_N M$, where we temporarily reinstated G_N.

The existence of superradiant solutions is determined by the imaginary part of the eigenvalue ω, which leads to growth of the occupation number of the mode $\psi_{\ell\mu}$. The superradiant rate is $\Gamma_{\text{SR}} \propto \alpha_{\text{eff}}^{4\ell+4} m$, and numerically it is found to be maximized around $\alpha_{\text{eff}} \sim 1$. This gives an approximate criterion for BHSR:

$$m \sim \frac{8\pi M_{pl}^2}{M} = 1.33 \times 10^{-10} \text{ eV} \left(\frac{1M_\odot}{M}\right). \qquad (3.57)$$

For BHSR to be effective, the superradiant timescale should be longer than any timescale of relevance for the BH, e.g., accretion. If BHSR is effective, then the BH will lose spin. Thus, large observed values of a_J will be disfavoured if a boson exists satisfying Eq. (3.57).

Astrophysical observations indicate the existence of BHs across a wide range of masses, from those formed by collapse of stars at the Chandrasekhar limit $M \approx 1.4M_\odot$, to the supermassive BHs (SMBHs) at the centres of galaxies. The spins of BHs can also be estimated, using X-ray spectroscopy of the accretion disk or by measurement of the gravitational waveform in the inspiral phase of binary systems. Detectable spins are generally large, $a_J \gtrsim 0.5$. Assuming that these large values would be disfavoured by a boson satisfying Eq. (3.57), we can estimate exclusions on UBDM. First, consider the stellar BHs, and assume a full spectrum of observations from the Chandrsekhar mass to the LIGO inspiral masses $M \approx 30 M_\odot$ [74]. This excludes UBDM for

$$4 \times 10^{-12} \, \text{eV} < m < 8 \times 10^{-11} \, \text{eV} \quad \text{(stellar BHs)}. \tag{3.58}$$

Next, consider SMBHs. The lightest currently known SMBH is in NGC4051, with mass $M \approx 1.9 \times 10^6 M_\odot$, while the *Event Horizon Telescope* has imaged the BH at the centre of M87 and determined the mass $M \approx 6.5 \times 10^9 M_\odot$. Again, assuming a continuous spectrum in between, we can exclude the range of UBDM masses:

$$2 \times 10^{-20} \, \text{eV} < m < 7 \times 10^{-17} \, \text{eV} \quad \text{(supermassive BHs)}. \tag{3.59}$$

These estimates agree somewhat favourably with more accurate treatments of BHSR modelling and BH population statistics (e.g., Ref. [75]).

To obtain the more accurate picture, the bosonic field equations on the Kerr background should be solved numerically. The oscillation timescale of the field is $\tau \sim 1/m$. For real scalar fields, the gravitational pressure oscillates with a frequency $2m$, sourcing oscillations of the metric potentials on a timescale faster than the superradiant timescale. This makes brute force numerical solution challenging, but many approximation methods are available.

BHSR also works for massive spin-one and spin-two fields (which are also UBDM candidates). The superradiant timescales can be vastly different, and specific treatments are necessary. Reference [76] considers spin-one vectors which have much smaller instability rates and thus weaker constraints. Reference [77] considers spin-two fields, which possess a particular mode mimicking the spin-zero case and thus have similar constraints. A significant difference occurs for complex fields. Due to the underlying $\mathbb{U}(1)$ symmetry and conserved particle number, the complex vector A_μ field does not source oscillations in the metric potentials with frequency m. This greatly simplifies the numerical task and has allowed direct simulation of superradiance with these so-called Proca fields [78]. The simulations are important because they include nonlinear back-reaction of the superradiant cloud on the Kerr spacetime and demonstrate that BHSR occurs in this more realistic setting.

One known "showstopper" for BHSR is the so-called "Bosenova" caused by attractive quartic self-interactions, which shut off the instability and prevent growth of the scalar cloud. The self-interaction term in the potential is $V_{\text{int}} = \lambda \phi^4 / 4!$, for some coupling constant λ. As the cloud grows, this term can become as large as the other terms in the energy budget. At this time, the scalar cloud collapses and super-radiance is shut off. This introduces a new timescale into the problem and practically gives rise to a maximum λ above which the superradiance rate is sub-dominant to the Bosenova rate and no spin extraction can occur. Numerical simulations [79] determine the maximum cloud occupation number before Bosenova occurs [80]:

$$N_{\text{Bose}} \sim 150 \frac{n^4}{\alpha_{\text{eff}} \lambda} = 5 \times 10^{78} \frac{n^4}{\alpha_{\text{eff}}^3} \left(\frac{M}{10 M_\odot} \right)^2 \left(\frac{f_a}{M_{pl}} \right)^2 , \qquad (3.60)$$

where n is the energy level of the occupied cloud and M_{pl} is the reduced Planck mass. In the last equality, we assumed that the scalar potential is of the ALP form $V(\phi) \propto -\cos(\phi/f_a)$, giving $\lambda = m^2/f_a^2$. Using this formula for stellar mass BHs, Ref. [80] finds that BHSR is shut off for $f_a \lesssim 10^{13}$ GeV; for SMBHs, this turns out to be $f_a \lesssim 10^{16}$ GeV.

Any UBDM interactions can compete with superradiance and possibly shut it off. Examples include interactions between the cloud and the Standard Model particles in the BH environment or the ALP interaction $g_{a\gamma\gamma}$, which leads to stimulated decay of the cloud [81, 82]. Of course, both "showstoppers" (Bosenova and axion–photon interactions) also predict new observables in the form of emission from BH regions for UBDMs of particular masses. Finally, we note that the superradiance phenomenon need not be limited strictly to BHs and can occur also near stars and neutron stars [83]—even though the astrophysical uncertainties are far greater.

? Problem 3.4 Estimating Superradiance Properties of UBDM

A simple way to estimate the relevance of BHSR is to inspect terms in the action,

$$S = \int d^4x \, \sqrt{-g} \left[\frac{M_{pl}^2}{2} R - \frac{1}{2}(\partial\phi)^2 - \frac{1}{2}m^2\phi^2 + \frac{\lambda}{4!}\phi^4 \right] , \qquad (3.61)$$

where R is the Ricci scalar of the Kerr background metric g and m, λ, and ϕ are the UBDM mass, (dimensionless) self-coupling, and field value, respectively. Note that it is useful to reinstate M_{pl} (or G_N) for this exercise. Assuming a suitable setup in which superradiance indeed occurs, estimate both the superradiance and Bosenova conditions, i.e., Eqs. (3.57) and (3.60). Note the similarity between your estimate of N_{Bose} and Eq. (3.60) when $\lambda = m^2/f_a^2$.

Solution on page 322.

UBDM Hints: LIGO and the QCD Axion

The exclusion estimates, Eqs. (3.58) and (3.59), assumed continuous BH distributions between the minimum and maximum values. In reality, the distributions are of course incomplete. In fact, this can serve as a discovery tool for UBDM. If light bosons with particular masses exist, then the observed BH mass and spin distribution should contain forbidden regions, and astrophysical BHs should cluster along superradiant "trajectories" in the (m, a_J) plane. Gravitational wave observations will, over time, provide a very complete survey of this plane. Furthermore, superradiant clouds emit their own gravitational waves due to level transitions and annihilation. From these effects, the LIGO observatory provides a discovery channel for UBDM with 10^{-13} eV $\lesssim m \lesssim 10^{-12}$ eV [84]. This region is disfavoured by current measurements of BH spins, but the excluded region is determined by the uncertainty on BH masses with a small number of measurements. Thus, there is the possibility to make discoveries with precise measurements and greater statistics. The accessible mass region for LIGO corresponds to the QCD axion with $f_a \sim M_{pl}$. For the proposed GW detectors in lower frequency bands corresponding to higher mass BHs (e.g., Laser Interferometer Space Antenna, LISA), discovery potential moves to lower UBDM masses.

3.2.5   Summary of Gravitational Constraints

Current constraints on UBDM mass and cosmic density from the CMB, galaxy formation, relaxation, and black hole superradiance are combined in Fig. 3.5, along with a selection of forecasts for upcoming surveys. They cover an astonishing 24 orders of magnitude in mass and place sub-percent constraints on the density parameter. We caution that the limits apply strictly only to scalar UBDM with $w_{\text{UBDM}} = -1$ in the early Universe and negligible self-interactions, e.g., ALPs and similar cases. However, the limits apply by order of magnitude to all UBDMs, particularly if they come from non-relativistic effects where model dependence is less important. In addition to the effects discussed in detail, we also show projections

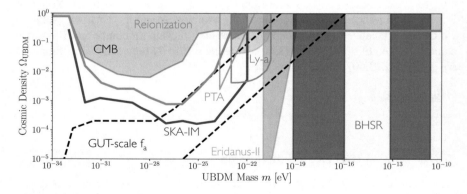

Fig. 3.5 Summary of gravitational constraints (shaded) on UBDM and forecasts (open) for upcoming surveys. Constraints assume a real scalar with potential $V(\phi) = m^2\phi^2/2$, see text for clarification on generalizing the bounds. CMB: cosmic microwave background [22, 85], PTA: pulsar timing array [32], BHSR: black hole superradiance [75], Ly-a: lyman alpha forest [86, 87], and SKA-IM: Square Kilometre Array intensity mapping [25]. Adapted from Ref. [88]

for the measurement of pulsar timing arrays (PTAs) with the *Square Kilometre Array* [32]. Current bounds from this technique [33] are not yet at the $O(1)$ level for Ω_{UBDM} and so do not appear.

3.3 Axion Compact Objects

ALP UBDM can form two different types of gravitationally bound objects which are distinct from ordinary DM galactic halos. These objects, miniclusters and axion stars, are interesting phenomenologically since they are far denser than galactic halos. They can thus have observational effects as sources of enhanced DM decay and conversion, gravitational lensing, or be observed in direct detection experiments if they happen to pass through the Earth.

3.3.1 Axion Stars

There exist several classes of (pseudo-)solitonic solutions to the Einstein–Klein–Gordon equations. These solutions go by many names and have been discovered and rediscovered many times. They date back to Wheeler's idea of a "geon": a wave confined to a finite region by gravity, thus mimicking a lump of matter. Ruffinni and Bonnazola [89] found explicit "boson stars" as time-independent fixed particle number state solutions for a complex scalar field coupled to general relativity: these are true solitons, stabilized by the existence of the conserved $\mathbb{U}(1)$ scalar field charge. Solutions also exist for a real scalar field. However, in this case, there is no conserved charge, and instead the solutions have a time-dependent metric and are known as "oscillitons" [90]. We could continue with the soliton bestiary for some time, but instead we will focus on the most well-motivated class of these objects: axion stars.[14]

First, consider the fully relativistic case. We are interested in time-dependent solutions for a scalar field coupled to general relativity. A public code is GRCHOMBO [91].[15] Like all stars, axion stars are stabilized by a balance between attraction (gravity and axion quartic self-interactions) and repulsion (gradient pressure and higher order interactions).[16] Initial conditions are found solving the boundary value problem on the initial spacetime volume (hypersurface) and evolved forward in time to investigate their stability. The solutions are a two-parameter

---

[14] To continue the bestiary just a little further, solutions are named for all scalar fields: inflaton stars, moduli stars, Higgs stars, etc.

[15] http://www.grchombo.org/.

[16] The axion potential is $V(\phi) = m^2 f_a^2 [1 - \cos \phi/f_a]$. Taylor expanding this we find that the ϕ^4 term is attractive, while higher order terms alternate in sign.

family in mass, M, and axion decay constant, f_a, giving a "phase diagram" that can be explored numerically [92].

The structure of the axion star phase diagram is easy to understand. As the mass of the star increases, the central value of the field ϕ_0 also increases. There are two possible instabilities, and which wins depends on f_a. For large f_a, the self-interactions can be neglected. Now, the ordinary GR lore applies: collapse to a BH at large mass. At low f_a, the axion has strong self-interactions, and these also drive collapse. Collapse increases ϕ_0 further until higher order repulsive interactions take over and expel relativistic axions from the collapsing core in an "axion nova" [93], which occurs at critical mass $M_{\text{nova}} = 10.4\, M_{pl} f_a / m g_4$, where g_4 is the coefficient of quartic interactions equal to unity for a cosine potential. For small f_a, the restoring interactions become important earlier during collapse and bring the star back to a stable configuration with only slightly lower mass than before the nova. As f_a increases, it takes more and more of the mass of the star to contract and reach the repulsive core, thus expelling a larger mass in the nova and reducing the mass of the stable remnant. The two types of instability are divided by a particular value of f_a. As $f_a \to \infty$, oscillatons and boson stars are found to be unstable when $\phi_0 \sim M_{pl}$ (this defines the "Kaup mass," $M \sim M_{pl}^2/m$), while self-interactions become important when $\phi_0 \sim f_a$, and so the boundary between the two unstable regions occurs for $f_a \sim M_{pl}$. A third phase boundary exists between the nova and BH regions, which simulations have found to be fractal in structure [94]. It is not clear this boundary could be reached by any astrophysical process, and so it is likely only a mathematical curiosity. The "triple point" between all three phases is found numerically to be near $(M, f_a) = (2.4 M_{pl}^2/m, 0.3 M_{pl})$, where M is the "Arnowitt–Deser–Misner" mass [95].[17]

Non-relativistic axion stars are far simpler to study: in the non-relativistic limit, the real scalar field possesses an effective conserved particle number. In this case, the solutions are simply referred to as solitons and the results apply generically to UBDM in the non-relativistic limit. Solitons are stationary waves of the form $\psi(r, t) = M_{pl} \chi(mr) e^{-i\gamma mt}$, where χ is a dimensionless function giving the radial profile, and γ is the energy eigenvalue. An important property of the SPEs (see Sect. 3.2.2) is their *scaling symmetry*:

$$(t, x, \psi, \Phi) \to (\lambda^{-2}t, \lambda^{-1}x, \lambda^2\psi, \lambda^2\Phi), \qquad (3.62)$$

where λ is the scale parameter (not to be confused with the quartic interaction strength in Sect. 3.2.4). The boundary value problem normalized to $\chi(0) = 1$, $\lambda = 1$, can be solved numerically and the results are fit by eigenvalue $\gamma = -0.692\lambda^2$ and radial density profile:

---

[17] Due to coordinate transformations, mass is not a straightforward quantity to define in general relativity (indeed, sometimes it is not defined). The Arnowitt–Deser–Misner mass is defined in the Hamiltonian formulation of general relativity and is essentially the conserved mass measured in the infinite future.

$$\frac{\rho_{\text{sol}}(r)}{m^2 M_{pl}^2} = \chi^2(mr) = \frac{1}{[1 + (0.230mr)^2]^8} \,. \tag{3.63}$$

These solutions are the ground state of the SPEs. They are a balance of the nonlinear and non-local gravitational force in the Poisson equation and the dispersive effect of the gradient energy term in the Schrödinger equation. Soliton dynamics can be studied using the numerical methods already discussed. In the limit of vanishing self-interactions, the soliton solutions are a one-parameter family given by the mass, M. Thanks to the scaling symmetry, we only need to find the solution once and then scale it using λ (see Problem 3.5).

How might axion stars form in astrophysical environments? Two mechanisms are seen in simulation of the SPEs. Which occurs depends on the scale, R, of the gravitational fluctuations compared to the de Broglie wavelength:

- Direct collapse: $R \sim \lambda_{\text{dB}}$ (e.g., Ref. [42])
- Kinetic condensation: $R \gg \lambda_{\text{dB}}$ (e.g., Ref. [41])

Direct collapse leads to rapid formation of axion stars on the gravitational free-fall time and by definition occurs in the smallest objects near to the cut-off scale of gravitational fluctuations, i.e., $M \sim M_0$, Eq. (3.43). This mechanism leads to an axion star in the centre of all DM halos close to the cut-off scale. If all mergers of this first generation are complete up to the largest scale of halos observed, then the numerically determined relationship between the star mass, $M_\star$, and the halo mass, M, is

$$M_\star \propto \left(\frac{M}{M_0} \right)^{1/3} M_0 \,, \tag{3.64}$$

where the constant of proportionality can be found in Ref. [96] and depends on the definition of M_0. This relationship is believed to derive from a combination of the virial theorem, equilibrium between the soliton and its gravitationally bound "atmosphere," and universal mass growth in the merger history of solitons [97]. Slow growth of solitons by accretion leads to significant scatter in the relation.[18]

The direct collapse mechanism is particularly relevant to the formation of solitonic cores in dwarf galaxies in the FDM regime (see hint box above) and the formation of axion stars in miniclusters (discussed below). Axion stars formed by this mechanism are in virial equilibrium with their environment for $t < t_{\text{relax}}$ and do not change appreciably in mass over such scales. The surrounding halo is a hot "atmosphere" for the star. The constant interaction with the halo causes the star to undergo radial oscillations at the normal mode frequencies [100].

Kinetic condensation gives rise to axion star formation in regions much larger than the de Broglie wavelength, for example, in the solar neighbourhood for

---

[18] Very recently some authors have even found a different best-fit exponent [98, 99], and numerical convergence may also play a part. The issue is not yet resolved at the time of writing.

the QCD axion. The scattering timescale thermalizes the distribution function on timescales of order τ_{gr}, Eq. (3.42), and at this time the local ground state is found in the form of an axion star which condenses spontaneously. Axion stars formed in this way continue to grow over time as they swallow up matter from the environment, with $M \propto (t/\tau_{\text{gr}})^p$. The index p is to be determined numerically and will evolve slowly in time with the wave distribution function. The growth process will eventually slow down when the star grows a gravitationally bound "atmosphere," at which point it should enter a local virial equilibrium solution close to Eq. (3.64).

Despite progress in our understanding of the formation and growth of axion stars, at the time of writing their abundance and galactic distribution is not fully understood even in benchmark models. The problem is partly one of scale: we do not know the mass above which the relation Eq. (3.64) breaks down and halos have no central soliton, but instead grow many small solitons in the kinetic regime.

Axion stars have a host of possible phenomenological consequences:

- **Galactic cores**: Solitons composed of Fuzzy DM with $m \sim 10^{-22}$ eV may help explain flat central densities in Milky Way dwarf satellites (tracer stars reside within the soliton) [42, 70] or central mass excesses (tracer stars outside the soliton). See hint box above for more details.
- **Direct detection**: The passage of axion stars through Earth, though rare, will greatly enhance the signal in a direct search and could be identified using a coordinated network of detectors like the Global Network of Optical Magnetometers for Exotic physics searches (GNOME) and GPS.DM [101, 102].
- **Indirect detection**: The high axion density creates a larger radio signal from decay and conversion of axions into photons (see Sect. 3.4.2). Cataclysmic signals could arise if the stars can reach the critical mass for an axion nova or stimulated decay due to interactions.
- **Relativistic axion stars**: If dense enough axion stars can be formed, they may show up as "Exotic Compact Objects" in gravitational wave detectors [103] and multi-messenger astronomy [104].

3.3.2 Miniclusters

A second special class of UBDM compact objects is formed by the process of spontaneous symmetry breaking, if this occurs during the normal course of thermal evolution of the Universe (as opposed to during the initial conditions epoch, inflation, or otherwise). The Peccei–Quinn (PQ) phase transition (see Chap. 2) occurs when the temperature of the Universe drops below approximately f_a. Recall that we write the complex PQ field as $\varphi = Re^{i\theta}$, and spontaneous symmetry breaking occurs when the field R takes on a vacuum expectation value. The following scenario applies specifically to ALPs where the field R is heavy and unstable (such that it decays at late times), while the field θ is initially massless but acquires a mass hierarchically smaller than the mass of R and at some time much later than the time of PQ symmetry breaking.

When PQ symmetry breaking occurs, R takes on a non-zero vacuum expectation value, and thus θ must also be specified. Since the axion field is massless at symmetry breaking, the only terms in the Lagrangian are proportional to $\partial\theta$, meaning there can be no preferred value for θ. The axion thus takes on a random value on essentially all scales. Imagine a pencil falling over from its point: in the absence of an external preference, the pencil falls in a random direction specified by an angle, θ, with the $\theta = 0$ axis arbitrary.

First, consider the simpler two-dimensional case, illustrated in Fig. 3.6. Because the PQ field is a continuous function (as all fields must be), for any random configuration of a complex field, there will be points in space around which θ makes a complete wrapping. At the wrapped point, the axion field θ is undefined (imagine shrinking the circle to a point: at the point, the circle must have zero size, and θ takes every value at once). The only way that this can be possible is if $R = 0$ at the wrapped point. The point in the complex field space where the radial coordinate is zero indeed has undefined phase. As long as the complete windings of θ persist, then at the centre of these windings R must remain at the origin, and thus symmetry breaking cannot happen. When the potential is $V(\varphi) = \lambda(|\varphi|^2 - f_a^2/2)^2$, this implies that the potential at the origin is $V(0) = \lambda f_a^4/4$, and this is the value of the potential at the centre of a point around which θ wraps.

In fact, in three dimensions, θ cannot wrap just a single point or else the field would be discontinuous. The field must wrap continuous one-dimensional lines (either infinitely long or in closed loops) known as cosmic strings, and in this particular case as axion strings, or global strings (since the symmetry breaking is of a global $\mathbb{U}(1)$).[19] This leads to the existence of PQ strings: continuous one-dimensional structures around which θ makes complete windings and where the radial field is pinned at $R = 0$. The strings are a class of topological soliton: localized field configurations stabilized by the topology of the field space. The formation mechanism is known as the *Kibble–Zurek mechanism* and is observed experimentally in condensed matter phase transitions with the same symmetries, for example, the transition from normal fluid to superfluid helium [4, 5].

What happens to the axion field? The equation of motion for the axion field in Fourier space is

$$\ddot{\tilde{\theta}}_k + 3H\dot{\tilde{\theta}}_k + (k^2/a^2 + m(T)^2)\tilde{\theta}_k = 0\,, \tag{3.65}$$

---

[19] This is generic for complex fields in three spatial dimensions. It is a topological property. Complex fields have symmetry group $\mathbb{U}(1)$ of rotations in the complex plane, i.e., loops. The mapping of $\mathbb{U}(1)$ onto $\mathbb{R}^3$ (Euclidean 3-space) is expressed by the first homotopy group $\pi_1(\mathbb{R}^3)$. This group is not the empty set, i.e., it is non-trivial, which can be seen by noting that $\mathbb{R}^3$ is the universal cover of T^3, the 3-torus, and $\pi_1(T^n) = \mathbb{Z}^n$. This last can be seen since one cannot shrink circles on tori to points continuously, and there are n distinct circles wrapping T^n.

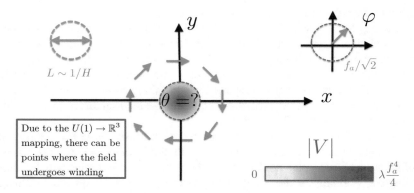

Fig. 3.6 Formation of axion strings from spontaneous symmetry breaking. The complex plane of the PQ field φ is illustrated in the upper right corner with the radius marked where R takes the vacuum expectation value and the circle along which the axion field is defined. When spontaneous symmetry breaking occurs, this circle is mapped onto the coordinates in the real plane, (x, y). The complex phase of the field (i.e., the value of the axion field) is shown by the orientation of the arrow. Complete wrappings of the field lead to defects where the phase is not defined. In this two-dimensional case, the defects are points, and in three dimensions, they are lines, i.e., strings. In the centre of the string, the PQ field has $\langle \varphi \rangle = 0$ and the potential takes the value indicated by the colour bar. Damping of sub-horizon modes $k > aH$ in Eq. (3.65) smooths fluctuations on length scales $L \sim H^{-1}$

where we are careful to distinguish between the field θ and the mode function $\tilde{\theta}_k$: $\tilde{\theta}_k = 0$ does not imply $\theta = 0$ as a preferred value, only that the mode k is absent from its spectrum and thus gradients of θ on the spatial scale $1/k$ are small. At early times, the QCD axion mass, $m(T)$, is vanishingly small compared to H and can be neglected.

Imagine initially that all modes are populated with some amplitude (for example, the inflationary fluctuations of the PQ field), and then the field configuration far from any string is allowed to evolve. Any mode "inside the horizon" has $k^2 \gg a^2 H^2$. These modes will undergo damped oscillation and decay in amplitude. Modes larger than the horizon, $k^2 \ll a^2 H^2$, remain pinned to their initial value by the friction term (coefficient of $\dot{\theta}_k$) in Eq. (3.65) given by $3H$ (Hubble friction). Thus, high frequency modes decay and low frequency modes remain static, smoothing the field on scales of order the horizon size.[20] Around any string, the axion field is wound $\theta \in (-\pi, \pi]$, and so we have $O(1)$ variation of the field on horizon size patches around the string. String formation is sketched in Fig. 3.6. Furthermore, numerical simulations indicate that string dynamics enter into a scaling solution with $O(1)$ strings per horizon volume.

---

[20] Note that the mode function $\tilde{\theta}_k$ decaying to zero does not imply a preference for the axion field θ to move to zero: in the massless limit, the shift symmetry prevents any such preference.

Strings decay when the axion mass becomes relevant to the mode evolution. Recall first that during cosmological expansion, temperature, T, always decreases as time, t, increases. Second, recall that the axion mass and the Hubble rate H are both decreasing functions of T. The mass term (coefficient of $\tilde{\theta}_k$) in Eq. (3.65) is comparable to the friction term when $m(T) \approx H(T)$. The axion mass term defines a preferred value for the field, $\theta = 0$, which is exactly why the PQ mechanism solves the strong-CP problem. Equation (3.65) is just a damped harmonic oscillator, and so the mode functions for all $k < aH$ (not just the short wavelength modes inside the horizon) will begin to oscillate when $m(T) \approx H(T)$, defining the special temperature T_{osc}. Now, everywhere, the axion field is making harmonic oscillations about zero. There are thus no longer regions around which it makes a complete and continuous winding. The axion field everywhere has an average value of zero (but importantly of course nonzero variance and energy density). This means that the radial mode is no longer required to take the value $R = 0$ along the strings. The axion field is everywhere defined, the radial field is not pinned, and it can undergo symmetry breaking at the string locations, i.e., the strings decay (or "unwind").

At this time, the axion field has a well specified distribution: in every horizon-size patch, it has $O(1)$ fluctuations, while on larger scales it is uncorrelated. The power spectrum, $P(k)$, is flat (white noise) for $k \ll a(T_{\mathrm{osc}})H(T_{\mathrm{osc}}) = k_J(T_{\mathrm{osc}})$ and cut off by the Jeans scale for $k \gg a(T_{\mathrm{osc}})H(T_{\mathrm{osc}})$. The normalization of the power spectrum is fixed by the variance, which should match the variance of the uniform distribution for θ, $\langle \theta^2 \rangle = \pi^2/3$. Just prior to $a(T_{\mathrm{osc}})$, the axion equation of state is $w \approx -1$, and so the fluctuations in θ do not source any curvature perturbations in the metric:[21] this is what is meant by the term *isocurvature*. It is this particular power spectrum (white noise isocurvature, with $O(1)$ variance, truncated at the horizon size at T_{osc}), which gives rise to the structures known as *axion miniclusters* as a remnant of string decay [105]. Figure 3.7 shows a snapshot from numerical simulation of the axion field after string decay [106], and miniclusters are located using a threshold based on spherical collapse under gravity [107].

The mass scale of miniclusters is the same as the mass scale of ordinary axion halos: it is fixed by the Jeans scale when field oscillations begin, Eq. (3.43), but now for very different values of the reference parameters. However, because of the large amplitude of these isocurvature fluctuations (variance of order unity), axion miniclusters begin to collapse earlier than ordinary DM halos and there is a significant nonlinear structure formation before matter–radiation equality.[22] The

---

[21] Intuitively, this can be understood because if $w = -1$ exactly, then this is equivalent to a cosmological constant, which is constant in space and time, and thus cannot source a spatially varying curvature.

[22] Contrast this to the evolution of large-scale inhomogeneities seen in the CMB and the inflationary "adiabatic" mode. These perturbations have small amplitude and a red slope in the power spectrum leading to smaller amplitude fluctuations on small scales. These fluctuations undergo logarithmic growth at early times in the radiation era, which is important to seed galaxy formation and is one of the pieces of evidence for DM discussed in Chap. 1. However, due to the existence of a free-streaming scale/Jeans scale for particle DM models, there is, in general, no

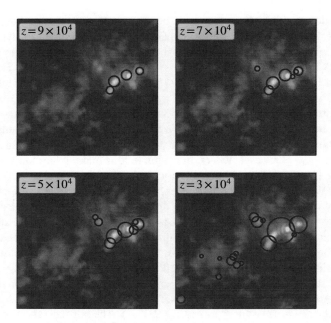

Fig. 3.7 Initial conditions for minicluster formation. After string decay, the axion field has large density perturbations, which subsequently collapse into the objects known as miniclusters. The figure shows a small patch of results from the numerical simulations of Ref. [106], which used lattice field theory methods to solve the equations of motion for the complex Peccei–Quinn field in the absence of gravity. The large hierarchies involved necessitate further approximations, and different simulation methods are not currently in precise agreement for the spectrum of perturbations extrapolated to physical values of the particle masses and couplings. Miniclusters are identified at different redshifts using a threshold derived from spherical collapse under gravity [107]

density of any DM halo is related to the background density at the time when it first collapses and leaves the Hubble flow. Thus, miniclusters are denser than ordinary halos and may survive repeated mergers up to the present day. Let us consider the phenomenology of these low-mass dense objects.

First, we need the minicluster mass which we estimate from the number density of axions within the comoving cosmological horizon at the time when field oscillations begin. We find T_{osc} in the usual way, setting $H(T_{\mathrm{osc}}) = A_1 m_a(T_{\mathrm{osc}})$.[23] Now, we need to calculate the horizon volume. Take the volume of rotation of the

---

nonlinear structure formation during the radiation era. For WIMPs with $m \lesssim 1$ TeV, structure formation in the adiabatic mode begins at $z \approx 500$ with the formation of Earth mass, $10^{-6} M_{\odot}$, halos [108].

[23] A_1 is simply a constant of proportionality to account for ambiguity defining T_{osc} and its later use in analytical formulae for the evolution of the energy density. Our earlier choices, e.g., Eq. (3.4), set $A_1 = 1$, physically assuming structure formation begs when the de Broglie wavelength is equal to the Hubble length, H^{-1}. Many authors choose $A_1 = 3$ when estimating the relic density.

spherical wave with comoving wavenumber $k_{osc} = a(T_{osc})H(T_{osc})$ over one-half wavelength:

$$V(k) = \int d^3x j_0(kr) = \frac{4\pi}{(ak)^3} \int_0^\pi dy\, y \sin(y) = \frac{4\pi^2}{(ak)^3} \equiv \frac{V_H}{(ak)^3}. \quad (3.66)$$

This defines the Hubble volume V_H. Alternative definitions are the cubic volume, $V(k) = (ak)^{-3}$, and the spherical volume of one-half wavelength $V(k) = \frac{4}{3}\pi(\pi/ak)^3$. The expected minicluster mass is simply $M_{MC} = V(a_{osc}k_{osc})\Omega_a \rho_{crit}$. We compute $M_{MC}(T_{osc})$ using the Friedmann equation to fix k_{osc}, $3H^2 M_{pl}^2 = (\pi^2/30)g_\star T^4$, and conservation of entropy to write $a(T) \propto T^{-1}g_{\star S}$ (normalized using the CMB measurement of z_{eq}). The result is well fit by

$$M_{MC}(T_{osc}) = 9.2 \times 10^{-13} V_H \left(\frac{T_{osc}}{2\,\text{GeV}}\right)^{-3} S\left(\log_{10}\left(\frac{T_{osc}}{\text{GeV}}\right)\right) M_\odot, \quad (3.67)$$

$$S(x) \equiv 0.5[1 + \tanh 4(x - 8.2)] + 1.3[1 + \tanh 4(8.2 - x)], \quad (3.68)$$

where $S(x)$ is an activation function that accounts for the behaviour of $g_\star$ and $g_{\star,S}$ in the Standard Model (dominantly the quark–hadron phase transition) and has only been roughly fit here using the results for $g_\star$ from Ref. [109]. Note that this expression is valid for any ALP with minicluster-like initial conditions.

From the QCD axion $m_{a,QCD}(T)$ dependence based on lattice QCD in Ref. [110], $T_{osc}(m_{a,QCD})$ is well fit by[24]

$$T_{osc}(m_{a,QCD}) = 2 \left(\frac{m_{a,QCD}}{100 A_1\,\mu\text{eV}}\right)^{0.165} \text{GeV}, \quad (3.69)$$

over the range of interest (broadly $10^{-5}\,\text{eV} \leq m_{a,QCD} \leq 10^{-3}\,\text{eV}$) for the relic density in this scenario (see also Fig. 3.1). For smaller axion masses, there is an $O(1)$ change in the constant at the front of Eq. (3.69), while the power law remains approximately the same.

Now, we would like to know the minicluster density profile. Kolb and Tkachev [111] wrote down the equation of motion for spherical collapse of an isolated top-hat density profile, with initial overdensity δ, in an expanding Universe dominated by radiation. The perturbation first grows in size as the Universe expands. It then turns around and collapses. Spherical collapse formally leads to a density singularity. However, in real collapses below the threshold for BH formation, aspherical perturbations lead to virialization (equilibrium between average kinetic

---

This ambiguity in the use of T_{osc} leads to significant uncertainty in analytical minicluster mass estimates, which can only be resolved by fitting results of numerical simulations.

[24] For the present purposes, a simple power law $m \propto T^{-4}$, matched to the zero-temperature result $m_{a,QCD} \equiv m_{a,QCD}(0) = 5.72\,\mu\text{eV}(10^{12}\text{GeV}/f_a)$ at $T_{QCD} = 140\,\text{MeV}$, is accurate enough.

and gravitational potential energy expressed by the virial theorem, see, e.g., Ref. [112]), and the collapsed object becomes self-supported with a finite average density. Virialization occurs when the radius of the perturbation is half of the turn-around radius. Using this information, one can compute the overdensity of the spherical system at the time of virialization. A numerical solution of the ordinary differential equation for spherical collapse gives the final average overdensity:[25]

$$\langle \rho_f \rangle = 140 \bar{\rho}_{\text{eq}} \delta^3 (1 + \delta). \tag{3.70}$$

Assuming the minicluster has a constant density, the radius is then calculated to be

$$R_{\text{MC}} = \left(\frac{3 M_{\text{MC}}}{4 \langle \rho_f \rangle} \right)^{1/3}. \tag{3.71}$$

Using $\bar{\rho}_{\text{eq}} = 2 \times 10^{14} M_\odot \text{kpc}^{-3}$,

$$R_{\text{MC}} = 3 \times 10^{-10} \text{ kpc} \frac{1}{\delta (1 + \delta)^{1/3}} \left(\frac{T_{\text{osc}}}{2 \text{GeV}} \right)^{-1}. \tag{3.72}$$

To make more use of this result, we need to know the minicluster radial profile, $\rho(r)$. Collapse of isolated density perturbations is self-similar and leads to a power law density profile with no preferred scale. Miniclusters are not isolated, and the formation proceeds much like ordinary DM halos, leading to Navarro–Frenk–White (NFW) profiles [114–116]. It is then natural to associate the radius R_{MC} with the NFW scale radius. If the initial distribution of δ could be measured, one would know the mass and size distribution of miniclusters.

In reality, the problem of miniclusters is far more complex than the simple story given here. Firstly, miniclusters do not have one fixed mass. Structure formation always proceeds hierarchically, and there is a mass function of miniclusters. This can be computed numerically via N-body simulation or semi-analytically from the initial power spectrum [107, 116–118]. The mass function takes on a power law spreading over many orders of magnitude around M_{MC}. Secondly, the distribution of δ in initial conditions is not uniquely defined. Numerical thresholding using the spherical collapse results allows some progress to be made [107], but the results still require calibration to N-body simulation. Unfortunately, N-body simulations cannot currently resolve the scale radius on all relevant scales and are not large enough to capture the rarest, densest, and thus most phenomenologically interesting miniclusters.

Finally, just like other UBDM halos, when the effects of the gradient energy (the UBDM de Broglie wavelength) are included, miniclusters have been shown to form

---

[25] For the more standard case during matter domination, which applies to ordinary DM halos, see, e.g., Ref. [113]. In the standard case, the equations can be solved analytically, leading to the well-known result that the virial overdensity $\langle \rho_f \rangle / \bar{\rho} \approx 200$, independent of δ.

central axion stars [119]. Axion stars in miniclusters follow approximately the same core–halo mass relation, Eq. (3.64), as an ordinary halo. For the QCD axion, the resulting axion stars are on very different mass scales for the UBDM particle mass and the halo mass than the reference FDM values in Eq. (3.64).

The minicluster power spectrum, mass function, size function, and central axion stars can all be used to constrain the QCD axion and ALPs in the post-inflation PQ symmetry breaking scenario. Some examples include:

- Microlensing and femtolensing [120–122] (see Problem 3.5 below)
- Radio signals from minicluster–neutron star collisions [123–125]
- The CMB isocurvature power spectrum and large-scale structure [126, 127]

? Problem 3.5 Microlensing Constraints on UBDM

Show that the scaling symmetry, Eq. (3.62), is a symmetry of the SPEs. Use this relationship and the profile, Eq. (3.63), to write down the mass–radius relation in units of solar masses ($M_\odot$) and kiloparsecs. The *Subaru Hyper Suprime Cam* (HSC) microlensing survey of M31 probes PBHs in the range of $10^{-11} M_\odot$ to $10^{-6} M_\odot$ [128]. The Einstein radius for gravitational microlensing is $R_E = 2[GM_\star x(1 - x)d_s]^{1/2}$, where d is the distance from the observer to the lens, d_s is the distance from the observer to the source, and $x = d/d_s$. Compare the axion star radius to R_E with $x = 1/2$ and $d_s = 770$ kpc and the distance to M31. What UBDM particle masses could be probed by HSC lensing due to axion stars?

Now, consider the mass–radius relation for miniclusters with initial overdensity δ, Eq. (3.72), and the minicluster mass relation, Eq. (3.68). What range of the (T_{osc}, δ) parameter space can be probed by microlensing?

Solution on page 323.

3.4   Indirect Detection of UBDM

3.4.1   Stellar and Supernova Energy Loss

In this section, we consider only constraints on axion-like couplings, i.e., pseudoscalar, anomalous, or shift symmetric (see Chap. 2). Analogous bounds can of course be derived for scalar and dilaton-like couplings. Axions are pseudoscalars, and their couplings to fermions depend on the orientation of the spin, while couplings of scalar particles do not. The spin dependence can lead to suppression of interactions, since macroscopic bodies are not in general strongly polarized. Being unsuppressed by spin effects, scalar constraints are often stronger. More details on some of the calculations are given in Chap. 5.

First and foremost, it is *extremely important* to remember that the constraints and effects we discuss in this section apply *independently* of whether the axion is (a large fraction of) the DM. The axions considered here are produced from Standard Model particles in stellar plasmas. They interact only very weakly and have a long mean-free path inside the plasma. Thus, stars are effectively transparent to axions, and the axions escape, allowing an additional cooling channel for the star. This changes the evolution of stars: in simple terms, it alters the progression of stars along the Hertzsprung–Russell (HR) diagram of stellar luminosity versus temperature. The relationship between the mass, age, and temperature of stars is thus different than in the Standard Model. Stellar physics is generally very well understood in terms of Standard Model physics alone and can be simulated using a code such as MESA [129], which can be modified to include axion-induced cooling [130]. For more details, see Refs [10, 131, 132].

Stars, including the Sun, can produce axions by the Primakoff process: photons inside the star convert into axions in the ambient magnetic and electric fields of the particles (electrons and nuclear ions) in the plasma. The rate for this process is

$$\Gamma_{\gamma \to a} = \frac{g_{a\gamma\gamma}^2 T \kappa_s^2}{32\pi} \left[\left(1 + \frac{\kappa_s^2}{4E^2} \right) \ln \left(1 + \frac{4E^2}{\kappa_s^2} \right) - 1 \right], \qquad (3.73)$$

where E is the photon energy, T is the temperature, and κ_s is the screening length. In the Debye–Hückel approximation, we have

$$\kappa_s^2 = \frac{4\pi\alpha}{T} \left(n_e + \sum_j Z_j^2 n_j \right), \qquad (3.74)$$

where n_e is the free electron density, and n_j is the density of the jth nuclear ion of charge Z_j. In a neutral medium with $n_e = n_j = 0$, the Primakoff rate goes to zero, since there is no background field to facilitate conversion.

Photon energies are distributed thermally, and the temperature varies with stellar radius. Kinematically, we require $E \geq m$ to produce an axion. Typical stellar interior temperatures are in the kiloelectron volt range, which gives the typical energy of the emitted axions, and approximately the maximum axion mass where this cooling channel is allowed. The luminosity in axions needs to be computed for a given stellar model. Applying this to the Sun gives $L_a = 1.85 \times 10^{-3} (g_{a\gamma\gamma}/10^{-10} \text{ GeV}^{-1})^2 L_\odot$. It is this solar luminosity in axions that *helioscope* experiments try to detect (see Chap. 5). We can derive a crude bound by demanding that the solar axion luminosity must be less than unity, since the evolution of the Sun is well described by emission dominantly in photons, i.e., $L_{\gamma,\text{Sun}} = 1 L_\odot$ by definition. Thus,

$$g_{a\gamma\gamma} < 2.3 \times 10^{-9} \text{ GeV}^{-1} \quad \text{(luminosity of the Sun)}. \qquad (3.75)$$

The bound in Eq. (3.75) can be improved by considering the statistics of populations of stars. The best understood case is for horizontal branch (HB) stars in globular clusters. Stars in globular clusters are all of a similar age and differ in their masses. The distribution of the stars gives an HR diagram that can be compared to models. The observable is the ratio of HB stars to red giant branch (RGB) stars, R, determined by placing stars on a colour–magnitude diagram. In the Standard Model, this ratio is a function of the primordial helium abundance, Y_{He}, stellar mass, and metallicity. Globular clusters are old systems, with ages in the range of 10 billion years. This gives a small range of available stellar masses and metallicities, which have a negligible effect on R. The value of Y_{He} can determined observationally by measurement of extragalactic H II regions which gives $Y_{He} = 0.2449 \pm 0.0040$ [133], which is consistent with the predictions of standard BBN and the CMB measurement of the baryon abundance [3]. Reference [134] reports a measured average value of $R = 1.38$ from 39 globular clusters, consistent with the Standard Model prediction.

With a given stellar evolution model, it is possible to compute the effect of the axion–photon coupling, $g_{a\gamma\gamma}$, and the axion–electron coupling, g_{aee}, on R. At present, there are two different models in the literature for the functional dependence, and each is presented in Ref. [135]. The specific forms are not enlightening, so we simply quote the bounds (derived in Ref. [136]). In both cases, the additional cooling channel lowers R compared to the Standard Model prediction leading to a degeneracy in the combined constraints, with the maximum value of one coupling only allowed when the other is strictly zero. Setting one coupling to zero, and fixing $Y_{He} = 0.25$, the individual bounds are

$$g_{a\gamma\gamma} < 4.95 \ (9.56) \times 10^{-11} \, \text{GeV}^{-1} \qquad (95\% \ \text{CL, HB/RGB stars}), \qquad (3.76)$$

$$g_{aee} < 2.95 \ (3.53) \times 10^{-13} \qquad (95\% \ \text{CL, HB/RGB stars}), \qquad (3.77)$$

where the number in brackets refers to the bound using the alternative model for R, which we see introduces an $O(1)$ shift in the bound on $g_{a\gamma\gamma}$. For the constraints in the combined parameter space, see Refs. [135, 136].

Supernova SN1987A provides an important bound on the axion nuclear couplings, g_{aNN} and g_d. During the core collapse process, a proto-neutron star is formed, the gravitational field of which traps neutrinos and causes them to be emitted over a delayed period of time. This model explains the duration of the burst of two dozen observed neutrinos coincident with SN1987A. Axion emission due to nuclear bremsstrahlung:

$$N + N \rightarrow N + N + a \qquad (g_{aNN} \ \text{coupling}), \qquad (3.78)$$

$$N + \gamma \rightarrow N + a \qquad (g_d \ \text{coupling}) \qquad (3.79)$$

would compete with neutrino emission and cool SN1987A too rapidly, shortening the neutrino burst, unless the total energy loss rate from either axion nuclear process obeys the bound $\varepsilon_a \lesssim 10^{19} \, \mathrm{erg \, g^{-1} s^{-1}} = 7.2 \times 10^{-18} \, \mathrm{eV}$.

For the first process, the cooling rate per unit mass is [132]

$$\varepsilon_a = \frac{1}{\rho} \left(\frac{C_N}{2f_a} \right)^2 \frac{n_N}{4\pi^2} \int_0^\infty d\omega \, \omega^4 S_\sigma = \left(\frac{C_N}{2f_a} \right)^2 \frac{T^4}{\pi^2 m_N} F \,, \qquad (3.80)$$

where ρ is the mass density of the supernova, n_N is the nucleon number density, ω is the axion angular frequency, and S_σ is the spin density structure function, which accounts for the fact that axions couple only to the nuclear spins. The coupling C_N is the nucleon coupling weighted as $C_N^2 = Y_n C_n^2 + Y_p C_p^2$, where Y_n and Y_p are the neutron and proton abundances, estimated as $Y_p = 0.3$ and $Y_n = 1 - Y_p = 0.7$ at the relevant epoch in the supernova.

The spin density structure function is non-trivial to compute but can be estimated using various approximations. The last equality in Eq. (3.80) defines the dimensionless function F from the integral of S_σ, which is estimated to be of order unity, and allows a simple estimate of the bound given the supernova internal temperature $T \approx 30 \, \mathrm{MeV}$. A recent analysis found the more accurate bound [137]

$$\frac{C_N}{2f_a} < 1.3 \times 10^{-9} \mathrm{GeV}^{-1} \quad \text{(SN1987A neutrino burst)} \,, \qquad (3.81)$$

a factor of approximately four weaker than the estimate with $F = 1$. The same analysis found an $O(1)$ effect from the modelling of supernova temperature and density profiles. The bound from SN1987A on C_N/f_a is particularly important for the QCD axion, since this couplings is always present, and so the bound can be cast as a model-independent constraint on the QCD axion mass. We use that $C_{KSVZ}^2 = 0.066 \Rightarrow C_{KSVZ} = 0.257$, leading to

$$m_{a,\mathrm{QCD}} \lesssim 0.06 \left(\frac{0.257}{C_N} \right) \, \mathrm{eV} \quad \text{(SN1987A neutrino burst)} \,. \qquad (3.82)$$

For the second process, ε_a is approximated by

$$\varepsilon_a = \frac{\Gamma}{V\rho} \approx \frac{\langle E_\gamma \rangle n_N n_\gamma \langle \sigma v \rangle}{\rho} \,, \qquad (3.83)$$

where n_i are the reactant number densities, $\langle E_\gamma \rangle$ is the average photon energy, ρ is the supernova mass density, and $\langle \sigma v \rangle$ is the thermally averaged cross section. The nuclear number density and supernova mass density are known, and the other parameters are fixed in terms of the internal temperature, T. To estimate the bound on g_d, Ref. [138] approximates the cross section as $\langle \sigma v \rangle = g_d^2 T^2$, leading to

$$g_d < 4 \times 10^{-9} \, \mathrm{GeV}^{-2} \,. \qquad (3.84)$$

UBDM Hints: Anomalous White Dwarf Cooling

White dwarfs (WDs) are stellar remnants whose electron-degenerate cores are supported by Fermi pressure against gravitational collapse. Their internal densities are relatively high as their masses are typically comparable to the mass of the Sun ($\sim 0.6\,M_\odot$), while their radii are of the order of the Earth's radius [139]. They cannot replenish their internal energy and therefore continuously cool down over time.

The evolution of WDs can be altered by introducing additional cooling channels. These can be provided by weakly interacting particles that efficiently carry away energy after being created in, and escaping from, the WD's core. A useful observable to infer the resulting additional cooling rate is the so-called period increase in variable WDs. These are WDs that periodically change in brightness over time as they pulsate due to non-radial excitations, called "gravity modes" (see, e.g., Ref. [140]), with potentially multiple pulsation periods associated with different coexisting sub-modes. For cooling WDs, the periods of their pulsations, Π, tend to increase over time with a rate $\dot{\Pi} = \mathrm{d}\Pi/\mathrm{d}t$ that can approximately be calculated via

$$\frac{\dot{\Pi}}{\Pi} \approx -\frac{1}{2}\frac{\dot{T}}{T} + \frac{\dot{R}}{R} \approx -\frac{1}{2}\frac{\dot{T}}{T}\,, \qquad (3.85)$$

where T and R are the internal temperature and radius of the WD, respectively [141]. The change in radius can usually be neglected for the observed low-luminosity dwarfs [142]. Axions and ALPs induce an energy loss that is proportional to g_{aee}^2 since axion–electron interactions dominate in the high-density electron-degenerate interior of the WDs [143, 144]. The resulting decrease in temperature, and therefore the additional contribution to the period increase in Eq. (3.85), is hence also proportional to g_{aee}^2.

Measurements of an anomalous period change can thus be used to estimate the associated axion–electron coupling. From the 250 known variable WDs [145], this has so far only been done for G117-B15A [146–150], R548 [148, 151], L19-2 [152], and PG 1351+489 [153]. The reason for this small fraction is that measuring the period change is very difficult: while the periods for the WDs listed here are of the order of a few minutes, their (inherently dimensionless) period changes, $\dot{\Pi}$, are less than about 10^{-13} in magnitude.

The left panel of Fig. 3.8 shows how the measured period increases in G117-B15A (blue line and shading) compared to theoretical prediction from simulating WD evolution with and without axions (red data points). To illustrate the dependence on the axion–electron coupling, we show the theoretical prediction as a function of g_{aee}^2.

The right panel of Fig. 3.8 shows the one-dimensional profile likelihoods for the four WDs listed above. Combining these likelihoods hints at an additional cooling channel corresponding to an axion–electron coupling of a few times 10^{-13} at more than 3σ confidence level [135, 136, 155].

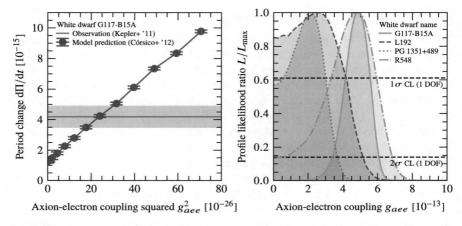

Fig. 3.8 Cooling hints in white dwarfs. *Left:* comparison of the predicted and measured period change as a function of axion–electron coupling squared for the WD variable G117-B15A (data from Refs. [150, 154]). *Right:* overview of the likelihood functions for different WD variables (reproduced from Ref. [136])

In addition to difficulties of observing the period change, there are a number of uncertainties involved in the modelling of WDs and their pulsations. Multiple challenges in quantifying the statistical and systematic uncertainties in WD modelling remain, such as the modelling of the transition from the main sequence to the WD phase. More details on WD modelling can be found in Refs. [150–153]. It is therefore not yet clearly established whether the cooling hints are due to systematics or indicate the presence of new physics—be it in form of a weakly interacting particle or a completely different astrophysical cooling channel.

A recent more general review of pulsating WDs can be found in Ref. [140]. Apart from the period increase discussed here, the WD luminosity function can also be used to probe the evolution of WDs and be seen as a hint for ALPs [156, 157].

3.4.2   Axion–Photon Conversion

In the presence of a magnetic field, axions convert into photons, and vice versa, by the Primakoff and inverse Primakoff process (see Chaps. 2, 4, and 5). This leads to constraints on the axion–photon coupling from any astrophysical environment penetrated by a magnetic field. In the following, we briefly mention some important instances.

Axions produced during supernova SN1987A can escape from the supernova event. These axions are subsequently converted back into visible photons in the form of gamma rays by the magnetic field of the Milky Way. This process would have led to a gamma ray burst coincident with SN1987A, which was not observed [158]. This places constraints on the axion mass and coupling at 95% CL [159]:

$$g_{a\gamma\gamma} < 5 \times 10^{-12} \, \text{GeV}^{-1} \quad (\text{for } m_a \lesssim 10^{-9} \, \text{eV}). \tag{3.86}$$

The bound gets rapidly worse at higher masses due to loss of coherence of the axion field on the scale of the galactic magnetic field and the resulting reduced photon fluence. The bound has an $O(1)$ dependence on the precise model of the galactic magnetic field.

Axion–photon conversion also occurs in the intergalactic medium and leads to modulation of the X-ray spectra of active galactic nuclei (AGN) and quasars (see, e.g., Refs [160, 161]). The modulation can be modelled statistically with a stochastic model for cluster magnetic fields. The strongest bound arises from the observation of a single source, NGC1275, by the Chandra satellite, which observed no modulations and sets the 3σ limit [162, 163]

$$g_{a\gamma\gamma} < 6 - 8 \times 10^{-13} \text{GeV}^{-1} \quad (\text{for } m_a \lesssim 10^{-12} \, \text{eV}). \tag{3.87}$$

Still further bounds can be derived from axion–photon conversion on cosmological scales. The conversion of CMB photons by Mpc scale primordial magnetic fields leads to CMB spectral distortions (i.e., departure from a black body spectrum) [164, 165]. Since the cosmic background explorer (COBE) satellite determined the CMB to be the most perfect black body in the Universe [166], any departures from perfection caused by axion–photon conversion are strongly constrained. On the other hand, the origin and spectrum of large-scale primordial cosmic magnetic fields is highly uncertain (e.g., Ref. [167]). Thus, bounds are given relative to the amplitude of the magnetic field power spectrum averaged on cosmic length scales, $A_B = \sqrt{\langle B^2 \rangle}$, as

$$g_{a\gamma\gamma} \lesssim 10^{-14} \, \text{GeV}^{-1} \left(\frac{1 \, \text{nG}}{A_B} \right) \quad (\text{for } m_a \lesssim 10^{-12} \, \text{eV}). \tag{3.88}$$

These bounds can be improved by up to two orders of magnitude by future CMB spectral measurements.

Acknowledgments We are grateful to Richard Brito and Jens Niemeyer for helpful discussions and to Jurek Bauer and David Ellis for producing Figs. 3.2 and 3.7. We are further indebted to Jurek Bauer and David Ellis for providing written solutions to problems 3.2, 3.3, and 3.5.

References

1. D.J.E. Marsh, Phys. Rep. **643**, 1 (2016)
2. M. Archidiacono, S. Hannestad, A. Mirizzi, G. Raffelt, Y.Y.Y. Wong, J. Cosmol. Astropart. Phys. **10**, 020 (2013)

3. N. Aghanim, Y. Akrami, M. Ashdown, J. Aumont, C. Baccigalupi, M. Ballardini, A. Banday, R. Barreiro, N. Bartolo, S. Basak, et al., Astron. Astrophys. **641**, A6 (2020)
4. T.W.B. Kibble, J. Phys. A Math. Theor. **9**, 1387 (1976)
5. W.H. Zurek, Nature **317**, 505 (1985)
6. T. Hiramatsu, M. Kawasaki, K. Saikawa, T. Sekiguchi, Phys. Rev. D **85**, 105020 (2012). [Erratum: Phys. Rev. D **86**, 089902 (2012)]
7. V.B. Klaer, G.D. Moore, J. Cosmol. Astropart. Phys. **1711**, 049 (2017)
8. M. Gorghetto, E. Hardy, G. Villadoro, J. High Energy Phys. **07**, 151 (2018)
9. E. Armengaud, et al., J. Cosmol. Astropart. Phys. **1906**, 047 (2019)
10. E.W. Kolb, M.S. Turner, *The Early Universe* (Addison-Wesley, Boston, 1990)
11. S. Dodelson, *Modern Cosmology* (Academic Press, Amsterdam, 2003)
12. V. Mukhanov, *Physical Foundations of Cosmology* (Cambridge University Press, Cambridge, 2005)
13. E.J. Copeland, M. Sami, S. Tsujikawa, Int. J. Mod. Phys. D **15**, 1753 (2006)
14. T. Clifton, P.G. Ferreira, A. Padilla, C. Skordis, Phys. Rep. **513**, 1 (2012)
15. D.S. Aguado, R. Ahumada, A. Almeida, S.F. Anderson, B.H. Andrews, B. Anguiano, E.A. Ortíz, A. Aragón-Salamanca, M. Argudo-Fernández, M. Aubert, et al., Astrophys. J., Suppl. Ser. **240**, 23 (2019)
16. J.R. Primack, SLAC Beam Line **31N3**, 50 (2001)
17. C.P. Ma, E. Bertschinger, Astrophys. J. **455**, 7 (1995)
18. A. Lewis, A. Challinor, A. Lasenby, Astrophys. J. **538**, 473 (2000)
19. J. Lesgourgues, arXiv:1104.2932 (2011)
20. R. Hložek, D. Grin, D.J.E. Marsh, P.G. Ferreira, Phys. Rev. D **91**, 103512 (2015)
21. A. Lewis, A. Challinor, Phys. Rep. **429**, 1 (2006)
22. R. Hložek, D.J.E. Marsh, D. Grin, Mon. Not. Roy. Astron. Soc. **476**, 3063 (2018)
23. B. Li, T. Rindler-Daller, P.R. Shapiro, Phys. Rev. D **89**, 083536 (2014)
24. J.C. Hwang, H. Noh, Phys. Lett. B **680**, 1 (2009)
25. J.B. Bauer, D.J. Marsh, R. Hložek, H. Padmanabhan, A. Laguë. Mon. Notices Royal Astron. Soc. **500**, 3162 (2021)
26. W. Hu, R. Barkana, A. Gruzinov, Phys. Rev. Lett. **85**, 1158 (2000)
27. L. Amendola, R. Barbieri, Phys. Lett. B **642**, 192 (2006)
28. D.J.E. Marsh, P.G. Ferreira, Phys. Rev. D **82**, 103528 (2010)
29. V. Poulin, T.L. Smith, D. Grin, T. Karwal, M. Kamionkowski, Phys. Rev. D **98**, 083525 (2018)
30. D.J. Bacon, R.A. Battye, P. Bull, S. Camera, P.G. Ferreira, I. Harrison, D. Parkinson, A. Pourtsidou, M.G. Santos, L. Wolz, et al., Publ. Astron. Soc. Aust. **37**, E007 (2020)
31. L.B. Newburgh, et al., Proc. SPIE Int. Soc. Opt. Eng. **9906**, 99065X (2016)
32. A. Khmelnitsky, V. Rubakov, J. Cosmol. Astropart. Phys. **1402**, 019 (2014)
33. N.K. Porayko, et al., Phys. Rev. D **98**, 102002 (2018)
34. F. Edwards, E. Kendall, S. Hotchkiss, R. Easther, J. Cosmol. Astropart. Phys. **1810**, 027 (2018)
35. V. Springel, Mon. Not. Roy. Astron. Soc. **364**, 1105 (2005)
36. L.M. Widrow, N. Kaiser, Astrophys. J. **416**, L71 (1993)
37. C. Uhlemann, M. Kopp, T. Haugg, Phys. Rev. D **90**, 023517 (2014)
38. P. Mocz, L. Lancaster, A. Fialkov, F. Becerra, P.H. Chavanis, Phys. Rev. D **97**, 083519 (2018)
39. R.E. Wyatt, *Quantum Dynamics with Trajectories* (Springer, Berlin, 2005)
40. L.E. Ballentine, *Quantum Mechanics* (World Scientific, Singapore, 1998)
41. D.G. Levkov, A.G. Panin, I.I. Tkachev, Phys. Rev. Lett. **121**, 151301 (2018)
42. H.Y. Schive, T. Chiueh, T. Broadhurst, Nat. Phys. **10**, 496 (2014)
43. P. Mocz, et al., Phys. Rev. Lett. **123**, 141301 (2019)
44. H.Y. Schive, T. Chiueh, T. Broadhurst, K.W. Huang, Astrophys. J. **818**, 89 (2016)
45. C. Schultz, J. Oñorbe, K.N. Abazajian, J.S. Bullock, Mon. Not. Roy. Astron. Soc. **442**, 1597 (2014)
46. R.J. Bouwens, et al., Astrophys. J. **803**, 34 (2015)

47. R.A. Windhorst, S.H. Cohen, R.A. Jansen, C. Conselice, H.J. Yan, New Astron. Rev. **50**, 113 (2006)
48. E. Leung, T. Broadhurst, J. Lim, J.M. Diego, T. Chiueh, H.Y. Schive, R. Windhorst, Astrophys. J. **862**, 156 (2018)
49. B. Bozek, D.J.E. Marsh, J. Silk, R.F.G. Wyse, Mon. Not. Roy. Astron. Soc. **450**, 209 (2015)
50. P.S. Corasaniti, S. Agarwal, D.J.E. Marsh, S. Das, Phys. Rev. D **95**, 083512 (2017)
51. M. Viel, G.D. Becker, J.S. Bolton, M.G. Haehnelt, Phys. Rev. D **88**, 043502 (2013)
52. N.Y. Gnedin, A.J.S. Hamilton, Mon. Not. Roy. Astron. Soc. **334**, 107 (2002)
53. M. Tegmark, M. Zaldarriaga, Phys. Rev. D **66**, 103508 (2002)
54. L. Hui, J.P. Ostriker, S. Tremaine, E. Witten, Phys. Rev. D **95**, 043541 (2017)
55. S. Chabanier, M. Millea, N. Palanque-Delabrouille, Mon. Not. Roy. Astron. Soc. **489**, 2247 (2019)
56. E. Armengaud, N. Palanque-Delabrouille, C. Yèche, D.J.E. Marsh, J. Baur, Mon. Not. Roy. Astron. Soc. **471**, 4606 (2017)
57. V. Iršič, M. Viel, M.G. Haehnelt, J.S. Bolton, G.D. Becker, Phys. Rev. Lett. **119**, 031302 (2017)
58. U.H. Zhang, T. Chiueh, Phys. Rev. D **96**, 063522 (2017)
59. K.H. Leong, H.Y. Schive, U.H. Zhang, T. Chiueh, Mon. Not. Roy. Astron. Soc. **484**, 4273 (2019)
60. B. Bar-Or, J.B. Fouvry, S. Tremaine, Astrophys. J. **871**, 28 (2019)
61. A.A. El-Zant, J. Freundlich, F. Combes, A. Halle, Mon. Notices Royal Astron. Soc. **492**, 877 (2020)
62. L. Lancaster, C. Giovanetti, P. Mocz, Y. Kahn, M. Lisanti, D.N. Spergel, J. Cosmol. Astropart. Phys. **2020**, 001 (2020)
63. D.J.E. Marsh, J.C. Niemeyer, Phys. Rev. Lett. **123**, 051103 (2019)
64. B.V. Church, P. Mocz, J.P. Ostriker, Mon. Not. Roy. Astron. Soc. **485**, 2861 (2019)
65. J. Binney, S. Tremaine, *Galactic Dynamics: Second Edition* (Princeton University Press, Princeton, 2008)
66. T.S. Li, et al., Astrophys. J. **838**, 8 (2017)
67. D. Crnojević, D.J. Sand, D. Zaritsky, K. Spekkens, B. Willman, J.R. Hargis, Astrophys. J. Lett. **824**, L14 (2016)
68. H.Y. Schive, T. Chiueh, T. Broadhurst, Phys. Rev. Lett. **124**, 201301 (2020)
69. K.K. Rogers, H.V. Peiris, Phys. Rev. Lett. **126**, 071302 (2021)
70. D.J.E. Marsh, A.R. Pop, Mon. Not. Roy. Astron. Soc. **451**, 2479 (2015)
71. S.M. Carroll, *Spacetime and Geometry. An Introduction to General Relativity* (Addison Wesley, Boston, 2004)
72. A. Arvanitaki, S. Dubovsky, Phys. Rev. D **83**, 044026 (2011)
73. R. Brito, V. Cardoso, P. Pani (eds.). *Superradiance*. Lecture Notes in Physics, vol. 906 (2015)
74. B.P. Abbott, et al., Phys. Rev. Lett. **116**, 061102 (2016)
75. M.J. Stott, D.J.E. Marsh, Phys. Rev. D **98**, 083006 (2018)
76. S.R. Dolan, Phys. Rev. D **98**, 104006 (2018)
77. R. Brito, V. Cardoso, P. Pani, Phys. Rev. D **88**, 023514 (2013)
78. W.E. East, F. Pretorius, Phys. Rev. Lett. **119**, 041101 (2017)
79. H. Yoshino, H. Kodama, Prog. Theor. Phys. **128**, 153 (2012)
80. A. Arvanitaki, M. Baryakhtar, X. Huang, Phys. Rev. D **91**, 084011 (2015)
81. J.G. Rosa, T.W. Kephart, Phys. Rev. Lett. **120**, 231102 (2018)
82. T. Ikeda, R. Brito, V. Cardoso, Phys. Rev. Lett. **122**, 081101 (2019)
83. F.V. Day, J.I. McDonald, J. Cosmol. Astropart. Phys. **10**, 051 (2019)
84. A. Arvanitaki, M. Baryakhtar, S. Dimopoulos, S. Dubovsky, R. Lasenby, Phys. Rev. D **95**, 043001 (2017)
85. R. Hložek, D.J.E. Marsh, D. Grin, R. Allison, J. Dunkley, E. Calabrese, Phys. Rev. D **95**, 123511 (2017)
86. T. Kobayashi, R. Murgia, A. De Simone, V. Iršič, M. Viel, Phys. Rev. D **96**, 123514 (2017)

87. A. Aghamousa, J. Aguilar, S. Ahlen, S. Alam, L.E. Allen, C.A. Prieto, J. Annis, S. Bailey, C. Balland, O. Ballester, et al., arXiv:1611.00036 (2016)
88. D. Grin, M.A. Amin, V. Gluscevic, R. HlČŠzek, D.J. Marsh, V. Poulin, C. Prescod-Weinstein, T.L. Smith, arXiv:1904.09003 (2019)
89. R. Ruffini, S. Bonazzola, Phys. Rev. **187**, 1767 (1969)
90. E. Seidel, W.M. Suen, Phys. Rev. Lett. **66**, 1659 (1991)
91. K. Clough, P. Figueras, H. Finkel, M. Kunesch, E.A. Lim, S. Tunyasuvunakool, Class. Quant. Grav. **32**, 245011 (2015)
92. T. Helfer, D.J.E. Marsh, K. Clough, M. Fairbairn, E.A. Lim, R. Becerril, J. Cosmol. Astropart. Phys. **1703**, 055 (2017)
93. D.G. Levkov, A.G. Panin, I.I. Tkachev, Phys. Rev. Lett. **118**, 011301 (2017)
94. F. Michel, I.G. Moss, Phys. Lett. B **785**, 9 (2018)
95. R. Arnowitt, S. Deser, C.W. Misner, Phys. Rev. **116**, 1322 (1959)
96. H.Y. Schive, M.H. Liao, T.P. Woo, S.K. Wong, T. Chiueh, T. Broadhurst, W.Y.P. Hwang, Phys. Rev. Lett. **113**, 261302 (2014)
97. X. Du, C. Behrens, J.C. Niemeyer, B. Schwabe, Phys. Rev. D **95**, 043519 (2017)
98. M. Mina, D.F. Mota, H.A. Winther, arXiv:2007.04119 (2020)
99. M. Nori, M. Baldi, Mon. Notices Royal Astron. Soc. **501**, 1539 (2021)
100. J. Veltmaat, J.C. Niemeyer, B. Schwabe, Phys. Rev. D **98**, 043509 (2018)
101. D.F. Jackson Kimball, D. Budker, J. Eby, M. Pospelov, S. Pustelny, T. Scholtes, Y.V. Stadnik, A. Weis, A. Wickenbrock, Phys. Rev. D **97**, 043002 (2018)
102. A. Derevianko, M. Pospelov, Nature Phys. **10**, 933 (2014)
103. G.F. Giudice, M. McCullough, A. Urbano, J. Cosmol. Astropart. Phys. **1610**, 001 (2016)
104. T. Dietrich, F. Day, K. Clough, M. Coughlin, J. Niemeyer, Mon. Not. Roy. Astron. Soc. **483**, 908 (2019)
105. C.J. Hogan, M.J. Rees, Phys. Lett. B **205**, 228 (1988)
106. A. Vaquero, J. Redondo, J. Stadler, J. Cosmol. Astropart. Phys. **04**, 012 (2019)
107. D. Ellis, D.J.E. Marsh, C. Behrens, Phys. Rev. D **103**, 083525 (2021).
108. A.M. Green, S. Hofmann, D.J. Schwarz, Mon. Not. Roy. Astron. Soc. **353**, L23 (2004)
109. O. Wantz, E.P.S. Shellard, Phys. Rev. D **82**, 123508 (2010)
110. S. Borsanyi, et al., Nature **539**, 69 (2016)
111. E.W. Kolb, I.I. Tkachev, Phys. Rev. D **50**, 769 (1994)
112. G. Sigl, *Astroparticle Physics: Theory and Phenomenology* (Springer, Berlin, 2017)
113. J. Peacock, *Cosmological Physics* (Cambridge University Press, Cambridge, 1999)
114. K.M. Zurek, C.J. Hogan, T.R. Quinn, Phys. Rev. D **75**, 043511 (2007)
115. M. Gosenca, J. Adamek, C.T. Byrnes, S. Hotchkiss, Phys. Rev. D **96**, 123519 (2017)
116. B. Eggemeier, J. Redondo, K. Dolag, J.C. Niemeyer, A. Vaquero, Phys. Rev. Lett. **125**, 041301 (2020)
117. J. Enander, A. Pargner, T. Schwetz, J. Cosmol. Astropart. Phys. **12**, 038 (2017)
118. M. Fairbairn, D.J.E. Marsh, J. Quevillon, S. Rozier, Phys. Rev. D **97**, 083502 (2018)
119. B. Eggemeier, J.C. Niemeyer, Phys. Rev. D **100**, 063528 (2019)
120. E.W. Kolb, I.I. Tkachev, Astrophys. J. Lett. **460**, L25 (1996)
121. A. Katz, J. Kopp, S. Sibiryakov, W. Xue, J. Cosmol. Astropart. Phys. **12**, 005 (2018)
122. M. Fairbairn, D.J.E. Marsh, J. Quevillon, Phys. Rev. Lett. **119**, 021101 (2017)
123. I. Tkachev, JETP Lett. **101**, 1 (2015)
124. A. Iwazaki, arXiv:1412.7825 (2014)
125. Y. Bai, Y. Hamada, Phys. Lett. B **781**, 187 (2018)
126. E. Hardy, J. High Energy Phys. **02**, 046 (2017)
127. M. Feix, S. Hagstotz, A. Pargner, R. Reischke, B.M. Schaefer, T. Schwetz, arXiv:2004.02926 (2020)
128. H. Niikura, et al., Nat. Astron. **3**, 524 (2019)
129. B. Paxton, L. Bildsten, A. Dotter, F. Herwig, P. Lesaffre, F. Timmes, Astrophys. J. Suppl. Ser. **192**, 3 (2011)
130. A. Friedland, M. Giannotti, M. Wise, Phys. Rev. Lett. **110**, 061101 (2013)

131. G.G. Raffelt, Phys. Rep. **198**, 1 (1990)
132. G.G. Raffelt, in *Axions* (Springer, Berlin, 2008), p. 51
133. E. Aver, K.A. Olive, E.D. Skillman, J. Cosmol. Astropart. Phys. **1507**, 011 (2015)
134. A. Ayala, I. Domínguez, M. Giannotti, A. Mirizzi, O. Straniero, Phys. Rev. Lett. **113**, 191302 (2014)
135. M. Giannotti, I. Irastorza, J. Redondo, A. Ringwald, J. Cosmol. Astropart. Phys. **1605**, 057 (2016)
136. S. Hoof, F. Kahlhoefer, P. Scott, C. Weniger, M. White, J. High Energy Phys. **03**, 191 (2019)
137. J.H. Chang, R. Essig, S.D. McDermott, J. High Energy Phys. **09**, 051 (2018)
138. P.W. Graham, S. Rajendran, Phys. Rev. D **88**, 035023 (2013)
139. H.L. Shipman, Astrophys. J. **228**, 240 (1979)
140. A.H. Corsico, L.G. Althaus, M.M. Miller Bertolami, S. Kepler, Astron. Astrophys. Rev. **27**, 7 (2019)
141. D.E. Winget, C.J. Hansen, H.M. van Horn, Nature **303**, 781 (1983)
142. S.D. Kawaler, D.E. Winget, C.J. Hansen, Astrophys. J. **295**, 547 (1985)
143. M. Nakagawa, Y. Kohyama, N. Itoh, Astrophys. J. **322**, 291 (1987)
144. M. Nakagawa, T. Adachi, Y. Kohyama, N. Itoh, Astrophys. J. **326**, 241 (1988)
145. A.D. Romero, L.A. Amaral, T. Klippel, D. Sanmartim, L. Fraga, G. Ourique, I. Pelisoli, G.R. Lauffer, S.O. Kepler, D. Koester, Mon. Not. R. Astron. Soc. **490**, 1803 (2019)
146. J. Isern, M. Hernanz, E. Garcia-Berro, Astrophys. J. **392**, L23 (1992)
147. T. Altherr, E. Petitgirard, T. del Rio Gaztelurrutia, Astropart. Phys. **2**, 175 (1994)
148. A. Bischoff-Kim, M.H. Montgomery, D.E. Winget, Astrophys. J. **675**, 1512 (2008)
149. J. Isern, E. Garcia-Berro, S. Torres, S. Catalan, Astrophys. J. **682**, L109 (2008)
150. A.H. Corsico, L.G. Althaus, M.M.M. Bertolami, A.D. Romero, E. Garcia-Berro, J. Isern, S.O. Kepler, Mon. Not. Roy. Astron. Soc. **424**, 2792 (2012)
151. A.H. Corsico, L.G. Althaus, A.D. Romero, A.S. Mukadam, E. Garcia-Berro, J. Isern, S.O. Kepler, M.A. Corti, J. Cosmol. Astropart. Phys. **1212**, 010 (2012)
152. A.H. Corsico, A.D. Romero, L.G. Althaus, E. Garcia-Berro, J. Isern, S.O. Kepler, M.M. Miller Bertolami, D.J. Sullivan, P. Chote, J. Cosmol. Astropart. Phys. **1607**, 036 (2016)
153. T. Battich, A.H. Corsico, L.G. Althaus, M.M. Miller Bertolami, M.M.M. Bertolami, J. Cosmol. Astropart. Phys. **1608**, 062 (2016)
154. S. Kepler, Publ. Astron. Soc. Pac.: Conf. Ser. **462**, 322 (2012)
155. M. Giannotti, I.G. Irastorza, J. Redondo, A. Ringwald, K. Saikawa, J. Cosmol. Astropart. Phys. **10**, 010 (2017)
156. M.M. Miller Bertolami, B.E. Melendez, L.G. Althaus, J. Isern, J. Cosmol. Astropart. Phys. **1410**, 069 (2014)
157. J. Isern, E. Garcia-Berro, S. Torres, R. Cojocaru, S. Catalan, Mon. Not. Roy. Astron. Soc. **478**, 2569 (2018)
158. E. Chupp, W. Vestrand, C. Reppin, Phys. Rev. Lett. **62**, 505 (1989)
159. A. Payez, C. Evoli, T. Fischer, M. Giannotti, A. Mirizzi, A. Ringwald, J. Cosmol. Astropart. Phys. **02**, 006 (2015)
160. M. Berg, J.P. Conlon, F. Day, N. Jennings, S. Krippendorf, A.J. Powell, M. Rummel, Astrophys. J. **847**, 101 (2017)
161. J.P. Conlon, F. Day, N. Jennings, S. Krippendorf, M. Rummel, J. Cosmol. Astropart. Phys. **07**, 005 (2017)
162. F. Day, S. Krippendorf, Galaxies **6**, 45 (2018)
163. C.S. Reynolds, M.D. Marsh, H.R. Russell, A.C. Fabian, R. Smith, F. Tombesi, S. Veilleux, Astrophys. J. **890**, 59 (2020)
164. A. Mirizzi, J. Redondo, G. Sigl, J. Cosmol. Astropart. Phys. **08**, 001 (2009)
165. H. Tashiro, J. Silk, D.J.E. Marsh, Phys. Rev. D **88**, 125024 (2013)
166. J.C. Mather, et al., Astrophys. J. **420**, 439 (1994)
167. R. Durrer, A. Neronov, Astron. Astrophys. Rev. **21**, 62 (2013)

Chapter 4
Microwave Cavity Searches

Maria Simanovskaia, Gianpaolo Carosi, and Karl van Bibber

Abstract The axion "haloscope" technique is a well-established method to search for dark matter axions with a resonant microwave cavity and has excluded axion models over several frequency ranges with unparalleled sensitivity. This chapter describes the basics of microwave cavity searches, including overviews of the main experimental components and details on the figure of merit for these searches.

4.1 Historical Introduction

By the early 1980s it was realized that a low mass axion would be a compelling dark matter candidate; the excellent agreement of theory and data for the neutrino signal seen from the Type-II supernova SN1987A a few years later providing an upper mass bound of $\sim 60\,\text{meV}$ [1].[1] Problematically, however, the axion-photon coupling for the QCD axion associated with those masses was so extremely small as to preclude conventional accelerator- or reactor-based searches by many orders of magnitude.

---

[1] Initially, a lower mass bound of $6\,\mu\text{eV}$ was set by overclosure arguments. Subsequent theoretical developments have eased that lower bound (see, for example, Refs. [2–4]), making the experimentalists' work harder; see the review of current theory in Chap. 3.

M. Simanovskaia (✉)
Department of Physics, Stanford University, Stanford, CA, USA
e-mail: simanovskaia@stanford.edu

G. Carosi
Lawrence Livermore National Laboratory, Livermore, CA, USA
e-mail: carosi2@llnl.gov

K. van Bibber
Department of Nuclear Engineering, University of California Berkeley, Berkeley, CA, USA
e-mail: karl.van.bibber@berkeley.edu

© The Author(s) 2023
D. F. Jackson Kimball, K. van Bibber (eds.), *The Search for Ultralight Bosonic Dark Matter*, https://doi.org/10.1007/978-3-030-95852-7_4

This conundrum was potentially solved by Pierre Sikivie in a seminal paper in 1983, where he showed that axions constituting the galactic halo dark matter could be detected by their resonant conversion to photons in a microwave cavity permeated by a magnetic field [5]. While the signal expected was extraordinarily weak, sensitivity estimates based on the technology of large-volume, high-field superconducting magnets, high-quality-factor cavities and ultralow noise amplifiers of that time appeared to make detection of the QCD axion very nearly within reach. Two early pilot experiments were soon mounted, one at the University of Florida (UF) [6] and the other at Brookhaven National Laboratory (BNL) by a Rochester-BNL-Fermilab (RBF) collaboration [7] providing experimental validation for that optimism, and setting limits on the axion-photon coupling $g_{a\gamma\gamma}$ within a factor of 10–100 of the model band. A watershed moment for the field was a workshop convened at BNL by Adrian Melissinos of the University of Rochester on April 13–14, 1989 [8] that brought together forty scientists and engineers, including experts in low-noise receivers, microwave resonators and superconducting magnets, to study whether projections of those technologies supported the idea of actually reaching the QCD model band and whether planning for a large-scale experiment was warranted at that time. The answer was unequivocally yes, and an R&D collaboration was formed from among the participants, opening the path to what has become a three-decade, world-wide effort on microwave cavity experiments and variations on the theme. These in turn have not only been beneficiaries but also drivers of technology development, particularly in quantum metrology.

4.2 Detection Principles

Microwave cavity searches rely on the axion's coupling to two photons through the inverse Primakoff effect (see Ref. [9] and Sect. 2.4.2). In a resonant microwave cavity immersed in a magnetic field, axions interact with the virtual photons of the magnetic field and convert to an oscillating electromagnetic field with a frequency ν_a corresponding to the axion mass m_a as $\nu_a \approx m_a c^2 / h$. The resonant conversion condition is that the axion mass is within the bandwidth of the microwave cavity at its resonance frequency. Since the axion mass is unknown, the cavity resonance frequency must be tuned to access a range of axion masses. As the resonance frequency of the cavity is tuned, the electromagnetic field inside the cavity is measured by a small probe antenna inserted in the cavity, which is in turn coupled to an ultralow noise preamplifier. There is an ongoing effort to maximize the axion signal power while reducing the background system noise in order to maximize the frequency search rate.

A standard detection schematic is illustrated in Fig. 4.1. The axion field a interacts with the virtual photons γ^* of the magnetic field and converts into a measurable oscillating electromagnetic field γ when the axion mass m_a is within the bandwidth of the cavity resonance frequency ν_c. The width of the axion signal is expected to be $\Delta\nu_a \leq 10^{-6}\nu_a$, and the bandwidth of the cavity resonance

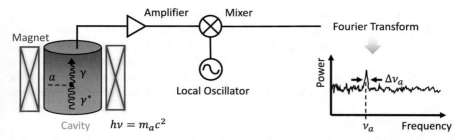

Fig. 4.1 Simplified axion detection schematic of microwave cavity searches. Much like an AM-radio, the high frequency axion signal is mixed down to audio frequencies by mixing with a local oscillator maintained at a fixed offset frequency from the cavity frequency. This allows for much lower digitization rates. The cartoon power spectrum shows a sample axion signal above the noise

frequency is determined by that resonance mode's quality factor. To measure the power inside the cavity, an inserted coaxial antenna probes the longitudinal electric field. Ultralow noise amplifiers boost the signal to a level where it can be properly mixed down to a lower frequency with a local oscillator and then the phase and power information can be digitally recorded. Finally, a Fourier transform is applied to the time-dependent signal resulting in a frequency-dependent power spectrum. Axion candidate frequencies are identified as signals above a target threshold and are revisited during a rescanning process to confirm if they are a persistent signal or statistical noise.

Several collaborations are implementing this axion detection method. These include, but are not limited to, the Axion Dark Matter eXperiment (ADMX), the Haloscope At Yale Sensitive To Axion Cold dark matter (HAYSTAC), the Center for Axion and Precision Physics (CAPP) Ultra Low Temperature Axion Search in Korea (CULTASK), and the CryOgenic Resonant Group Axion CoNverter (ORGAN). These modern experiments, derived from the early pilot experiments of the RBF and UF collaborations, use similar detection techniques but have unique designs and mostly operate over different frequency ranges.

4.2.1 Signal Power

The signal power in a microwave cavity search can be derived by solving the equations of motion for the electromagnetic field coupled to the axion in the case of a resonant microwave cavity permeated by a static magnetic field and the axion field [10]. It is determined by a combination of theoretical parameters describing axion physics and measurable parameters describing the experimental apparatus in the equation

$$P_{\text{sig}} = \left(g_{a\gamma\gamma}^2 \frac{\hbar^3 c^3 \rho}{m_a^2} \right) \times \left(\frac{1}{\mu_0} B_0^2 \omega_c V C_{mn\ell} Q_0 \frac{\beta}{(1+\beta)^2} \frac{1}{1 + (2\Delta\nu_a/\Delta\nu_c)^2} \right),$$

(4.1)

where the factors in the first set of parentheses involve theoretical parameters set by nature and the factors in the second set of parentheses are experimental parameters. Theoretical parameters include the model-dependent coupling constant $g_{a\gamma\gamma}$, local dark matter density $\rho \approx 0.45\,\text{GeV}\,\text{cm}^{-3}$ (commonly used in axion searches [11] and consistent with recent measurements [12]), and the axion mass m_a. The coupling constant itself has units of GeV^{-1} and can be further expressed as $g_{a\gamma\gamma} = g_\gamma \alpha/(\pi f_a) = g_\gamma \alpha m_a/(\pi \cdot 0.006\,\text{GeV}^2)$ where the dimensionless g_γ changes between classes of models. Representative values are $g_\gamma = -0.97$ for the Kim–Shifman–Vainshtein–Zakharov (KSVZ) [13, 14] family of models and $g_\gamma = 0.36$ for the Dine–Fischler–Srednicki–Zhitnitsky (DFSZ) [15, 16] family of models. The relevant experimental parameters are external magnetic field strength B_0, cavity resonance frequency $\omega_c = 2\pi \nu_c$, cavity volume V, mode-specific cavity form factor $C_{mn\ell}$ (often, C_{010}), unloaded quality factor Q_0, cavity coupling parameter β ($\beta = 1$ corresponds to critical coupling, $\beta < 1$ is undercoupled, and $\beta > 1$ is overcoupled), and cavity linewidth $\Delta \nu_c$.

Typical values for the HAYSTAC detector are $B_0 = 9\,\text{T}$, $\omega_c = 2\pi\,5\,\text{GHz}$, $V = 1.5\,\text{L}$, $C_{010} = 0.5$, $Q_L = 10^4$, $\beta = 2$, $\Delta \nu_a \leq 5\,\text{kHz}$, and $\Delta \nu_c = \nu_c/Q_L = \omega_c/(2\pi Q_L)$, where Q_L is the cavity loaded quality factor defined by $Q_L = Q_0/(1 + \beta)$. Altogether, the expected power for these parameters at the axion–photon coupling set by the KSVZ family of models is $P_{\text{sig}} \approx 10^{-24}\,\text{W}$.

? Problem 4.1 Axion to Photon Production Rate

The ADMX experiment searches for dark matter axions with DFSZ scale coupling to photons and consists of a $B_0 = 7.6\,\text{T}$ magnet, a cavity system with $V = 150\,\text{L}$, $C_{010} = 0.45$, $Q_0 = 180,000$, and typical $\beta = 2$. Assuming the existence of a $3.3\,\mu\text{eV}$ dark matter axion with DFSZ coupling ($g_\gamma = 0.36$) at a typical local density of $\rho = 0.45\,\text{GeV}\,\text{cm}^{-3}$ what is the average number of photons emitted from the cavity every second when it is tuned to the correct frequency?

Solution on page 326.

4.2.2 Noise Considerations

For any phase-insensitive linear receiver the system noise temperature T_{sys} may be written

$$k_B T_{\text{sys}} = h\nu N_{\text{sys}} = h\nu \left(\frac{1}{e^{h\nu/k_B T} - 1} + \frac{1}{2} + N_A \right), \qquad (4.2)$$

where the three additive contributions correspond, respectively, to a blackbody photon gas in equilibrium with the cavity at temperature T, the zero-point fluctuations of the photon field, and the input-referred added noise of the receiver. The latter two terms combine to form the standard quantum limit (SQL), with $N_A \geq 1/2$ [17].

The Dicke radiometer equation [18] combines the expected signal power with the system noise temperature to form the signal-to-noise ratio Σ:

$$\Sigma = \frac{P_{\text{sig}}}{k_B T_{\text{sys}}} \sqrt{\frac{\tau}{\Delta \nu_a}}, \tag{4.3}$$

where τ is the integration time, and $\Delta \nu_a$ is the expected linewidth of the axion.

There is an active effort in the microwave cavity search community working to increase expected signal power and decrease system noise temperature rather than integrating for longer to improve Σ. Increasing the applied magnetic field or improving cavity performance (volume, quality factor, or form factor) increases the expected signal power, as suggested by Eq. (4.1). To decrease system noise temperature, the experiments are cooled to temperatures as low as possible and state-of-the-art amplifier technologies are implemented.

4.2.3 Scan Rate

Because the mass, and hence the oscillation frequency, of the axion is unknown, resonant microwave cavity searches must scan over a wide range of frequencies. Therefore, the ultimate figure of merit is the scan rate, which incorporates the expected signal power and noise considerations, and quantifies how quickly searches can scan through different frequencies at a given sensitivity

$$R \equiv \frac{d\nu}{dt} \approx \frac{4}{5} \frac{Q_L Q_a}{\Sigma^2} \left(g_{a\gamma\gamma}^2 \frac{\hbar^3 c^3 \rho_a}{m_a^2} \right)^2 \times \left(\frac{1}{\hbar \mu_0} \frac{\beta}{1+\beta} B_0^2 V C_{mn\ell} \frac{1}{N_{\text{sys}}} \right)^2. \tag{4.4}$$

Most of these terms are recognizable from the expression of the signal power in Eq. (4.1). Using typical values for HAYSTAC, to achieve the benchmark KSVZ sensitivity g_{KSVZ}, the scan rate of the first run would have been approximately 40 MHz/yr. Since the scan rate scales as the fourth power of the coupling constant, the scan rate would have been 640 MHz/yr to achieve twice the KSVZ sensitivity $2g_{\text{KSVZ}}$.

Improving the scan rate allows us to search through a mass range more quickly, but we are still limited by the tuning range of the cavity and amplifier electronics. The frequency range we can probe depends on the resonance frequency of the microwave cavity. In general, higher-frequency cavities have a smaller volume and therefore suffer from a smaller expected signal power as well as an increase in operational complexity due to a higher resonance mode density. New cavity designs are being developed that expand the accessible frequency range while improving sensitivity. This requires investigating various geometries using electromagnetic simulations, prototypes, and microwave testing.

4.3 Resonant Microwave Cavities

A resonant microwave cavity supports many modes with various electric and magnetic field profiles. Microwave cavity searches generally focus on one resonant mode and use a cavity design that optimizes the mode of interest for the figure of merit within a frequency tuning range while preserving mode purity.

The figure of merit of a cavity resonant mode is determined by the scan rate R (Eq. 4.4), which is partially composed of cavity geometry and resonant mode characteristics. The components include the quality factor Q, which quantifies losses, the form factor $C_{mn\ell}$, which describes the alignment of the resonant mode electric field to the external magnetic field, and the cavity volume V. The scan rate depends on these quantities as

$$R \equiv \frac{dv}{dt} \propto Q\, C_{mn\ell}^2\, V^2. \tag{4.5}$$

The volume in the cavity figure of merit involves the internal cavity space through which electric fields penetrate. The volume generally decreases with increasing frequency, which is one of the main challenges in designing higher-frequency cavities. The HAYSTAC cavity, shown in Fig. 4.2, is a closed cylindrical volume of 5.08 cm radius and 25.4 cm length with one tuning rod of 2.54 cm radius that can be rotated off center.

4.3.1 Resonant Cavity Modes

The resonant modes present in a cavity are characterized by their electromagnetic field profiles that obey the Maxwell equations and the usual boundary conditions. The Maxwell equations in vacuum, assuming that the electric and magnetic fields have time dependence $e^{-i\omega t}$, reduce to

$$\begin{aligned}
\nabla \times \boldsymbol{E}\,(\rho, \phi, z) &= i\frac{\omega}{c}\boldsymbol{B}\,(\rho, \phi, z) \\[4pt]
\nabla \times \boldsymbol{B}\,(\rho, \phi, z) &= -i\mu\epsilon\frac{\omega}{c}\boldsymbol{E}\,(\rho, \phi, z) \\[4pt]
\nabla \cdot \boldsymbol{E}\,(\rho, \phi, z) &= 0 \\[4pt]
\nabla \cdot \boldsymbol{B}\,(\rho, \phi, z) &= 0,
\end{aligned} \tag{4.6}$$

Fig. 4.2 HAYSTAC copper-coated stainless steel resonant microwave cavity tunable over the frequency range 3.6–5.8 GHz. The top cover is not shown

where E is the electric field and B is the magnetic field. At a perfect electric conductor (PEC) surface, the tangential electric fields and normal magnetic fields vanish, the tangential component of magnetic field is related to the surface current density j_s, and the normal component of electric field is related to the surface charge density ρ_s. These boundary conditions are, respectively, summarized in the following equations:

$$\hat{n} \times E = 0$$
$$\hat{n} \cdot B = 0$$
$$\hat{n} \times H = j_s$$
$$\hat{n} \cdot D = \rho_s,$$

(4.7)

where $\hat{n}$ is the vector normal to the PEC surface s.

The solutions to these equations and boundary conditions describe the resonant modes present in a cavity. The modes include transverse magnetic (TM), transverse electric (TE), and transverse electric and magnetic (TEM) modes. TM modes are characterized by a transverse magnetic field and longitudinal electric field, TE modes are characterized by a transverse electric field and longitudinal magnetic field, and TEM modes are characterized by a transverse electric and magnetic field. TEM modes only exist in structures with a central conductor. Characteristics of the three types of resonant modes are summarized in Table 4.1.

Each TE and TM mode is classified by three numbers m, n, and ℓ that describe the variation in the azimuthal, radial, and longitudinal directions, respectively. For example, the TM_{020} and TM_{011} modes have one extra radial node and longitudinal

Table 4.1 Resonant mode descriptions, assuming that the $\hat{z}$ direction is along the length of the cavity

Mode type	Description
Transverse magnetic (TM)	$B_z = 0$ everywhere; boundary condition $E_z\,(r = s) = 0$
Transverse electric (TE)	$E_z = 0$ everywhere; boundary condition $\frac{\partial B_z}{\partial n}\,(r = s) = 0$
Transverse electromagnetic (TEM)	$E_z = 0$ and $B_z = 0$ everywhere

node, respectively, in the electric field compared to the TM_{010} mode. For an empty cylindrical cavity of radius r_{cavity} and height h_{cavity}, the resonant frequencies of the $TM_{mn\ell}$ modes are given by

$$\omega TM_{mn\ell} = \frac{c}{\sqrt{\mu \epsilon}} \sqrt{\frac{x_{mn}^2}{r_{\text{cavity}}^2} + \frac{\ell^2 \pi^2}{h_{\text{cavity}}^2}}, \tag{4.8}$$

where $m, \ell = 0, 1, 2, \ldots$, $n = 1, 2, 3, \ldots$, c is the speed of light, μ is the permeability, ϵ is the permittivity, and x_{mn} is the nth root of the Bessel function of the first kind $J_m\,(x)$. The resonant frequencies of the $TE_{mn\ell}$ modes are given by

$$\omega TE_{mn\ell} = \frac{c}{\sqrt{\mu \epsilon}} \sqrt{\frac{(x'_{mn})^2}{r_{\text{cavity}}^2} + \frac{\ell^2 \pi^2}{h_{\text{cavity}}^2}}, \tag{4.9}$$

where $m = 0, 1, 2, \ldots$, $n = 1, 2, 3, \ldots$, $\ell = 1, 2, 3, \ldots$ and x'_{mn} is the nth root of the derivative of the Bessel function of the first kind $J'_m\,(x)$.

Some of the resonant modes can be tuned in frequency by moving rods inside cylindrical cavities and some have approximately stationary resonance frequencies upon changing the position of the rod. Since the TE frequencies are primarily determined by the length of the cavity, they do not change significantly when the central conductor moves or changes in size.

? Problem 4.2 Cavity Resonance Frequencies

For an empty cylindrical cavity of radius 5.0 cm, what is the frequency of the TM_{010} mode? Find the number of TE modes within 1.0 GHz of the TM_{010} mode for cavity heights of 5.0 cm, 10.0 cm, and 20.0 cm.

Solution on page 326.

4.3.2   Quality Factor

As suggested in Eq. (4.5), the cavity figure of merit scales as the quality factor Q. The quality factor is a property of each resonant mode in the cavity and is given by the ratio of the stored energy U to the dissipated power P_d in the cavity, multiplied by the resonant mode frequency ω:

$$Q = \omega \frac{U}{P_d}. \tag{4.10}$$

The stored energy in the cavity is proportional to the square of the electric field integrated over the cavity volume filled with material of dielectric constant ϵ:

$$U = \frac{1}{2}\epsilon \int_{\text{cavity volume}} |E|^2 \, dV. \tag{4.11}$$

The power loss in the cavity is proportional to the square of the magnetic field integrated over the metallic surfaces inside the cavity:

$$P_d = \frac{\omega \mu \delta}{4} \int_{\text{cavity surfaces}} |H|^2 \, dA, \tag{4.12}$$

where μ is the magnetic permeability of the metallic cavity surfaces, and the skin depth δ is the distance that electric fields are allowed to penetrate into the metallic surfaces. The classical skin depth is given by

$$\delta = \sqrt{\frac{2}{\omega \mu \sigma}}, \tag{4.13}$$

where σ is the conductivity of the metallic surface. Conductivity improves with decreasing temperature, and the classical skin depth is expected to improve as well. However, at sufficiently low temperatures, the skin depth reaches an asymptote. In HAYSTAC, cooling the cavities from room temperature to 4 K gives an improvement of the quality factor by approximately a factor of four at a frequency around 1 GHz. For comparison, the conductivity improves by a factor of over a hundred in that temperature range. The classical description becomes invalid when the skin depth decreases below the electron's mean free path. In this regime, the skin depth depends on the electron density instead of the normal conductivity. This anomalous skin depth [19] is given by

$$\delta_a = \left(\frac{\sqrt{3}\,c^2 m_e v_F}{8\pi^2 \omega n e^2} \right)^{1/3}, \tag{4.14}$$

where m_e is the electron mass, v_F is the Fermi velocity, n is the conduction electron density, and e is the electron charge [20].

The quality factor is determined primarily by the material and geometry of the cavity and can also vary greatly between resonant modes. Materials with a higher conductivity typically have smaller skin depths and therefore higher quality factors. The desire to have higher conductivity motivates making or plating the cavities with oxygen-free high-conductivity (OFHC) copper and annealing them for further conductivity improvement. Superconducting materials are appealing but the thickness of the superconductor must be kept small enough to prevent lossy vortices from forming in the strong magnetic field. This technology is being explored by the ADMX, CAPP, and QUAX experiments (the latter of which is further described in Chap. 8 [21, 22]),

4.3.3 Form Factor

The form factor is a measure of how well the electric field of a resonant mode aligns with the applied external magnetic field. It is given by

$$C_{mn\ell} = \frac{\left(\int_{\text{cavity volume}} \boldsymbol{E} \cdot \boldsymbol{B}_0 \, dV\right)^2}{B_0^2 \, V \int_{\text{cavity volume}} \epsilon_r \, |\boldsymbol{E}|^2 \, dV}, \tag{4.15}$$

where V is the cavity volume not occupied by a metallic object and filled with material of dielectric constant $\epsilon = \epsilon_0 \epsilon_r$. Note that if the cavity is partially filled with a dielectric material, ϵ_r varies in space.

The form factor is maximized when $\boldsymbol{E} \cdot \boldsymbol{B}_0$ integrated over the volume is maximized. Since the applied magnetic field for microwave cavity searches is commonly in the $\hat{z}$ direction, all resonant modes without electric field components in the $\hat{z}$ direction have form factors that are identically zero, and many TM modes have portions of their field that cancel out, thereby lowering their integrated form factor to near zero.

? Problem 4.3 Form Factor of an Annular Cavity

Compare the form factor of the TM_{010}-like mode and the form factor of the TM_{030}-like mode in an annular cavity. Assume that the magnetic field $\boldsymbol{B}_0$ is in the $\hat{z}$ direction and the electric field of the TM_{0n0}-like mode in an annular cavity with cavity radius r_{cavity}, rod radius r_{rod}, and height h_{cavity} can be described by

$$E_z(\rho) \sim \sin k_0 (\rho - r_{\text{rod}}), \text{ where } k_0 = \frac{\pi n}{r_{\text{cavity}} - r_{\text{rod}}}. \tag{4.16}$$

Solution on page 327.

4.3.4   Tuning and Mode Density

Quality factor, form factor, and volume quantitatively describe the behavior of interest of a single resonant mode at a given frequency. These quantities give a general sense of performance across a tuning range, but if the range is full of intruder modes, it will be interrupted. A TM mode resonance frequency decreases when a rod rotates away from the center of the cavity. In comparison, the TE mode resonance frequency does not change significantly. When the TE and TM mode frequencies approach each other, the two modes mix, producing two hybrid modes, in analogy with two-level mixing in quantum mechanics [23]. If the mode of interest hybridizes significantly, it will be difficult or impossible to interpret the results of the experiment, thus leading to a notch in frequency coverage of the experiment. Mode density is difficult to quantify, but it is a key consideration for cavity design. The problem of mode density worsens for cavities of too large an aspect ratio h_{cavity}/r_{cavity}; practically one is constrained to stay with cavity designs of $h_{cavity}/r_{cavity} \sim 5$ or lower.

4.3.5   Multiple Cavity Systems

A cavity's TM_{010} frequency scales inversely to the radius and thus the volume, assuming a constant length-to-radius aspect ratio, decreases as $V \propto \nu^{-3}$ in going to higher frequencies. For a fixed magnet solenoid volume one can simply increase the number of cavities N each with their own independent receiver chains which can then combine their powers statistically for a $\sqrt{N}$ improvement to the $ of a single cavity. However, one can also take advantage of the coherent nature of the axion signal to recover this volume more efficiently by co-adding the in-phase voltage signals of multiple frequency-locked cavities. The axion signal, though it has an unknown global phase, will generate the same, in-phase, signal in each cavity. The voltages from each cavity can thus be combined in phase to provide $N \times V_a$ output voltage. The noise power from each cavity would be added incoherently providing an added noise level of $\sqrt{N} \times V_{noise}$. Squaring these to get power we see that we can get a signal-to-noise enhancement of $N \times \$_{single\ cavity}$ [24, 25]. The price that one pays in such a scheme is the added complexity of controlling all the cavity systems so that they are within a linewidth of the other cavities.

4.3.6   Testing Cavities

Before incorporating the resonant microwave cavity in the detector, it must be thoroughly studied and characterized. Changing the cavity geometry (for example,

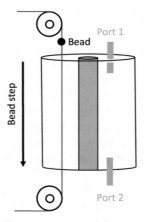

Fig. 4.3 Bead perturbation technique setup

by moving a tuning rod) changes the mode frequencies. Mode maps track these changes by showing mode frequencies at each cavity geometry change.

A vector network analyzer measures reflection and transmission of microwave signals in the frequency range of interest. When the coaxial antennas couple to a resonant mode, more signal is transmitted. By measuring the scattering parameters between two ports either through transmission (S_{12}, S_{21}) or reflection (S_{11}, S_{22}), the frequencies and quality factors of resonant modes in the cavity can be measured. All measurements are done with weak coupling to the coaxial antennas to minimize perturbations due to the antenna presence in the cavity. Scattering parameter measurements give information on the frequencies and quality factors of resonant modes, but not on their electric field distribution. To get insight on the resonant mode electric field distribution, the cavity is probed by pulling a relatively (compared to the cavity size) small bead through the length of the cavity and measuring the resonance frequency at each step. An example setup and bead pull measurement are shown in Fig. 4.3. The presence of the bead inside the cavity perturbs the electromagnetic field and shifts the resonance frequency by a magnitude proportional to the square of the strength of the electric field at the bead location [26]. If the bead is only slightly perturbing the electric field, the expected frequency shift is given by

$$\frac{\Delta\omega}{\omega} = \frac{-(\epsilon - 1)}{2} \frac{V_{\text{bead}}}{V_{\text{cav}}} \frac{E(r)^2}{\langle E(r)^2 \rangle_{\text{cav}}}, \tag{4.17}$$

where V_{bead} and V_{cav} are the volumes of the bead and cavity, respectively [27]. This bead perturbation technique, which is commonly used in the microwave engineering community, allows the resonant mode to be identified, to determine whether the mode is significantly hybridized, and to ensure that the cavity is properly aligned. This is essential to confirming that the cavity form factor corresponds to its calculated value.

4.4 Amplifiers

The current microwave cavity axion experiments are establishing limits on the axion that correspond to signals with powers on the order of 10^{-24} W, or equivalently on the order of one axion-to-microwave-photon conversion per second. Achieving sensitivity to such small signals depends critically on the equivalent system noise temperature T_{sys} and thus requires the state-of-the-art ultralow noise detectors.

The first-generation RBF and UF experiments, and the ADMX experiment for its first several years of operation, utilized High Electron Mobility Transistor amplifiers (HEMTs). Noise added by HEMTs decreases with temperature down to a minimum of a few Kelvin when cooled to liquid helium temperatures, but plateaus before reaching the SQL. Since the scan rate in Eq. (4.4) is inversely proportional to T_{sys}^2, decreasing the system noise temperature would significantly decrease the amount of time it would take to scan through the axion parameter space accessible by resonant cavity searches.

? Problem 4.4 The Standard Quantum Limit

What is the noise temperature of the standard quantum limit for 700 MHz? What about for 6 GHz?

Solution on page 327.

Amplifiers presently in use are operating at or near a system noise temperature corresponding to the SQL, and recently a squeezed-vacuum state receiver has been employed to circumvent the SQL [28].

4.4.1 Quantum-Limited Amplifiers

Unlike the HEMT noise temperature that plateaus at a few Kelvin, the noise temperature of amplifiers based on DC superconducting quantum interference devices (SQUIDs) decreases roughly linearly as the physical system temperature decreases to around 0.1 K [29]. SQUIDs are naturally applied to low frequencies, but replacing the input coil with a tunable microstrip resonator enables operation of SQUIDs up to 1 GHz [30]. Using SQUIDs provides the ability to drive the system noise temperature close to the SQL while introducing the challenge of magnetic shielding. Since they are sensitive to magnetic flux, experiments must magnetically shield them from the high magnetic fields permeating the resonant microwave cavity.

SQUIDs can operate up to a few GHz, but the axion parameter space extends to higher frequencies. Josephson parametric amplifiers (JPAs) are naturally resonant devices designed to operate in the 2–12 GHz range. A JPA is a nonlinear LC circuit, with the inductance provided by an array of SQUIDs [31]. Like the SQUID, the JPA must be carefully magnetically shielded. To give the scale of shielding, in HAYSTAC, the magnetic shielding is composed of a second superconducting magnet coil to negate the field from the main magnetic in the region of the quantum amplifiers, passive persistent coils, and ferromagnetic and superconducting shields [32]. Ultimately successful operation of JPAs in HAYSTAC required that the remnant field be reduced to a level corresponding to much less than one flux quantum in the region of the device. Although these amplifiers offer improved noise performance at a range of frequencies, they are limited by the SQL.

4.4.2 Sub-quantum Limited Amplifiers

During initial operation, HAYSTAC operated the JPA as a low-noise phase-insensitive linear amplifier subject to the SQL. To overcome the SQL, experiments can borrow from developments in quantum measurement technology to manipulate the noise in the system. In microwave cavity experiments, an antenna measures a voltage that is proportional to the electric field inside the cavity and can be decomposed into components

$$\hat{V} = \hat{X} \cos \omega_c t + \hat{Y} \sin \omega_c t, \qquad (4.18)$$

where ω_c is the cavity resonance frequency, and $\hat{X}$ and $\hat{Y}$ are quadratures of the cavity field. The variances of the quadratures have a minimum uncertainty limit of $\sigma_{\hat{X}}^2 \sigma_{\hat{Y}}^2 \geq 1/4$. JPAs can squeeze the vacuum state to increase uncertainty in one quadrature while decreasing it in the other. To decrease the system noise temperature below the SQL, experiments can implement vacuum squeezing by operating two JPAs in a phase-sensitive mode (the amplifier applies different gains to the two quadratures). In this operation, one JPA prepares the microwave cavity in a squeezed-vacuum state before amplifying with a second JPA that is 90° out of phase. Then, the noise is decreased in the measured quadrature increasing the signal-to-noise ratio [33]. Implementing a squeezed-state receiver (SSR) allowed HAYSTAC to enhance the scan rate by a factor of 1.9 [28]. The benefit of using an SSR is limited by the cable transmissivity and lossy microwave components, so future efforts will work to improve the connections in the system.

4.5 Operational Experiments

Since the initial experiments in the 1980s, several microwave cavity searches have excluded axion parameter space while updating their cavity and amplifier designs

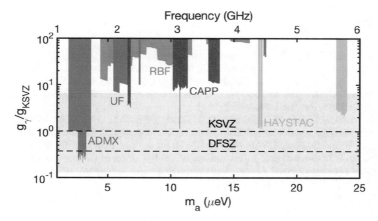

Fig. 4.4 Excluded parameter space by microwave cavity searches. The dashed lines represent the KSVZ and DFSZ models and the yellow band describes their uncertainty

Table 4.2 Summary of representative values from selected microwave cavity searches

Experiment	Time	Frequency	Volume	Amplifier	Magnet	T_{SYS}/T_{SQL}
RBF [7]/UF [34]	1985–1990	2.5 GHz	5 L	HEMT	6 T/8 T	100–200
ADMX @ LLNL [35]	1995–2010	0.6 GHz	200 L	HEMT, SQUID	7.6 T	50–100
ADMX @ UW [36]	2016-present	0.8 GHz	150 L	SQUID, JPA	7.6 T	10
CAPP [37]	2019-present	1.6 GHz	3.5 L	HEMT	8 T	12
HAYSTAC I [38]	2015–2018	6 GHz	1.5 L	JPA	9 T	2
HAYSTAC II [28]	2019-present	4 GHz	1.5 L	SSR	8 T	<1

to probe various axion mass ranges. The parameter space that has been excluded by the various microwave cavity searches is shown in Fig. 4.4.

Table 4.2 compares representative values of microwave searches. The ratio of system noise temperature to the SQL has improved with developments in cooling and amplifier technology over the years. Although the technological advances have been impressive, more innovation is needed to probe the vast axion parameter space available for microwave cavity searches.

Acknowledgments This work was performed under the auspices of the U.S. Department of Energy by Lawrence Livermore National Laboratory under Contract DE-AC52-07NA27344, and the National Science Foundation under grant No. 1914199.

References

1. J.H. Chang, R. Essig, S.D. McDermott, J. High Energy Phys. **09**, 051 (2018)
2. P.W. Graham, A. Scherlis, Phys. Rev. D **98**, 035017 (2018)
3. M. Tegmark, A. Aguirre, M.J. Rees, F. Wilczek, Phys. Rev. D **73**, 023505 (2006)
4. P. Agrawal, G. Marques-Tavares, W. Xue, J. High Energy Phys. **2018**, 1 (2018)
5. P. Sikivie, Phys. Rev. Lett. **51**, 1415 (1983)
6. C. Hagmann, P. Sikivie, N.S. Sullivan, D.B. Tanner, Phys. Rev. D **42**, 1297 (1990)
7. S. DePanfilis, A.C. Melissinos, B.E. Moskowitz, J.T. Rogers, Y.K. Semertzidis, W.U. Wuensch, H.J. Halama, A.G. Prodell, W.B. Fowler, F.A. Nezrick, Phys. Rev. Lett. **59**, 839 (1987)
8. C. Jones, A. Melissinos (eds.). *Proceedings of the Workshop of Cosmic Axions: Brookhaven National Laboratory, April 13–14, 1989* (World Scientific, Singapore, 1990)
9. H. Primakoff, Phys. Rev. **81**, 899 (1951)
10. L. Krauss, J. Moody, F. Wilczek, D.E. Morris, Phys. Rev. Lett. **55**, 1797 (1985)
11. H. Peng, et al., Nucl. Instrum. Meth. **A444**, 569 (2000)
12. J.I. Read, J. Phys G Nucl. Part. Phys. **41**, 063101 (2014)
13. J.E. Kim, Phys. Rev. Lett. **43**, 103 (1979)
14. M. Shifman, A. Vainshtein, V. Zakharov, Nucl. Phys. B **166**, 493 (1980)
15. A.R. Zhitnitsky, Sov. J. Nucl. Phys. **31**, 260 (1980). [Yad. Fiz.31,497(1980)]
16. M. Dine, W. Fischler, M. Srednicki, Phys. Lett. B **104**, 199 (1981)
17. C.M. Caves, Phys. Rev. D **26**, 1817 (1982)
18. R.H. Dicke, Rev. Sci. Inst. **17**, 268 (1946)
19. A.B. Pippard, Proc. R. Soc. Lond. A. **191**, 385 (1947)
20. C. Kittel, *Quantum Theory of Solids* (Wiley, Hoboken, 1987)
21. D. Di Gioacchino, C. Gatti, D. Alesini, C. Ligi, S. Tocci, A. Rettaroli, G. Carugno, N. Crescini, G. Ruoso, C. Braggio, P. Falferi, C.S. Gallo, U. Gambardella, G. Iannone, G. Lamanna, A. Lombardi, R. Mezzena, A. Ortolan, R. Pengo, E. Silva, N. Pompeo, IEEE Trans. Appl. Superconduct. **29**, 1 (2019)
22. D. Alesini, C. Braggio, G. Carugno, N. Crescini, D. D'Agostino, D. Di Gioacchino, R. Di Vora, P. Falferi, S. Gallo, U. Gambardella, C. Gatti, G. Iannone, G. Lamanna, C. Ligi, A. Lombardi, R. Mezzena, A. Ortolan, R. Pengo, N. Pompeo, A. Rettaroli, G. Ruoso, E. Silva, C.C. Speake, L. Taffarello, S. Tocci, Phys. Rev. D **99**, 101101 (2019)
23. I. Stern, G. Carosi, N. Sullivan, D. Tanner, Phys. Rev. Appl. **12**, 044016 (2019)
24. D.S. Kinion, First results from a multiple-microwave-cavity search for dark-matter axions. Ph.D. Thesis (2001)
25. J. Jeong, S. Youn, S. Ahn, C. Kang, Y.K. Semertzidis, Astropart. Phys. **97**, 33 (2018)
26. L.C. Maier, J.C. Slater, J. Appl. Phys. **23**, 68 (1952)
27. J.C. Slater, Rev. Mod. Phys. **18**, 441 (1946)
28. K.M. Backes, D.A. Palken, S.A. Kenany, B.M. Brubaker, S.B. Cahn, A. Droster, G.C. Hilton, S. Ghosh, H. Jackson, S.K. Lamoreaux, A.F. Leder, K.W. Lehnert, S.M. Lewis, M. Malnou, R.H. Maruyama, N.M. Rapidis, M. Simanovskaia, S. Singh, D.H. Speller, I. Urdinaran, L.R. Vale, E.C. van Assendelft, K. van Bibber, H. Wang, Nature **590**, 238 (2021)
29. F.C. Wellstood, C. Urbina, J. Clarke, Phys. Rev. B **49**, 5942 (1994)
30. M. Mück, M.O. André, J. Clarke, J. Gail, C. Heiden, Appl. Phys. Lett. **72**, 2885 (1998)
31. M.A. Castellanos-Beltran, K.W. Lehnert, Appl. Phys. Lett. **91**, 083509 (2007)
32. S.A. Kenany, M. Anil, K. Backes, B. Brubaker, S. Cahn, G. Carosi, Y. Gurevich, W. Kindel, S. Lamoreaux, K. Lehnert, S. Lewis, M. Malnou, D. Palken, N. Rapidis, J. Root, M. Simanovskaia, T. Shokair, I. Urdinaran, K. van Bibber, L. Zhong, Nucl. Instrum. Methods. Phys. Res. A **854**, 11 (2017)
33. M. Malnou, D.A. Palken, B.M. Brubaker, L.R. Vale, G.C. Hilton, K.W. Lehnert, Phys. Rev. X **9**, 021023 (2019)
34. C. Hagmann, P. Sikivie, N. Sullivan, D.B. Tanner, S.I. Cho, Rev. Sci. Inst. **61**, 1076 (1990)

35. S.J. Asztalos, G. Carosi, C. Hagmann, D. Kinion, K. van Bibber, M. Hotz, L.J. Rosenberg, G. Rybka, J. Hoskins, J. Hwang, P. Sikivie, D.B. Tanner, R. Bradley, J. Clarke, Phys. Rev. Lett. **104**, 041301 (2010)
36. T. Braine, R. Cervantes, N. Crisosto, N. Du, S. Kimes, L.J. Rosenberg, G. Rybka, J. Yang, D. Bowring, A.S. Chou, R. Khatiwada, A. Sonnenschein, W. Wester, G. Carosi, N. Woollett, L.D. Duffy, R. Bradley, C. Boutan, M. Jones, B.H. LaRoque, N.S. Oblath, M.S. Taubman, J. Clarke, A. Dove, A. Eddins, S.R. O'Kelley, S. Nawaz, I. Siddiqi, N. Stevenson, A. Agrawal, A.V. Dixit, J.R. Gleason, S. Jois, P. Sikivie, J.A. Solomon, N.S. Sullivan, D.B. Tanner, E. Lentz, E.J. Daw, J.H. Buckley, P.M. Harrington, E.A. Henriksen, K.W. Murch, Phys. Rev. Lett. **124**, 101303 (2020)
37. J. Choi, S. Ahn, B.R. Ko, S. Lee, Y.K. Semertzidis, Nucl. Instrum. Methods. Phys. Res. A **1013**, 165667 (2021)
38. L. Zhong, S. Al Kenany, K.M. Backes, B.M. Brubaker, S.B. Cahn, G. Carosi, Y.V. Gurevich, W.F. Kindel, S.K. Lamoreaux, K.W. Lehnert, S.M. Lewis, M. Malnou, R.H. Maruyama, D.A. Palken, N.M. Rapidis, J.R. Root, M. Simanovskaia, T.M. Shokair, D.H. Speller, I. Urdinaran, K.A. van Bibber, Phys. Rev. D **97**, 092001 (2018)

Chapter 5
Solar Production of Ultralight Bosons

Julia K. Vogel and Igor G. Irastorza

Abstract This chapter will spotlight axions produced in the core of the Sun. A first focus will be put on the production mechanism for axions in the solar interior through coupling of axions to photons via the Primakoff effect as well as their interactions with electrons. In addition to the axion production, the axion-to-photon conversion probability is a crucial quantity for solar axion searches (also referred to as helioscopes) and determines the expected number of photons from solar axion conversion that are detectable in a ground-based search. After these basic considerations, the helioscope concept will be detailed, and past, current, and future experimental realizations of axion helioscopes will be discussed. This includes the analysis used to aim at axion detection and upper limit calculations in case no signal above background is detected in experimental data. For completeness, alternative approaches other than traditional helioscopes to search for solar axions are discussed.

5.1 Production of Axions in the Sun

5.1.1 Solar Models and the Origin of Solar Axions

Axions can be produced in the core of stars via the Primakoff effect [1], which converts axions to photons and vice versa in strong electromagnetic fields as shown in the Feynman graphs of Fig. 5.1. In the extremely hot and dense core of the Sun—the closest celestial axion source to Earth—the two-photon coupling of pseudoscalars allows for the conversion of blackbody (BB) photons into axions.

J. K. Vogel (✉)
Lawrence Livermore National Laboratory, Livermore, CA, USA
e-mail: vogel9@llnl.gov

I. G. Irastorza
Center for Astroparticle and High Energy Physics (CAPA), University of Zaragoza, Zaragoza, Spain
e-mail: igor.irastorza@unizar.es

© The Author(s) 2023
D. F. Jackson Kimball, K. van Bibber (eds.), *The Search for Ultralight Bosonic Dark Matter*, https://doi.org/10.1007/978-3-030-95852-7_5

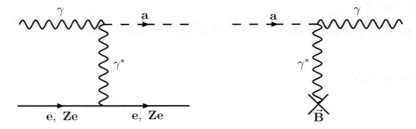

Fig. 5.1 Left: Feynman diagram of the Primakoff effect in the solar interior. Photons can be converted into axions in the electric field of the charged particles in the plasma. Right: in a laboratory magnetic field, the axion couples to a virtual photon from the transverse component of the magnetic field via the inverse Primakoff effect

The BB photons in this case have energies in the keV range. The virtual photon is hereby provided by the strong electromagnetic field, originating from the charged particles in the plasma. In nonrelativistic conditions, the Primakoff effect is relevant, since in this case, electrons and nuclei can be considered heavy in comparison to the energies of the surrounding photons. Therefore, the differential cross section here (not taking into account recoil effects) is given by [2]

$$\frac{d\sigma_{\gamma \to a}}{d\Omega} = \frac{g_{a\gamma\gamma}^2 Z^2 \alpha}{8\pi} \frac{|\boldsymbol{p}_\gamma \times \boldsymbol{p}_a|^2}{q^4}, \tag{5.1}$$

where the axion and photon energies are considered equal and the momentum transfer is given by $q = \boldsymbol{p}_\gamma - \boldsymbol{p}_a$ with $\boldsymbol{p}_\gamma$ and $\boldsymbol{p}_a$ being the momentum of the photon and the axion, respectively. The axion-to-photon coupling constant is represented by $g_{a\gamma\gamma}$, Z is the atomic number, and α denotes the fine-structure constant. The cutoff of the long-range Coulomb potential in vacuum for massive axions is given by the minimum required momentum transfer

$$q_{\min} = \frac{m_a^2}{2E_a}, \tag{5.2}$$

for the axion mass being small compared to its energy ($m_a \ll E_a$), yielding a total cross section of

$$\sigma_{\gamma \to a} = Z^2 g_{a\gamma\gamma}^2 \left[\frac{1}{2}\ln\left(\frac{2E_a}{m_a}\right) - \frac{1}{4} \right]. \tag{5.3}$$

The cutoff of the long-range Coulomb potential in a plasma is the consequence of screening effects resulting in an additional factor in the differential cross section such that

$$\frac{d\sigma_{\gamma \to a}}{d\Omega} = \frac{g_{a\gamma\gamma}^2 Z^2 \alpha}{8\pi} \frac{|\boldsymbol{p}_\gamma \times \boldsymbol{p}_a|^2}{q^4} \frac{q^2}{\kappa^2 + q^2}. \tag{5.4}$$

The Debye–Hückel scale κ represents screening effects via [3]

$$\kappa^2 = \frac{4\pi\alpha}{T_\odot} \sum_j Z_j^2 n_j, \tag{5.5}$$

with $T_\odot$ describing the temperature in the solar core plasma and n_j is the number density of charged particles carrying the charge $Z_j e$. Near the center of the Sun, the Debye–Hückel scale $\kappa \approx 9$ keV. The total scattering cross section taking into account this modification was calculated by Raffelt [3, 4]. Under the assumption of a nonrelativistic medium and negligible recoil effects, an expression for the transition rate $\Gamma_{\gamma \to a}$ can be obtained by summing over all target species of the medium

$$\Gamma_{\gamma \to a} = \frac{T_\odot \kappa^2 g_{a\gamma\gamma}^2}{32\pi^2} \frac{|\boldsymbol{p}_\gamma|}{E_\gamma} \int d\Omega \, \frac{|\boldsymbol{p}_\gamma \times \boldsymbol{p}_a|^2}{q^2 (q^2 + \kappa^2)}. \tag{5.6}$$

Angular integration then yields [4]

$$\Gamma_{\gamma \to a} = \frac{T_\odot \kappa^2 g_{a\gamma\gamma}^2}{32\pi} \frac{p_\gamma}{E_\gamma} \left\{ \frac{\left[(p_\gamma + p_a)^2 + \kappa^2 \right] \left[(p_\gamma - p_a)^2 + \kappa^2 \right]}{4 p_\gamma p_a \kappa^2} \right.$$

$$\left. \times \ln \left[\frac{(p_\gamma + p_a)^2 + \kappa^2}{(p_\gamma - p_a)^2 + \kappa^2} \right] - \frac{\left(p_\gamma^2 - p_a^2 \right)^2}{4 p_\gamma p_a \kappa^2} \ln \left[\frac{(p_\gamma + p_a)^2}{(p_\gamma - p_a)^2} \right] - 1 \right\}, \tag{5.7}$$

where $p_\gamma = |\boldsymbol{p}_\gamma|$ and $p_a = |\boldsymbol{p}_a|$ are the absolute values of the photon and axion momenta, respectively. For the Sun, the effective mass of the photon in the medium, i.e., the plasma frequency ω_p, is small. Typically, it is around 0.3 keV, while the solar core temperature is $T_\odot = 15.6 \times 10^6$ K $= 1.3$ keV, leading to typical photon energies of about $3T_\odot \approx 4$ keV. We therefore neglect the plasma frequency in the following, and photons will be treated as massless. Recoil effects can be ignored, such that $E_\gamma = E_a$ in the photon-to-axion conversion and $p_\gamma = E_\gamma = E_a$ and $p_a = \sqrt{E_a^2 - m_a^2}$ can be assumed, simplifying Eq. (5.7) to

$$\Gamma_{\gamma \to a} = \frac{T_\odot \kappa^2 g_{a\gamma\gamma}^2}{32\pi} \left\{ \frac{(m_a^2 - \kappa^2)^2 + 4E_a^2 \kappa^2}{4 E_a p_a \kappa^2} \ln \left[\frac{(E_a + p_a)^2 + \kappa^2}{(E_a - p_a)^2 + \kappa^2} \right] \right.$$

$$\left. - \frac{m_a^4}{4 E_a p_a \kappa^2} \ln \left[\frac{(E_a + p_a)^2}{(E_a - p_a)^2} \right] - 1 \right\}. \tag{5.8}$$

For axion masses small compared to the axion energy, i.e., $p_a \approx E_a$, the next to last term tends to zero and the above equation reduces further to

$$\Gamma_{\gamma \to a} = \frac{T_\odot \kappa^2 g_{a\gamma\gamma}^2}{32\pi} \left[\left(1 + \frac{\kappa^2}{4E^2} \right) \ln \left(1 + \frac{4E^2}{\kappa^2} \right) - 1 \right]. \tag{5.9}$$

The differential axion flux expected at Earth is then simply the convolution of the transition rate with the distribution of blackbody photons of the Sun followed by an integration using a standard solar model

$$\frac{d\Phi_a (E_a)}{dE_a} = \frac{1}{4\pi d_\odot^2} \int_0^{R_\odot} d^3 r \, \frac{1}{\pi^2} \frac{E_a^2}{e^{E_a/T} - 1} \Gamma_{\gamma \to a}, \tag{5.10}$$

where an average distance to the Sun of $d_\odot \approx 1.50 \times 10^{13}$ cm can be assumed. Van Bibber et al. [5] were the first to derive an approximate formula by including the standard solar model developed by Bahcall et al. [6] in 1982. Raffelt and Serpico revised the early results using an updated solar model [7] by fitting the following function to the solar data:

$$\frac{d\Phi_a(E_a)}{dE_a} = A \left(\frac{E_a}{E_0} \right)^\beta e^{-(\beta+1)E_a/E_0}, \tag{5.11}$$

where A is a normalization factor, E_0 corresponds to the average axion energy $\langle E_a \rangle$, and β is related to higher moments of energy. The best fit is obtained as

$$\frac{d\Phi_a(E_a)}{dE_a} = 6.020 \times 10^{10} \, (g_{10})^2 \, \frac{(E_a/\text{keV})^{2.481}}{e^{((E_a/\text{keV})/1.205)}} \left[\text{cm}^{-2}\text{s}^{-1}\text{keV}^{-1} \right], \tag{5.12}$$

with an accuracy at the 1% level for energies in the 1–11 keV range and g_{10} defined as

$$g_{10} = \frac{g_{a\gamma\gamma}}{10^{-10} \, \text{GeV}^{-1}}. \tag{5.13}$$

The average axion energy is $\langle E_a \rangle = 4.2$ keV, and the maximum of the axion energy distribution is expected to be around 3 keV. Note that this is the case for KSVZ axions (hadronic axions, proposed by Kim [8], Shifman et al. [9]), for which the Primakoff production mechanism dominates. For the Dine–Fischler–Srednicki–Zhitnitsky (DFSZ) model axions [10, 11], with axion-electron interaction present at tree level, for which "ABC processes" (axio-recombination, bremsstrahlung, and Compton scattering) dominate, the peak is shifted toward lower energies.[1] The total axion flux for hadronic models is then proportional to g_{10}^2 as

---

[1] See Chap. 2 and Sect. 5.1.2 for details.

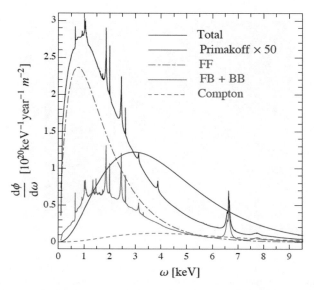

Fig. 5.2 Solar axion flux on Earth. The coupling of axions to photons here is assumed to be $g_{a\gamma\gamma} = 10^{-12}$ GeV^{-1}, the interaction strength with electrons $g_{aee} = 10^{-13}$. For a typical KSVZ (hadronic model, see Chap. 2 for details on axion models), the Primakoff effect is the dominant component and the differential axion flux is represented by the blue line, while for the DFSZ model, in which axions and electrons interact at tree level, the various components of the ABC flux take over (red lines): FF = free-free (bremsstrahlung), FB = free-bound (axio-recombination), and BB = bound-bound (axio-deexcitation). The black line is the total ABC flux. Please note that to show ABC and Primakoff spectra in the same plot, the latter has been multiplied by a factor of 50. Figure from Ref. [12]

$$\Phi_a = 3.75 \times 10^{11} (g_{10})^2 \text{ cm}^{-2} \text{ s}^{-1}. \tag{5.14}$$

The axion luminosity for the standard solar model is

$$L_a = 1.85 \times 10^{-3} (g_{10})^2 L_\odot, \tag{5.15}$$

where $L_\odot$ refers to the solar photon luminosity. No major updates to the solar models have been made since then, so these predictions still hold. Figure 5.2 shows the differential solar axion flux for hadronic and non-hadronic axion models.

When using an imaging device—such as an X-ray telescope—to detect photons from axion-to-photon conversion, as common in solar axion search experiments (referred to as *helioscopes*), a helpful approach is to consider the differential axion flux as an apparent surface luminosity $\varphi_a(E_a, r)$ of the solar disk. This implies that the flux (for $g_{10} = 1$) is calculated per unit surface area of the apparent 2-dimensional solar disk. It is a function of the axion energy E_a and a dimensionless radial coordinate r ($0 \leq r \leq 1$), representing the radius normalized to the solar radius $R_\odot$. The apparent surface luminosity $\varphi_a(E_a, r)$ can be formulated as [13]

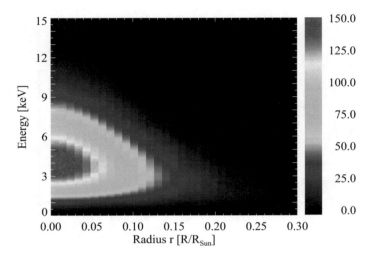

Fig. 5.3 Contour plot of the axion surface luminosity of the Sun resulting from the Primakoff effect as a function of energy and dimensionless radial coordinate r. The units, in which the flux is given, are axions/(cm^2·s·keV) per unit surface area on the solar disk. Here, $g_{10} = 1$ has been assumed

$$\varphi_a\left(E_a, r\right) = \frac{R_\odot^3}{2\pi^3 d_\odot^2} \int_r^1 ds \frac{s}{\sqrt{s^2 - r^2}} E_a\, k\, f_B \Gamma_{a\to\gamma} \qquad (5.16)$$

and is given in units of cm^{-2} s^{-1} keV^{-1} per unit surface area; $d_\odot$ is the average distance of Earth from the Sun as in Eq. (5.10), s represents the radial position in the Sun, determining physical quantities, such as temperature and density, and k is the wavenumber. $f_B = \left(e^{E_a/T_\odot} - 1\right)^{-1}$ denotes the Bose-Einstein distribution. In Fig. 5.3, the axion surface luminosity as seen on Earth is shown as a function of axion energy E_a and radial coordinate r. Only Primakoff conversion has been taken into account here (hadronic models), but a similar plot can also be derived for non-hadronic models, in which axions also significantly interact with electrons. The color scale is given in units of axions/(cm^2·s·keV) per unit surface area on the solar disk. This shows that most axions originate from the inner 20% of the solar radius. Furthermore, the axion flux is expected to be largest at energies around 3 keV for hadronic axions. Figure 5.4 illustrates the energy dependence of the axion surface luminosity for several radial coordinates obtained by integration up to the corresponding values of r. The total axion flux at Earth can be obtained from the apparent surface luminosity $\varphi_a\left(E_a, r\right)$ using

$$\Phi_a = 2\pi \int_0^1 dr\, r \int_{\omega_p}^\infty dE\, \varphi_a\left(E_a, r\right), \qquad (5.17)$$

where ω_p is again the plasma frequency.

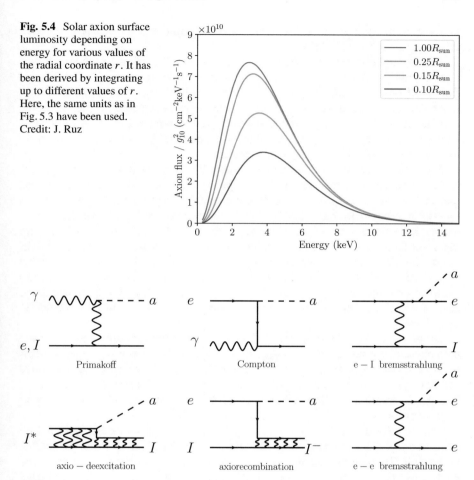

Fig. 5.4 Solar axion surface luminosity depending on energy for various values of the radial coordinate r. It has been derived by integrating up to different values of r. Here, the same units as in Fig. 5.3 have been used. Credit: J. Ruz

Fig. 5.5 Different processes responsible for axion production in the Sun, including both the Primakoff process and the ABC processes. Figure from Ref. [12]

5.1.2 Non-Primakoff Solar Axions

In non-hadronic models, like the DFSZ models (see Chap. 2), axions couple with electrons at tree level. This coupling allows for additional mechanisms of axion production in the Sun, namely: atomic axio-recombination and axion-deexcitation, axio-bremsstrahlung in electron-ion or electron-electron collisions, and Compton scattering with emission of an axion. Figure 5.5 shows the Feynman diagrams of all these processes. Collectively, solar axions from the flux generated by all these channels are referred to as ABC (or BCA) solar axions, from the initials of the aforementioned processes. The most up-to-date computations of these production channels can be found in [12].

The spectral distribution of ABC solar axions, as well as of each of the individual components, is shown in Fig. 5.2. Although the relative strength of ABC and Primakoff fluxes depends on the particular values of the g_{aee} and $g_{a\gamma\gamma}$ couplings, and therefore on the details of the axion model being considered, for non-hadronic models the ABC flux is expected to dominate. For example, for the representative values taken to produce Fig. 5.2, the Primakoff spectrum has been multiplied by 50 to make it comparable with the ABC spectrum.

Although all processes contribute substantially, free-free (bremsstrahlung) processes constitute the most important component and are responsible for the fact that ABC axions are of somewhat lower energies than Primakoff axions, with a spectral maximum around ~1 keV. This is because the axio-bremsstrahlung cross section increases for lower energies and, in the hot solar core, electrons are more abundant than photons, and their energies are high with respect to atomic orbitals. In addition, the axio-deexcitation process is responsible for the presence of several narrow peaks, each one associated with different atomic transitions of the species present in the solar core. These two features would be of crucial importance in the case of a positive detection to confirm an axion discovery, as will be discussed in Sect. 5.4.2.3.

Despite the above, due to the fact that g_{aee} is more strongly bounded from astrophysical considerations than $g_{a\gamma\gamma}$ (see Chap. 3), the sensitivity of experiments to ABC axions has so far not been sufficiently high to reach and study unconstrained values of g_{aee}. This may change with the next generation of solar axion helioscopes, like the International AXion Observatory (IAXO), that will enjoy sensitivity to values down to $g_{aee} \sim 10^{-13}$ (Sect. 5.4).

For the sake of completeness, we should mention that the existence of axion-nucleon couplings g_{aNN} also allows for additional mechanisms of axion production in the Sun. These emissions are monoenergetic and are associated with particular nuclear reactions in the solar core. Some examples of the emissions that have been searched for experimentally are 14.4 keV axions emitted in the M1 transition of ^{57}Fe nuclei and MeV axions from ^{7}Li and $D(p, \gamma)^{3}$He nuclear transitions or ^{169}Tm (see Ref. [14] for details and references).

Note that while the above considerations are mainly focusing on axions, axion-like particles (ALPs, [15–17], see also Chap. 2) share—to a large extent—the same theory and phenomenology as axions. Interestingly, most of the experiments searching for the effects of axion couplings to standard model particles (photons, electrons, nucleons) are therefore also sensitive to these more generic axionlike particles. Generally speaking, ALPs are pseudo-Nambu-Goldstone bosons with small masses and rather weak interaction strength originating from the spontaneous breaking of a symmetry at very high energy scales (Chap. 2, Sect. 2.3). They generally mix with photons similarly to axions but do not exhibit the axion-typical relation between axion mass m_a and coupling constant $g_{a\gamma\gamma}$, i.e., they are not a part of the Peccei-Quinn (PQ) mechanism [18, 19] for quantum chromodynamics (QCD) axions and do not acquire their masses from effects in QCD but rather through corresponding dynamics that explicitly break a global symmetry. These more generic particles (every axion is an ALP, but not every ALP is an axion) are invoked in various scenarios, theoretically well motivated at the low-energy frontier of particle physics (see Ref. [14] and also Chap. 2, Sect. 2.5). They are sometimes also referred to as non-QCD or non-PQ axions, which is why the term *axions* is often used to refer to both QCD axions and ALPs.

5.1.3 Constraints on the Solar Axion Flux

The solar axion flux expectation can be constrained by using known solar properties. First, an additional energy loss channel via axion emission would increase the consumption of nuclear fuel, and since the Sun has lived through about half of its helium burning phase, its solar axion luminosity should not exceed the solar photon luminosity. This consideration can, for example, rule out apparent "signals" of the PVLAS-type [20], since these would require an axion luminosity of $L_a > 10^6 \times L_\odot$, such that the solar lifetime would be about 1000 years [21]. Indeed, for $g_{10} \gtrsim 20$, it becomes basically impossible to construct a self-consistent solar model due to excessive axion losses [22].

Precision helioseismology and the measured solar neutrino flux are another avenue to constrain the axion-photon coupling strength. In an updated statistical analysis [23], these two observations were combined to provide a stringent upper limit on the coupling constant of $g_{10} < 4.1$ at the 4σ confidence level. Helioscope upper limits on $g_{a\gamma\gamma}$ are consistent with these solar constraints, in that the solar axion flux, which corresponds to the published limits, is too small to significantly affect the abovementioned observations. A similar argument holds to constrain the coupling of axions to electrons g_{aee} [24]. Axion losses can thus be seen as minor perturbations of solar models.

5.1.4 Do Axions Escape from the Sun?

In order to be detected in an experiment, solar axions first need to escape the Sun. Their mean free path (MFP) ℓ_a must therefore be larger than the solar radius. In natural units, the photon-axion conversion rate given in Eq. (5.9) and the inverse MFP of a photon of energy E_γ considering the Primakoff effect are identical. Thus, the MFP can be obtained from Eq. (5.9) in the static limit (no recoil, screening included). With a temperature $T \approx 1.3$ keV and $\kappa \approx 9$ keV at the solar center, ℓ_a for 4 keV axions is

$$\ell_a \approx 6 \times 10^{24} g_{10}^{-2} \text{cm} \approx 8 \times 10^{13} g_{10}^{-2} R_\odot. \quad (5.18)$$

Thus, the coupling constant $g_{a\gamma\gamma}$ would have to be larger than the best solar axion limits as observed by the CERN Axion Solar Telescope (CAST, [13, 25–31]) by a factor of 10^7 in order to have reabsorption of axions in the Sun.

In the extreme case of such a strong coupling, axions would influence the solar structure. They would be responsible for the bulk of the energy transport within the Sun, which is otherwise carried by the photons. In order to be trapped in the Sun, axions would have to interact strongly enough to have an MFP smaller than that of photons, which is ≈ 1 mm in the solar interior. Thus, the solar structure will only remain unaffected if the MFP ℓ_a of axions is not much larger than a millimeter. Otherwise, the energy transfer rate in the Sun would be extremely accelerated and the solar structure would be dramatically altered. This condition is so stringent that reabsorption is not a possibility worth considering for axions or axionlike particles [21].

5.2 Axion-to-Photon Conversion Probability for Solar Axions

To detect solar axions in a laboratory experiment, helioscopes employ magnetic fields to convert axions into X-ray photons via the inverse Primakoff effect (see the right part of Fig. 5.1). The virtual photon in this interaction is provided by the transverse component of the magnetic field. The conversion process works in a manner analogous to neutrino oscillations [2].

Although the photon has spin-one and axions are spin-zero particles, mixing is possible in an external magnetic or electric field that enables matching of the missing quantum numbers. The conversion from a free photon into a spin-zero axion requires a change in the azimuthal quantum number of angular momentum (J_z) which for photons equals $J_z = \pm 1$ and $J_z = 0$ for axions. Therefore, a longitudinal field, i.e., a field with azimuthal symmetry, does not allow for these transitions given the fact that it cannot change J_z. A transverse field, however, does allow for mixing between photons and axions.

The determining wave equation for particles propagating perpendicular to a transverse magnetic field B has been derived by Raffelt and Stodolsky [32] as

$$\left[\begin{pmatrix} \omega - \dfrac{m_\gamma^2}{2\omega} - i\dfrac{\Gamma}{2} & g_{a\gamma\gamma}\dfrac{B}{2} \\ g_{a\gamma\gamma}\dfrac{B}{2} & \omega - \dfrac{m_a^2}{2\omega} \end{pmatrix} - i\partial_z \right] \begin{pmatrix} A_\parallel \\ a \end{pmatrix} = 0. \tag{5.19}$$

$A_\parallel$ is the amplitude of the photon field component parallel to the magnetic field B, a is the amplitude of the axion field, ω represents the frequency, m_γ is the effective photon mass in the gas, and m_a is the axion mass. Damping is included via the inverse absorption length Γ of photons. Up to a global phase, a first-order solution can be found by using a perturbative approach as

$$\langle A_\parallel(z)|a(0)\rangle = \frac{1}{2}g_{a\gamma\gamma}\exp\left(-\int_0^z dz'\,\frac{\Gamma}{2}\right)$$

$$\times \int_0^z dz'\,B\,\exp\left(i\int_0^{z'} dz''\left[\frac{m_\gamma^2 - m_a^2}{2\omega} - i\frac{\Gamma}{2}\right]\right). \quad (5.20)$$

The conversion probability $P_{a\to\gamma}$ of axions into photons at a length $z = L$ of the magnetic field can be obtained in Lorentz–Heaviside units (see Problem 5.1) as

$$P_{a\to\gamma} = |\langle A_\parallel(z)|a(0)\rangle|^2 = \left(\frac{Bg_{a\gamma\gamma}}{2}\right)^2 \frac{1}{q^2 + \Gamma^2/4}\left[1 + e^{-\Gamma L} - 2e^{-\Gamma L/2}\cos(qL)\right],$$
$$(5.21)$$

where q is the absolute momentum transfer between the real photon in the medium and the axion (see Problem 5.2) given by

$$q = \left|\frac{m_\gamma^2 - m_a^2}{2E_a}\right|, \quad (5.22)$$

where E_a is the energy of the axion.

? Problem 5.1 Natural Lorentz–Heaviside Units

The conversion probability of axions into photons in the presence of a transverse magnetic field as given in Eq. (5.21) uses natural Lorentz–Heaviside units for which the dimensions GeV and (T · m) are equivalent, due to the fact that charge is dimensionless and natural units are used ($c = 1$). Show how 1 GeV = 1.010 T · m and therefore $P_{a\to\gamma}$ is dimensionless.

Solution on page 328.

? Problem 5.2 Momentum Transfer

Derive the equation for the axion-to-photon momentum transfer as shown in Eq. (5.22).

Solution on page 329.

There are two cases to consider for the probability of conversion in experimental solar axion searches: (1) an evacuated conversion region and (2) a conversion volume filled with a low-Z buffer gas. Both scenarios will be discussed in the following.

5.2.1 Coherence Condition and Conversion Probability in Vacuum

Using Eq. (5.21) in the limit of $m_\gamma \to 0$, the probability of axion-to-photon conversion in a magnetic field in vacuum can be derived. Assuming negligible absorption ($\Gamma \approx 0$) results in

$$P_{a\to\gamma} = \left(\frac{BLg_{a\gamma\gamma}}{2}\right)^2 \left(\frac{\sin\left(\frac{qL}{2}\right)}{\left(\frac{qL}{2}\right)}\right)^2, \qquad (5.23)$$

again with magnetic field strength B and length L. The momentum transfer q as given in Eq. (5.22) simplifies to

$$q = \frac{m_a^2}{2E_a}. \qquad (5.24)$$

This enables a coherence condition for which photon and axion waves are in phase and nonzero conversion probability can be obtained. This coherence condition can be expressed as

$$\frac{qL}{2} \ll \pi, \qquad (5.25)$$

which is shown in Fig. 5.6 where the $(\sin(x)/x)^2$ term of Eq. (5.23) with $x = qL/2$ is plotted as a function of x and it nicely illustrates that the largest contributions are found for values $x \lesssim \pi$. In terms of axion mass, the condition can be written as

$$m_a \ll \sqrt{\frac{4\pi E_a}{L}}, \qquad (5.26)$$

such that the coherence condition is fulfilled for axion masses smaller than 0.02 eV for a realistic example of a 10 m long magnet and a typical solar axion energy of $E_a \approx 4.2$ keV. In the limit of $x \to 0$, the $\sin^2(x)/x^2$ term tends to 1 and Eq. (5.23) reduces to

$$P_{a\to\gamma} = \left(\frac{BLg_{a\gamma\gamma}}{2}\right)^2, \qquad (5.27)$$

i.e., the conversion probability in vacuum.

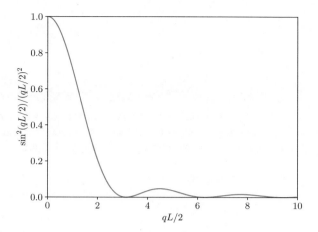

Fig. 5.6 Dependence of the coherence term on different values of its argument $qL/2$. The major contributions to the probability function result from values of $qL/2$ which are smaller than π

5.2.2   *Coherence Condition and Conversion Probability in a Buffer Gas*

Accessing higher axion masses than possible in a conversion volume under vacuum in a given experiment can be achieved by filling the magnetic field region with a buffer gas. In order to minimize the absorption of X-ray photons in the buffer gas, elements with low atomic number Z are strongly preferred. Additional constraints due to operation of the magnet as a superconductor often require low operating temperatures of a few Kelvin and therefore usually only helium and hydrogen are good options since others are not gaseous given the required operating pressure. In the case of buffer gas use, Eqs. (5.21) and (5.22) no longer simplify as in the vacuum case.

5.2.2.1   Effective Mass of the Photon

While photons in vacuum can be considered massless and travel at the speed of light c, they acquire an effective mass when passing through a transparent medium at a speed $v < c$. In the classical wave picture, the slowdown can be explained as a delay of the photon wave due to interference of the incident light with photons coming from matter polarized by the original photons. Considering the situation from the particle view, it can be understood as a mixing effect between initial photon and quantum excitations of the traversed matter, resulting in a particle with effective mass. The photon energy is given by $E_\gamma^2 = m_\gamma^2 = \hbar^2 \omega_p^2$, where ω_p is the plasma frequency, and the effective photon mass in helium can be derived (see Problem 5.3) as

$$m_\gamma = \sqrt{0.020 \frac{p/\text{mbar}}{T/\text{K}}} \, \text{eV}. \tag{5.28}$$

? Problem 5.3 Effective Photon Mass in a Buffer Gas

In a buffer gas, photons acquire an effective mass, which enables the study of higher mass axions as compared to the use of an evacuated magnetic field region in axion helioscope experiments. Derive Eq. (5.28) for helium.

Solution on page 330.

5.2.2.2 Momentum Transfer

With the effective photon mass, the momentum transfer between an axion and a (real) photon can be calculated. q will be minimal for axion masses close to the corresponding effective photon mass of the considered buffer gas pressure. Since the momentum transfer has to be small in order to fulfill the coherence condition of Eq. (5.25), only a narrow range of axion masses can be studied at a specific gas pressure.

5.2.2.3 The Absorption of Photons in a Buffer Gas

The absorption of the photons originating from axions via the Primakoff effect in a buffer gas is another important factor influencing the conversion probability. In general, the absorption Γ of these photons is defined as the inverse of the absorption length ℓ:

$$\Gamma(E_a) = \frac{1}{\ell(E_a)} = \rho\mu(E_a), \tag{5.29}$$

where ρ is the density of the gas and $\mu(E_a)$ represents the energy-dependent mass absorption coefficient, which is given by

$$\mu(E_a) = \frac{N_A}{A}\sigma_A(E_a), \tag{5.30}$$

with Avogadro's constant N_A and mass number A. The scattering cross section σ_A takes into account photoelectric, coherent, and incoherent contributions. In practice, the magnetic field region of an axion helioscope will be filled with a low-Z buffer gas at a certain pressure p_{gas} and temperature T_{gas}. It is therefore useful to consider that at standard temperature and pressure (STP), the ideal gas equation yields

$$\ell_{\text{STP}} = \ell \times \left(\frac{T_{\text{STP}}}{T_{\text{He}}} \right) \times \left(\frac{p_{\text{He}}}{p_{\text{STP}}} \right),$$ (5.31)

for helium gas, and thus Eq. (5.29) is

$$\Gamma(E_a) = \mu(E_a) \rho_{\text{STP}} \frac{T_{\text{STP}} p_{\text{He}}}{p_{\text{STP}} T_{\text{He}}}.$$ (5.32)

The density under standard conditions ρ_{STP} for ^{4}He is 0.1786 g/L.

5.2.2.4 Mass Range of Coherence

Restoring coherence by means of a buffer gas in the magnetic field region makes small axion mass ranges around the effective photon mass accessible as can be seen from Eq. (5.22) and the coherence condition of Eq. (5.25), i.e.,

$$\sqrt{m_\gamma^2 - \frac{4\pi E_a}{L}} < m_a < \sqrt{m_\gamma^2 + \frac{4\pi E_a}{L}}.$$ (5.33)

Since the effective photon mass depends on the pressure and the axion mass range that can be explored varies with axion energy and length of the magnetic field region in a given experiment, the accessible axion mass range changes. For example, axion masses around a photon mass of 0.43 eV can be scanned with ^{4}He at 1.8 K, since the maximum operating pressure before the ^{4}He gas liquefies at these temperatures is 16.4 mbar. If an axion helioscope is operated at room temperature (293 K) instead, a similar photon mass is obtained for an operating gas pressure of 2.7 bar. Depending on the buffer gas, the magnet length, and the operating temperature, different solar axion experiments will be able to access slightly different mass ranges around the calculated photon mass.

5.3 Expected Number of Photons from Solar Axion Conversion

The expected number of photons N_γ from axion-to-photon conversion in a magnetic field can be obtained as a function of axion mass and the coupling constant (for a given pressure of the buffer gas) as

$$N_\gamma = \int_E \frac{d\Phi(E_a, g_{a\gamma\gamma}^2)}{dE_a} P_{a \to \gamma}(E_a, m_a, g_{a\gamma\gamma}^2) \, \epsilon(E_a) \, \Delta t \, A \, dE_a,$$ (5.34)

with detection area A and detection efficiency $\epsilon(E_a)$. The exposure time, i.e., the time an axion helioscope is able to point at the solar core while tracking the Sun, is

Fig. 5.7 Expected number of photons from axion-to-photon conversion in a magnetic field for a typical axion helioscope as described in the text (namely, CAST). The pressure of the buffer gas (here, ^{4}He) is given at 1.8 K and therefore 5.49 mbar corresponds to a density of 0.147 kg/m^3

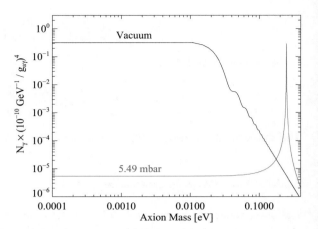

Δt. Since the expected solar axion flux and the conversion probability each depend quadratically on $g_{a\gamma\gamma}$, the number of expected photons relates to the axion-photon coupling constant as

$$N_\gamma \propto g_{a\gamma\gamma}^4 \ . \tag{5.35}$$

The number of expected photons from conversion of axions in vacuum and at a particular pressure p of a buffer gas (^{4}He is used as an example) at 1.8 K in the magnetic field region is shown in Fig. 5.7 for the experimental conditions of a typical axion helioscope (i.e., CAST). An exposure time of 90 min and 100% efficiency of the detector ($\epsilon = 1$) have been assumed for this plot along with a magnetic field of 9 T throughout a 10 m long region. The sensitive area included is 14.52 cm^2, corresponding to the size of the magnet bore for the current leading axion helioscope (CAST).

5.4 Axion Helioscope Experiments

As discussed in Sect. 5.2, axion helioscopes employ strong transverse magnetic fields B over a length L to convert solar axions into photons. Due to the fact that axions have a mass, axion and photon waves will be out of phase after a certain distance which determines the coherence condition (Eq. 5.25). For typical solar axion energies and a magnet length of ≈ 10 m, coherence is conserved for axion masses up to about 10^{-2} eV, while for higher masses coherence in vacuum is lost and the experimental sensitivity decreases. It can be restored by the use of a buffer gas in the magnetic field region for higher axion masses due to the photon acquiring an effective mass in a medium. Thus, the coherence condition will be fulfilled for axion masses close to the effective photon mass. By changing the pressure of the gas inside the magnetic field region systematically, the photon mass can be increased in a controlled manner and higher masses can be scanned via pressure-step scanning.

5.4.1   Concept of Axion Helioscopes

A typical axion helioscope requires at least two key components: a powerful magnet and one or more high-sensitivity, ultralow background X-ray detectors. In latest implementations of the concept, as shown in Fig. 5.8, an X-ray focusing device is added at the end of the magnet to concentrate the signal photons and increase the signal-to-noise ratio. Such an X-ray telescope also enables the use of large cross-sectional magnets (to boost conversion probability) and simultaneously small-area detectors (to enable ultralow background levels), which in combination boost helioscope experiments to the next level. By aligning the magnet with the core of the Sun and tracking its movement, an excess of X-rays at the end of the magnet is expected as compared to background measurements when the magnet is not pointing at the Sun. The helioscope detection concept was first proposed in the 1980s [5, 33] and initially experimentally implemented by Lazarus et al. with a few hours of data acquired [34]. Later, the second-generation helioscope SUMICO [35] was built at the University of Tokyo, providing the first self-consistent limit to solar axions compatible with solar physics. During the last two decades, the helioscope principle has been advanced by the CERN Axion Solar Telescope (CAST [13, 25–31]) pushing the sensitivity to solar axions significantly due to innovative concepts employed by the experiment: a superior magnet, X-ray optics, and enhanced detectors. The next generation of axion helioscopes, the

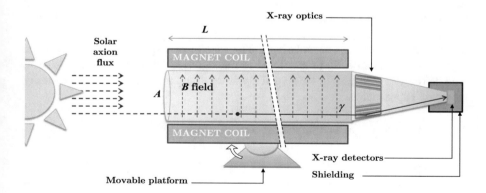

Fig. 5.8 A conceptual setup of an axion helioscope using X-ray focusing to enhance the experimental sensitivity. Axions from the Sun are converted in a strong transverse magnetic field and the emerging photons are then focused by an X-ray telescope into a small focal spot. A low background detector is located in the focal plane of the optics to capture an image of the photons produced by conversion from axions. Figure adapted from Ref. [36]

International AXion Observatory (IAXO, [36–39]) and its intermediate-scale phase BabyIAXO, will build on these improvements and further enhance solar axion searches by pushing sensitivities far beyond the ones reached by CAST.

Generally speaking, so far each generation of axion helioscopes improved the sensitivity to g_{10} by a factor of ≈ 7, mostly by successfully recycling existing magnets and other equipment. Improving over the current state of the art provided by CAST requires purpose-designed components for the key helioscope pieces (magnet, detectors) as well as focusing devices without which the use of the full potential of these new components would be impossible. To maximize the figure of merit (FOM) f of a helioscope

$$f = f_M \times f_O \times f_D \times f_T, \tag{5.36}$$

a global optimization of the FOM for magnet f_M, optics f_O, detectors f_D, and exposure f_T is needed. These are defined as

$$f_M = B^2 L^2 A, \tag{5.37}$$

$$f_O = \frac{\epsilon_o}{\sqrt{a}}, \tag{5.38}$$

$$f_D = \frac{\epsilon_d}{\sqrt{b}}, \tag{5.39}$$

$$f_T = \sqrt{\epsilon_t t}, \tag{5.40}$$

where B, L, and A are the magnet parameters (field strength, field length, and cross-sectional area), and ϵ_o, ϵ_d, and ϵ_t efficiencies of optics, detectors, and data acquisition, respectively. Furthermore, a is the total focal spot area of the telescopes and b the detector background normalized to unit area and time, while t is the total exposure time for observations of the solar disk center. It is worth noting that in order to maximize f, all components need to be optimized simultaneously in a multi-parameter process considering the expected axion spectrum. The helioscope figure of merit is directly proportional to the signal-to-noise (S/N) enhancement, and therefore a measure for the sensitivity to the coupling that can be used to easily compare experiments.

Tutorial: Figure of Merit for Helioscopes

According to Eqs. (5.37)–(5.40), the figure of merit for a helioscope can most easily be boosted by improving the magnet if optics and detectors can be built to fully enable the use of these improvements. As Eq. (5.37) clearly shows, increasing the magnetic field strength B and the length L of the magnet would be the most efficient ways. Why do next-generation helioscopes opt to increase the cross-sectional area A instead?

Since the magnet parameters (B, L, and A) are all interconnected, they need to be optimized together. While magnets with larger fields ($>$ 10 Tesla) have been previously built, they are usually much shorter than the current state of the art (CAST, with magnet length $\approx$ 10 m) or cannot be tilted sufficiently to track the Sun without impairing the cooling needed for superconductive operation. Increasing the length while keeping a (relatively) lower magnetic field B would be technologically feasible. However, the length L feeds into the coherence condition (see Eq. 5.25), i.e., for efficient conversion the product of momentum transfer q and length L must be small. For large L, the accessible axion masses become small as can be seen from Eqs. (5.26) and (5.33). Therefore, the best approach is to increase the cross-sectional area A of the magnet. Note that this in turn, however, requires the use of large focusing optics covering the complete magnet bore in order to focus the putative signal onto a small spot such that small-area detectors can be used. This is necessary since axion searches are by definition rare-event searches and the detector background needs to be as low as possible (zero background is the goal), which is generally only achievable with small-area detectors. Increasing the exposure time increases the sensitivity, but the upper limit on $g_{a\gamma\gamma}$ (in the absence of an axion signal) goes with the 8th root of time following Eqs. (5.34) and (5.35), i.e., in order to improve the limit by a factor of 2, one would need to measure a factor of 2^8 times longer. Considering that scanning axion masses with a buffer gas in the magnetic field region needs many pressure steps and each step is usually measured for one solar tracking, one would have to spend 256 days instead of 1 to achieve a factor 2 improvement in the upper limit on the coupling.

End of Tutorial

5.4.2   Current and Future Axion Helioscopes

5.4.2.1   The CERN Axion Solar Telescope (CAST)

To date, the CAST experiment has been the most powerful axion helioscope ever built. The magnet drives the sensitivity of any helioscope due to the $B^2 L^2$ dependence of the conversion probability as seen from Eq. (5.37): a 9 Tesla, 9.3 m superconducting magnet is the primary element at the heart of the CAST experiment. With its two magnet bores of 14.5 cm^2 each, it was originally built as one of the early prototypes for the Large Hadron Collider (LHC) at CERN, which had straight bores as opposed to the bent ones fabricated later on and eventually used in the LHC, and then it was repurposed for axion searches with CAST. The magnet itself boosts the conversion probability by 2 orders of magnitude [26] compared to the predecessor helioscope [35]. CAST is equipped with an elevation and azimuthal drive, such that the experiment is able to follow the Sun twice a day during sunrise and sunset for 90 min each. When not tracking the Sun, CAST

Fig. 5.9 Experimental setup of the CERN Axion Solar Telescope (CAST). The magnet (blue) is installed on a movable platform (green), and detectors are mounted on either end of the magnet. On the right side, the original telescope used at CAST is visible (silver). Also shown is the cryo cooling tower of the experiment with the helium supply lines. Credit: CERN/CAST

acquires background data in a magnet parking position. Right from the start, this helioscope employed an X-ray telescope in combination with a Charge Coupled Device (CCD) as a focal plane detector [40] on one of its four magnet bore exits. A second optic was installed later on [41]. CCDs are highly sensitive, pixelated photon detectors based on semiconductor technology (less sensitive versions can be found in many digital cameras and imaging devices). The CCD of CAST is a spare flight detector from the European Space Agency's XMM Newton mission and has greatly enhanced the sensitivity of the helioscope. A variety of other detectors have been used over the course of the experiment, including several generations of novel MICROMEsh GAseous Structure (MICROMEGAS, MM, [42]) and microbulk MM detectors (i.e., a more advanced version of MM detectors), a time projection chamber (TPC, [43]) and other more specialized equipment (see [31] and the references therein for further details). Both TPCs and MMs are gaseous, low background particle detectors combining elements of Multiwire Proportional Chambers (MWPC) and conventional drift chambers. While TPC and MM share the same detection principles of amplifying charges that are created by ionization in the gas volume of the detectors, the MM detectors represent a more recent evolution that includes a metallic micro-mesh positioned in close proximity to the readout electrode dividing the gas region into two. This is a key feature that enables high gain as well as the ability to detect fast signals and also allows for ultralow background performance benefiting from the use of low-radioactivity materials used to build the detectors. Figure 5.9 shows the CAST experiment including all its main components: magnet, optics, and detectors.

The experiment was divided into two main phases: (1) CAST Phase I (vacuum) and (2) Phase II (gas phase with ^{4}He and ^{3}He in the magnet bores). After completion of both phases, the experiment revisited some vacuum measurements to make use of improved detection techniques and dedicated some time to chameleon[2] searches [47–49], which are candidates for dark energy, and the use of microwave cavities within the CAST magnet. During its initial observational program (Phase I), CAST operated with evacuated magnet bores studying axion masses $m_a <$ 0.02 eV yielding an upper limit result of $g_{a\gamma\gamma} < 8.8 \times 10^{-11}$ GeV^{-1} at the 95% confidence level [13, 26]. During CAST Phase II operations, the magnet bores were filled with ^{4}He and ^{3}He to extend the search range up to axion masses of $m_a = 1.17$ eV with average limits of $g_{a\gamma\gamma} \lesssim 2.3$–3.3 $\times 10^{-10}$ GeV^{-1} at 95% C.L. for masses larger than 0.02 eV [27, 28]. For Phase II, the exact values depend on the individual pressure setting of the buffer gases. In recent years, improvements of the previous vacuum data results have been enabled with upgraded MM detector systems coupled to a novel X-ray telescope, the IAXO pathfinder system [41]. This approach using improved instrumentation resulted in a new benchmark limit for $m_a < 0.02$ eV of $g_{a\gamma\gamma} < 0.66 \times 10^{-10}$ GeV^{-1} (95% C.L.) [31]. One of the main goals of the CAST experiment has been to supersede the most stringent limits from astrophysical observations of horizontal branch stars (see Chap. 3) at the level of $g_{a\gamma\gamma} \lesssim 0.8 \times 10^{-10}$ GeV^{-1} [50]. CAST has studied both QCD and non-QCD axions, but most notably excluded KSVZ axions (see Chap. 3) around the 1 eV axion mass as shown in Fig. 5.10. Furthermore, the experiment has delivered results on more exotic physics cases of solar axions from M1 transition of Fe-57 nuclei [51], high-energy (MeV) axions from ^{7}Li and D$(p, \gamma)^3$He nuclear decays [52], axion–electron coupling constants for solar axions [53], and other ALP searches, such as for chameleons [47–49].

Tutorial: Understanding Helioscope Exclusion Plots

   Figure 5.10 is a typical example of a helioscope exclusion plot, showing the upper limit obtained by a solar axion search in the case where no signal was detected above background. Why does the red line (measurements in vacuum) sharply rise at $m_a \approx 10^{-2}$ eV and why does the black upper limit (combined previous vacuum and buffer gas phase results) display a "wiggly" structure for the higher axion masses?
   The sharp rise of both curves at around 10^{-2} eV is the result of a loss of coherence when operating with vacuum in the magnetic field region. Note that the value of m_a for which coherence is lost depends on the length L of the magnetic field, as seen from Eq. (5.25), and therefore depends on the specific helioscope. The black line is a combination of vacuum and buffer gas measurements, i.e., here coherence is restored—as can be seen from Eq. (5.33)—by scanning through small pressure steps with a buffer gas (^{4}He and ^{3}He, in the case of CAST). Each "wiggle" in the upper limit corresponds roughly to one specific pressure setting, i.e., a specific narrow

---

[2] Chameleons [44, 45] are hypothetical scalar particles postulated as candidates for dark energy and interact less strongly with matter than with gravity. Depending on the energy density of their surrounding environment, these particles have a variable effective mass. A more detailed description of chameleons and their expected properties can be found in Ref. [46].

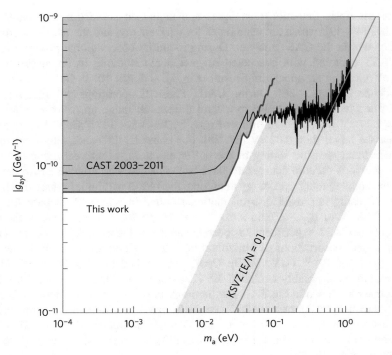

Fig. 5.10 CAST exclusion plot showing the recent benchmark result [31] of the experiment obtained with the IAXO pathfinder system [41]. The exclusion limits for $g_{a\gamma\gamma}$ at 95% C.L. are shown for previous data (black) and the latest results (red). QCD axions are expected to live in the yellow model band and the green line indicates the standard KSVZ axion model with $E/N = 0$. This ratio is the quotient of electromagnetic anomaly E and color anomaly N of the axion current [54, 55] and can acquire various values depending on the different axion models (see Chap. 2). Figure taken from Ref. [31], in which the interested reader will also be able to find additional details and references

axion mass range. The exact values depend on the number of actually observed (background) counts during the tracking, the exact time spent at the respective pressure setting, detectors active during tracking, and so on. Note also that the factor of roughly 2–3 between the vacuum and buffer gas measurements is due to the different exposure times (as well as improved detection systems): the vacuum phase includes about 2 years of tracking data, while 2+ pressure steps were measured per day during the gas phase.

End of Tutorial

5.4.2.2 The International Axion Observatory (IAXO)

The most straightforward way to improve over current helioscope designs (see the tutorial on helioscope figures of merit earlier in this chapter) is to boost the cross-sectional area of the magnet, equip all magnet bores with X-ray focusing devices, and utilize ultralow background detectors—these improvements are the basis for the next-generation axion helioscope IAXO [37]. The expected gain of IAXO over CAST is a factor of 10^4–10^5 in signal-to-noise ratio, which corresponds to an improvement in sensitivity to the coupling constant $g_{a\gamma\gamma}$ by $\approx 30\times$. These advances promise sensitivities to discover axionlike particles with a coupling to photons as small as $g_{a\gamma\gamma} \approx 10^{-12}$ GeV^{-1} or to electrons down to $g_{aee} \approx 10^{-13}$. IAXO also has the potential to find QCD axions in the 1 meV–1 eV mass range where these particles are able to solve the strong *CP* problem. Figure 5.11 shows the envisioned layout of the IAXO experiment.

Currently being designed, IAXO represents the next generation of axion helioscopes and builds on technologies with a proven track record in CAST as well as other particle physics experiments [56, 57] and astronomy missions [58, 59]. The key piece of IAXO is a 25 m long magnet with a 2.5 T (5.1 T) average (peak) field. For the first time, a helioscope will use a toroidal multibore configuration [60] with 8 coils of 70 cm diameter each and a total diameter of 5.1 m resulting in an intense

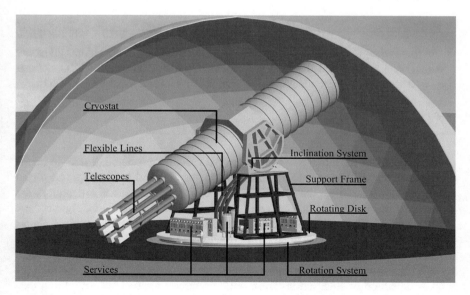

Fig. 5.11 Schematic view of the IAXO experiment. The 25 m long magnet with its 8 bores is shown along with eight X-ray optics and detectors, the flexible service lines, cryogenics, power service units, and the horizontal and vertical drive system. A lifesize person has been added for comparison. Figure taken from Ref. [39]

field over a large conversion volume. The magnet figure of merit f_M alone provides a $300\times$ improvement over CAST.

Each of IAXO's 8 bores will be covered by a 70 cm diameter telescope adopted from space science and optimized for axion searches. Telescopes for X-rays are based on the principle of total external reflection of light at grazing incidence. Therefore, the angle of incoming photons in the keV range needs to be below a critical angle (≈ 1 deg) in order for the X-rays to be reflected rather than absorbed. Reflectivity can be further enhanced by making use of Bragg's law, resulting in constructive interference via coating the mirror substrates with a "multilayer." Multilayers consist of periodic or non-periodic structures of alternating thin film layers of two or more materials (absorbers and spacers) deposited on an optical substrate. The focusing devices for IAXO will be built based on segmented glass technology originally developed for NASA's Nuclear Spectroscopic Telescope Array (NuSTAR, [58]) mission and replicated optics similar to those flown on the JAXA/NASA satellites Hitomi (ASTRO-H, see [59]) and XRISM [61]. While segmented glass optics are assembled out of thousands of individual mirror pieces, replicated optics are built up from multiple full revolution shells. For IAXO, the number and position of the substrates as well as the exact prescription for the coating are being carefully designed to optimize the throughput of the optics and match both the axion spectrum and the detector responses [62], making use of the IAXO pathfinder results [41].

? Problem 5.4 Estimating the Focal Spot Size for Solar Axion Observations

Given that the region of the Sun from which most axions are expected has an extent of $s_{\text{object}} = 3$ arcmin and the imaging capability of an envisioned IAXO optic is $s_{\text{optic}} = 2$ arcmin, estimate the expected focal spot area for a focal length of 5 m.

Solution on page 331.

The focal plane detectors will be ultralow background, pixelated devices to image the focused signal. In order to achieve the low background levels required for an efficient axion search ($\lesssim 10^{-7}$–10^{-8} counts/keV^{-1}cm^{-2}s^{-1}), these detectors are fabricated from radiopure materials and require sophisticated shields. The baseline technology for IAXO will use small gaseous detectors with pixelized readout planes (Microbulk MICROMEGAS [42, 63]) as previously developed and tested at CAST. In addition, other detector technologies are being studied [39] to reach higher sensitivities, lower energy thresholds, and better energy resolution for applications such as detection of solar axions from ABC processes via their g_{aee} coupling (see Sect. 5.1.2). Just like CAST before, IAXO will have a gas phase extending the helioscope's sensitivity to QCD axions at the higher axion masses.

As an intermediate step toward IAXO, BabyIAXO [64] is being designed and is just moving into the beginnings of its construction phase. BabyIAXO will be a scaled-down version of IAXO to test all IAXO components while simultaneously delivering first significant science results to supersede CAST's latest benchmark results. BabyIAXO will feature a 10 m long magnet with 2 bores of 60–70 cm diameter and two optic detector systems similar in design and layout to the ones of the full-scale IAXO experiment. Most likely, the experiment will be equipped with custom-designed optics close to the final IAXO telescope specifications and a telescope that is a flight spare from ESA's XMM Newton mission [65]. The baseline detectors for BabyIAXO are envisioned to be building on MICROMEGAS microbulk technology.

5.4.2.3   Physics Prospects of IAXO

The physics prospects for IAXO and BabyIAXO in comparison to current best limits from CAST and astrophysical hints (see Chap. 3) are shown in Fig. 5.12: at the high-mass end of the axion mass range large parts of QCD axion model space (KSVZ [8, 9] and DFSZ [10, 11]) can be tested, including viable dark matter models. Furthermore, the *ALP miracle* [66] parameter space in which ALPs simultaneously

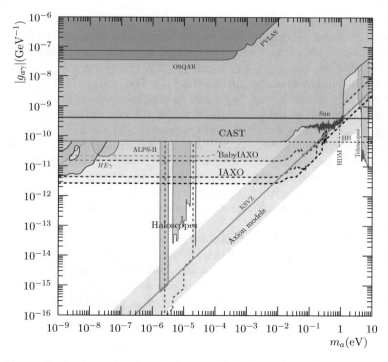

Fig. 5.12 Sensitivity prospects for BabyIAXO and IAXO in the context of other experimental and observational bounds. The interested reader is referred to Ref. [14] for a detailed recent review of axion researches, their results, and future prospects. Figure taken from Ref. [36]

solve dark matter and inflation can be studied. Also, at the high-mass end, IAXO will test non-hadronic models (axion–electron coupling) that would be able to explain the stellar cooling anomaly [67]. At the lower end of the axion masses, ALP parameter space invoked by observed hints of the anomalous transparency of the Universe to ultrahigh-energy (UHE) photons will be accessible to IAXO (partially with BabyIAXO), and in the intermediate axion mass range, there is a large region of parameter space that BabyIAXO and IAXO can probe for the very first time and in which viable ALP cold dark matter could exist.

In the case of a discovery, the study of the spectral features of the signal would provide additional insight into the nature of the new axionlike particle. With sufficient energy resolution and statistics, the measured spectrum could be decomposed as a sum of an ABC contribution and a Primakoff component. For adequate parameter ranges, it has been shown that independent determination of $g_{a\gamma\gamma}$ and g_{aee} should then be possible [68]. In particular, the narrow lines of Fig. 5.2 could first be used to unambiguously identify an ABC component and, eventually, they could be studied as a probe for solar metallicity [69]. Moreover, for values of m_a at the onset of loss of coherence, the measured spectrum gets distorted (depleted) at low energies. Again, with sufficient energy resolution and statistics, this effect can be used to determine the mass of the axion for a certain region of parameters [70].

Although this chapter is devoted to axions from the Sun, it is worth mentioning that there is another astrophysical source whose axions could potentially be detected with the help of helioscopes: namely, a nearby supernova (SN) explosion. In the first 10 s after the bounce of a core-collapse SN, axions are copiously produced via nucleon–nucleon axion-bremsstrahlung [71]. The energy of these axions can be several tens of MeVs, and they could in principle convert back into photons in a helioscope and be detected there, provided the experiment is equipped with appropriate high-energy detectors. According to [72], if the SN explosion occurs within a few hundred parsecs from Earth, the axions arrive in sufficiently high numbers, and one can expect a detectable, even though potentially small, signal in future helioscopes like BabyIAXO or IAXO. Apart from being equipped with gamma-ray detectors, the helioscope should point to the SN in advance of the actual explosion, something that could be accomplished with the help of a pre-SN neutrino alert.

In summary, IAXO and BabyIAXO are the next generation of axion helioscopes and will dramatically increase the sensitivity to solar axions compared to CAST, the currently most powerful axion helioscope. Furthermore, these novel, large-scale experiments also have the potential to serve as a multi-purpose facility for generic axion and ALP research in the coming decade, e.g., by incorporating microwave cavities and functioning as a haloscope (amongst other options). Together helioscopes, haloscopes, and laboratory searches provide complementary approaches to finally close in on QCD axions, axionlike particles, and other dark matter candidates to either discover these elusive particles or strongly constrain and potentially rule out their existence. For an instructive and detailed recent review of experimental axion and ALP searches, see Ref. [14].

? Problem 5.5 Calculating an Exclusion Plot Using the Maximum Likelihood Method

While the goal of every axion search is obviously to find the (still hypothetical) particle, if no signal above background is detected, one can still extract useful knowledge from the observations and exclude previously viable parameter space via an upper limit calculation for the axion–photon coupling constant $g_{a\gamma\gamma}$. Using a helioscope operated with buffer gas in the magnetic field region as an example (e.g., CAST's Phase II with ^{4}He), i.e., low statistics per pressure step, outline how an upper limit at 95% confidence level can be obtained using an extended maximum likelihood method under the assumption that uncertainties in the background can be neglected. Keep in mind that neighboring pressure settings can provide additional information about a specific density step. Here, the expected number of photons from axion-conversion N_{ik} at a specific gas density setting p_k (kth setting) in the energy bin E_i (ith energy bin) is given as

$$N_{ik} = \int_{E_i}^{E_{i+1}} \frac{d\Phi_i}{dE_a} P_{a\rightarrow\gamma,ik}\,\epsilon_i\,\Delta t_k\,A\,dE_a. \tag{5.41}$$

How can results from multiple pressure settings, detectors, and even experimental phases be combined to a single upper limit in the end?

Solution on page 331.

5.5 Alternative Experiments to Search for Solar Axions

5.5.1 Stationary Helioscopes

While conventional axion helioscopes are constructed to point at and follow the Sun, other approaches have also been considered. One such novel modulation helioscope technique uses a stationary setup in which a gaseous time projection chamber (TPC) is installed in a strong magnetic field, such as the Axion Modulation hELIoscope Experiment (AMELIE, see [73]). Since solar axions are most efficiently converted when their incidence direction is perpendicular to the magnetic field, a modulation signal varying during the day due to Earth's rotation is expected. Given that the signal would furthermore vary over the course of the year, axions would leave a distinct temporal signature in addition to their usual spectral one. This novel helioscope technology is not competitive with the standard helioscope technique for low axion masses ($m_a \lesssim 0.1$ eV) due to the fact that coherence of conversion is lost because of the short range of the X-rays from axion conversion in the high-pressure or high-Z gas used for large photon absorption. For higher axion masses, however, the approach might prove useful since the scanning of axion masses with a buffer gas in a conventional axion helioscope to study model-compatible axions at the higher end of the mass range remains challenging.

5.5.2 Crystalline Detectors Using Primakoff–Bragg Conversion

Instead of an external magnetic field, axion–photon conversion can also take place in the electromagnetic field at the atomic level inside materials. Therefore, crystalline detectors can be used to coherently convert solar axions into photons, which is the case when the angle of incidence of the axion fulfills the Bragg condition with the plane of the crystal [74, 75]. Pioneering results investigating these Bragg patterns were achieved with the SOLAX, COSME, and DAMA experiments. SOLAX used a Germanium spectrometer to study axion masses $m_a \lesssim 1$ keV and derived an upper limit on the coupling constant of $g_{a\gamma\gamma}(95\%$ CL$) < 2.7 \times 10^{-9}$ GeV^{-1} for this range [76]. COSME provided a similar result with its Ge detector yielding $g_{a\gamma\gamma}(95\%$ CL$) < 2.78 \times 10^{-9}$ GeV^{-1}, independent of the axion mass [77]. The best result so far was achieved by DAMA with a NaI(Tl) crystal [78]: $g_{a\gamma\gamma}(90\%$ CL$) < 1.7 \times 10^{-9}$ GeV^{-1}, again independent of the axion mass. It is worth noting that these bounds are not as strong as those derived from solar physics. Even though future experiments like CUORE [79] are expected to provide improved sensitivity to $g_{a\gamma\gamma}$, they will not be able to compete with axion helioscopes for axion masses below 1 eV. While for higher masses they become more competitive, these heavy axions are disfavored by cosmology and astrophysics [80] (see also Chap. 3).

5.5.3 Non-Primakoff Effect Conversions

While the axion–photon coupling constant is the preferred parameter to study for most axion searches, since it is generic to all axion models, axions could also interact with matter via their coupling to electrons and nucleons. WIMP searches using liquid xenon detectors [81–83] have looked for potential signals due to the axio-electric effect [84] in their ionization detectors [85, 86]. The axio-electric effect is similar to the photo-electric effect, but, instead of a photon, an axion hits the electron and ionizes the target atom (e.g., xenon). The advantage here is that the final signals depend directly on g_{aee} rather than on a product of the axion–electron and the axion–photon coupling. LUX sets the most competitive limit at a 90% C.L. as $g_{aee} < 3.5 \times 10^{-12}$, which is, however, not yet able to compete with limits from astrophysics [67]. Recently, the XENON collaboration reported a 3.5σ excess compatible with a potential axion signal [87] but cautioned that tritium background could explain the observed feature and cannot be excluded as the real cause for the excess at present time. Measurements with the next-generation XENONnT experiment will enable further studies of the observed feature. (Editor's note: indeed, after this chapter was written, the XENONnT experiment ruled out axions as the cause of this excess, see arXiv:2207.11330.)

To probe axion–nucleon couplings, monochromatic solar axions emitted in M1 nuclear transitions can be searched for with detectors containing the same nuclide (see [14] and the references therein for a more detailed discussion). These experiments are however not able to compete with astrophysical limits.

However, for the time being, helioscopes remain the most promising approach to find solar axions and ALPs.

Acknowledgments Parts of this chapter were adapted from Ref. [88]. Part of this work was performed under the auspices of the U.S. Department of Energy by Lawrence Livermore National Laboratory under Contract DE-AC52-07NA27344 with support from the LDRD program through grant 17-ERD-030. I.G.I. acknowledges support from the European Research Council (ERC) under the European Union's Horizon 2020 research and innovation programme, grant agreement ERC-2017-AdG 788781 (IAXO+) as well as from the Spanish "Agencia Estatal de Investigación" under grant PID2019-108122GB-C31.

References

1. H. Primakoff, Phys. Rev. **81**, 899 (1951)
2. G.G. Raffelt, *Stars as Laboratories for Fundamental Physics: The Astrophysics of Neutrinos, Axions, and Other Weakly Interacting Particles* (University of Chicago Press, Chicago, USA, 1996)
3. G.G. Raffelt, Phys. Rev. D **37**, 1356 (1988)
4. G.G. Raffelt, Phys. Rev. D **33**, 897 (1986)
5. K.V. Bibber, P.M. McIntyre, D.E. Morris, G.G. Raffelt, Phys. Rev. D. **39**, 2089 (1989)
6. J.N. Bahcall, W.F. Huebner, S.H. Lubow, P.D. Parker, R.K. Ulrich, Rev. Mod. Phys. **54**, 767 (1982)
7. J.N. Bahcall, M.H. Pinsonneault, Phys. Rev. Lett. **92**, 121301 (2004)
8. J.E. Kim, Phys. Rev. Lett. **42**, 103 (1979)
9. M.A. Shifman, A.I. Vainshtein, V.I. Zakharov, Nucl. Phys. B **166**, 493 (1980)
10. A.P. Zhitnitskiĭ, Sov. J. Nucl. Phys. **31**, 260 (1980)
11. M. Dine, W. Fischler, M. Srednicki, Phys. Lett. B **104**, 199 (1981)
12. J. Redondo, JCAP **12**, 008 (2013)
13. S. Andriamonje, et al., JCAP **0704**, 010 (2007)
14. I.G. Irastorza, J. Redondo, Prog. Part. Nucl. Phys. **102**, 89 (2018)
15. E. Masso, R. Toldra, Phys. Rev. D **52**, 1755 (1995)
16. E. Masso, Nucl. Phys. Proc. Suppl. **114**, 67 (2003)
17. A. Ringwald, Phys. Dark Univ. **1**, 116 (2012)
18. R.D. Peccei, H.R. Quinn, Phys. Rev. Lett. **38**, 1440 (1977)
19. R.D. Peccei, H.R. Quinn, Phys. Rev. D **16**, 1791 (1977)
20. E. Zavattini, et al., Phys. Rev. Lett. **96**, 110406 (2006)
21. G.G. Raffelt, Lect. Notes Phys. **741**, 51 (2008)
22. H. Schlattl, A. Weiss, G.G. Raffelt, Astropart. Phys. **10**, 353 (1999)
23. N. Vinyoles, et al., J. Cosmol. Astropart. Phys. **1510**, 015 (2015)
24. P. Gondolo, G.G. Raffelt, arXiv:0807.2926 (2008)

25. K. Zioutas, et al., Nucl. Instrum. Meth. **A425**, 480 (1999)
26. K. Zioutas, et al., Phys. Rev. Lett. **94**, 121301 (2005)
27. E. Arik, et al., JCAP **0902**, 008 (2009)
28. E. Arik, et al., Phys. Rev. Lett. **107** (2011)
29. M. Arik, et al., Phys. Rev. Lett. **112**, 091302 (2014)
30. M. Arik, et al., Phys. Rev. D **92**, 021101 (2015)
31. V. Anastassopoulos, et al., Nature Phys. **13**, 584 (2017)
32. G.G. Raffelt, L. Stodolsky, Phys. Rev. D **37**, 1237 (1988)
33. P. Sikivie, Phys. Rev. Lett **51**, 1415 (1983)
34. D.M. Lazarus, et al., Phys. Rev. Lett. **69**, 2333 (1992)
35. S. Moriyama, et al., Phys. Lett. B **434**, 147 (1998)
36. E. Armengaud, et al., J. Cosmol. Astropart. Phys. **1906**, 047 (2019)
37. I.G. Irastorza, et al., J. Cosmol. Astropart. Phys. **1106**, 013 (2011)
38. I.G. Irastorza, et al., The international axion observatory IAXO. Letter of intent to the CERN SPS Committee. Tech. Rep. CERN-SPSC-2013-022. SPSC-I-242, CERN, Geneva (2013)
39. E. Armengaud, et al., JINST **9**, T05002 (2014)
40. M. Kuster, et al., New J. Phys. **9**, 169 (2007)
41. F. Aznar, et al., J. Cosmol. Astropart. Phys. **1512**, 008 (2015)
42. S. Aune, et al., JINST **9**, P01001 (2014)
43. D. Autiero, et al., New J. Phys. **9**, 171 (2007)
44. J. Khoury, A. Weltman, Phys. Rev. D **69**, 044026 (2004)
45. J. Khoury, A. Weltman, Phys. Rev. Lett. **93**, 171104 (2004)
46. C. Burrage, J. Sakstein, JCAP **11**, 045 (2016)
47. V. Anastassopoulos, et al., Phys. Lett. B **749**, 172 (2015)
48. V. Anastassopoulos, et al., J. Cosmol. Astropart. Phys. **01**, 032 (2019)
49. V. Anastassopoulos, et al., Phys. Dark Univ. **26**, 100367 (2019)
50. A. Friedland, M. Giannotti, M. Wise, Phys. Rev. Lett. **110**, 061101 (2013)
51. S. Andriamonje, et al., J. Cosmol. Astropart. Phys. **0912**, 002 (2009)
52. D. Miller, et al., J. Cosmol. Astropart. Phys. **1003**, 032 (2010)
53. K. Barth, et al., J. Cosmol. Astropart. Phys. **1305**, 010 (2013)
54. D.B. Kaplan, Nucl. Phys. B **260**, 215 (1985)
55. M. Srednicki, Nucl. Phys. B **260**, 689 (1985)
56. H.H.J. Ten Kate, et al., Physica C **468**, 2137 (2008)
57. H.H.J. Ten Kate, et al., IEEE Trans. Appl. Supercond. **18**, 352 (2008)
58. F.A. Harrison, et al., ApJ **770**, 103 (2013)
59. T. Takahashi, et al., J. Astron. Telesc. Instrum. Syst. **4**, 021402 (2018)
60. I. Shilon, A. Dudarev, H. Silva, U. Wagner, H.H.J. ten Kate, IEEE Trans. Appl. Supercond. **24**, 1 (2013)
61. B. Williams, et al., arXiv:2003.04962 (2020)
62. A.C. Jakobsen, M.J. Pivovaroff, F.E. Christensen, Proc. SPIE **8861**, 886113 (2013)
63. I.G. Irastorza, et al., J. Cosmol. Astropart. Phys. **1601**, 034 (2016)
64. I.G. Irastorza, et al., in preparation for submission to JINST (2020)
65. F. Jansen, et al., Astron. Astrophys. **365**, L1 (2001)
66. R. Daido, F. Takahashi, W. Yin, J. High Energy Phys. **2018**, 104 (2018)
67. M. Giannotti, I.G. Irastorza, J. Redondo, A. Ringwald, K. Saikawa, J. Cosmol. Astropart. Phys. **1710**, 010 (2017)
68. J. Jaeckel, L.J. Thormaehlen, JCAP **03**, 039 (2019)
69. J. Jaeckel, L.J. Thormaehlen, Phys. Rev. D **100**, 123020 (2019)
70. T. Dafni, C.A. O'Hare, B. Lakić, J. Galán, F.J. Iguaz, I.G. Irastorza, K. Jakovčic, G. Luzón, J. Redondo, E. Ruiz Chóliz, Phys. Rev. D **99**, 035037 (2019)
71. G.G. Raffelt, J. Redondo, N. Viaux Maira, Phys. Rev. D **84**, 103008 (2011)
72. S.F. Ge, K. Hamaguchi, K. Ichimura, K. Ishidoshiro, Y. Kanazawa, Y. Kishimoto, N. Nagata, J. Zheng, *Supernova-Scope for the Direct Search of Supernova Axions* (2020)
73. J. Galan, et al., JCAP **12**, 012 (2015)

74. E.A. Paschos, K. Zioutas, Phys. Lett. B **323**, 367 (1994)
75. R.J. Creswick, et al., Phys. Lett. B **427**, 235 (1998)
76. F.T. Avignone III, et al., Phys. Rev. Lett. **81**, 5068 (1998)
77. A. Morales, et al., Astropart. Phys. **16**, 325 (2002)
78. R. Bernabei, et al., Phys. Lett. B **515**, 6 (2001)
79. D. Li, R.J. Creswick, F.T. Avignone, Y. Wang, J. Cosmol. Astropart. Phys. **1510**, 065 (2015)
80. S. Cebrian, et al., Astropart. Phys. **10**, 397 (1999)
81. K. Abe, et al., Phys. Lett. B **724**, 46 (2013)
82. C. Fu, et al., Phys. Rev. Lett. **119**, 181806 (2017)
83. D.S. Akerib, et al., Phys. Rev. Lett. **118**, 261301 (2017)
84. A. Derevianko, V.A. Dzuba, V.V. Flambaum, M. Pospelov, et al., Phys. Rev. D **82**, 065006 (2010)
85. A. Ljubicic, D. Kekez, Z. Krecak, T. Ljubicic, Phys.Lett. B **599**, 143 (2004)
86. A. Derbin, A. Kayunov, V. Muratova, D. Semenov, E. Unzhakov, Phys. Rev. D **83**, 023505 (2011)
87. E. Aprile, et al., Phys. Rev. D **102**, 072004 (2020)
88. J.K. Vogel, Searching for Solar Axions in the eV-Mass Region with the CCD Detector at CAST. Ph.D. thesis

Chapter 6
Magnetic Resonance Searches

John W. Blanchard, Alexander O. Sushkov, and Arne Wickenbrock

Abstract Ultralight bosonic dark matter (UBDM), such as axions and axionlike particles (ALPs), can interact with Standard Model particles via a variety of portals. One type of portal induces electric dipole moments (EDMs) of nuclei and electrons and another type generates torques on nuclear and electronic spins. Several experiments search for interactions of spins with the galactic dark matter background via these portals, comprising a new class of dark matter haloscopes based on magnetic resonance.

6.1 Searching for Axionlike Dark Matter via Nuclear Magnetic Resonance

Searches for ultralight bosonic dark matter (UBDM) based on magnetic resonance rely on the interaction between the UBDM field and spin. The possible forms of these interactions are detailed in Chap. 2 (see Table 2.1 and surrounding discussion). Here we focus on axions and axionlike particles (ALPs) and write these interactions in a format that aids the description of relevant experiments. We also assume that the axion or ALP field is the dominant component of the dark matter energy density. Chapter 8 covers searches that do not make this assumption.

J. W. Blanchard
Helmholtz-Institut Mainz, GSI Helmholtzzentrum für Schwerionenforschung GmbH, Mainz, Germany

NVision Imaging Technologies GmbH, Ulm, Germany
e-mail: john@nvision-imaging.com

A. O. Sushkov (✉)
Department of Physics, Boston University, Boston, MA, USA
e-mail: asu@bu.edu

A. Wickenbrock
Helmholtz Institut Mainz, Johannes Gutenberg-Universität, Mainz, Germany
e-mail: wickenbr@uni-mainz.de

© The Author(s) 2023
D. F. Jackson Kimball, K. van Bibber (eds.), *The Search for Ultralight Bosonic Dark Matter*, https://doi.org/10.1007/978-3-030-95852-7_6

6.1.1 Interactions with Nuclear Spins

The electric dipole moment (EDM) interaction of the axion field a with nuclear spin I is described by the non-relativistic Hamiltonian:

$$\mathcal{H}_{\text{EDM}} = g_d a E^* \cdot I / I, \tag{6.1}$$

where g_d is the coupling constant and E^* is the effective electric field, see below. The gradient interaction is described by the non-relativistic Hamiltonian (see discussion in Sect. 2.4.3):

$$\mathcal{H}_{\text{gr}} = \hbar c g_{\text{aNN}} \nabla a \cdot I, \tag{6.2}$$

where g_{aNN} is the coupling constant. Both $\mathcal{H}_{\text{EDM}}$ and $\mathcal{H}_{\text{gr}}$ have the same form as the Zeeman Hamiltonian, $\mathcal{H}_Z = -\hbar \gamma B^* \cdot I$, where γ is the nuclear spin gyromagnetic ratio, and B^* is an effective magnetic field, proportional either to $a E^*$ or to ∇a. To a first approximation, the axionlike dark matter field $a(t) = a_0 \cos(\omega_a t)$ oscillates at the Compton angular frequency $\omega_a = m_a c^2 / \hbar$, where c is the speed of light in vacuum, $\hbar$ is the reduced Planck constant, and m_a is the unknown ALP mass. Therefore the effective magnetic field B^* also oscillates at this angular frequency.

We consider here magnetic resonance experiments that search for spin energy shifts or for spin precession induced by these oscillating interactions. The energy shift is quantified by the expectation value of the relevant Hamiltonian, $\langle \mathcal{H} \rangle$, where $\mathcal{H}$ could be either $\mathcal{H}_{\text{EDM}}$ or $\mathcal{H}_{\text{gr}}$. To quantify the rate of spin precession, it is convenient to define the Rabi frequency Ω_a. Assuming that the effective magnetic field B^* is linearly polarized, in the rotating wave approximation this Rabi frequency is given by

$$\Omega_a = \frac{\langle \mathcal{H} \rangle}{2\hbar}. \tag{6.3}$$

6.1.1.1 The EDM Interaction with P,T-odd Moments of Nucleons and Nuclei

As discussed in Chap. 2, the axion concept was invented as a solution of the strong CP problem of quantum chromodynamics (QCD). It is the QCD axion interaction with the gluon field that achieves this goal, by relating the θ parameter of the QCD Lagrangian to the axion field: $\theta = a/f_a$. In the presence of a dynamical background axionlike field $a(t)$, the oscillating $\theta(t)$ can lead to experimentally observable effects. For an isolated nucleon, θ induces an electric dipole moment (EDM):

$$d_n = g_d a = 2.4 \times 10^{-16} \theta \, \text{e} \cdot \text{cm} = 2.4 \times 10^{-3} \theta \, \text{e} \cdot \text{fm}, \tag{6.4}$$

calculated with 40% accuracy [1, 2]. We note the unit conversion: $1\,e \cdot cm = 1.5 \times 10^{13}\,GeV^{-1}$. The neutron EDM experiment discussed in Ref. [3] and the proton storage ring experiment described in Ref. [4] search for these oscillating nucleon EDMs.

When the nucleon is bound inside an atomic nucleus, this EDM is screened. According to the Schiff theorem, the EDM of a point-like nucleus is completely screened by atomic electrons in the low-frequency limit [5]. Taking into account finite nuclear size, the nuclear Schiff moment is given by

$$S = \frac{e}{10}\left(\langle r^2 r \rangle - \frac{5}{3Z}\langle r^2 \rangle \langle r \rangle\right), \tag{6.5}$$

where e is the elementary electric charge, Z is the atomic number, and $\langle r^k \rangle = \int r^k \rho(r) d^3 r$ are the moments of nuclear charge density $\rho(r)$. The Schiff moment sources the (P)arity- and (T)ime-odd electrostatic potential

$$\varphi(r) = 4\pi (S \cdot \nabla)\delta(r). \tag{6.6}$$

Let us consider two contributions to the nuclear Schiff moment: (1) the permanent nucleon EDM d_n, and (2) P,T-odd nuclear forces.

1. The contribution due to d_n arises because of non-coincident densities of nuclear charge and dipole moment. It can be estimated using Eq. (8.76) from Ref. [6]:

$$4\pi S_{EDM} \approx d_n \times \frac{4\pi}{25}\frac{(K+1)I}{I(I+1)}r_0^2, \tag{6.7}$$

where $K = (\ell - 1)(2I + 1) = 1$ and $r_0 = 1.25 A^{1/3}$, ℓ being the orbital angular momentum of the valence nucleon. Note that the definition of the Schiff moment in Ref. [6] differs from ours by a factor of 4π, which appears on the left-hand side.

2. The P,T-odd nuclear interaction of a non-relativistic nucleon with nuclear core is parametrized by strength η [7]:

$$W = \frac{G_F}{\sqrt{2}}\frac{\eta}{2m_n}\sigma \cdot \nabla\rho(r), \tag{6.8}$$

where $G_F \approx 10^{-5}\,GeV^{-2}$ is the Fermi constant, m_n is the nucleon mass, σ is its spin, and $\rho(r)$ is the density of core nucleons. A vacuum θ angle gives rise to this interaction via the P,T-odd pion-nucleon coupling constant [6, 8]:

$$\eta = 1.8 \times 10^6\,\theta. \tag{6.9}$$

Nuclear physics calculations express the nuclear Schiff moment of a particular nucleus in terms of parameter η, see, for example, Ref. [9].

Once the nuclear Schiff moment is expressed in terms of θ, atomic calculations are used to connect physical observables to the value of the Schiff moment. For example, the observable can be the value of the energy shift of a nuclear spin state in an applied electric field. The connection can then be made to the effective electric field E^* defined in Eq. (6.1). This calculation is performed for ^{207}Pb nuclear spins in ferroelectric crystals in Refs. [10, 11]. For ferroelectrically poled PMN-PT (lead magnesium niobate-lead titanate) it is found that the effective electric field is $E^* = 340\,\mathrm{kV/cm}$ with estimated uncertainty $\approx 50\%$ [12]. The effect on nuclear spins is equivalent to that of an effective magnetic field

$$B^*_{\mathrm{EDM}} = -\frac{g_d a E^*}{\hbar \gamma I},\tag{6.10}$$

where γ is the spin gyromagnetic ratio.

6.1.1.2 The Gradient Interaction

In order to calculate the gradient of the axionlike field a that appears in Eq. (6.2), it is necessary to consider the integral over the velocity distribution of the axionlike galactic dark matter field. Importantly, there are contributions from both the lab velocity with respect to the galactic rest reference frame, and from the spread of the dark matter virial velocity distribution. The effect on nuclear spins is equivalent to that of an effective magnetic field

$$B^*_{\mathrm{aNN}} = -(c g_{\mathrm{aNN}}/\gamma)\nabla a,\tag{6.11}$$

where γ is the spin gyromagnetic ratio.

6.1.2 Interactions with Electron Spins

Electron spins can also couple to the ALP field via the derivative fermion coupling. This gradient interaction generates the electron spin Hamiltonian with the same form as Eq. (6.11). For an electron spin the coupling constant in the Lagrangian is often written as $g_{\mathrm{aee}}/(2m_e)$, where m_e is the electron mass, and g_{aee} is unitless. However, the physics is exactly the same—an electron spin experiences a torque due to the gradient ∇a, which acts as an effective magnetic field, whose magnitude is proportional to g_{aee}. There are stringent astrophysical limits on the coupling constant g_{aee}, and the QUest for AXions experiment (QUAX) is a laboratory search for this interaction at frequencies near 10 GHz [13].

6.2   Basics of NMR

Nuclear magnetic resonance (NMR) encompasses a broad and versatile set of techniques that have found application in a wide range of disciplines. A typical NMR experiment involves measurement of nuclear spin dynamics in an applied bias magnetic field (Fig. 6.1).

In pulsed magnetic resonance experiments, the spins are perturbed by resonant radiofrequency (RF) magnetic field pulses, and the subsequent spin evolution is detected. Since the introduction of digital fast Fourier transform algorithms, most modern applications of NMR utilize the pulsed scheme [14, 15]. Searches for permanent electric dipole moments are an example of pulsed NMR experiments in the fundamental-physics context [16]. In continuous wave (CW) magnetic resonance experiments, the excitation field is present continuously and the bias field is varied. Spin-based dark matter haloscope experiments usually employ the CW scheme [17]. Here we provide a basic introduction to NMR, in order to help readers make sense of nuclear-spin-based dark matter searches. For a more thorough treatment, the reader is referred to Refs. [18–20].

6.2.1   Nuclear Magnetism

In virtually all cases,[1] detection of an NMR signal involves measurement of the magnetic field produced by nuclear spins in a sample. The magnetic moment of a

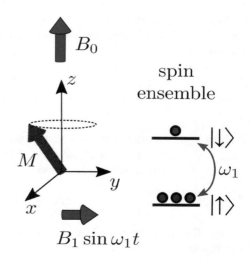

Fig. 6.1 A schematic of a typical NMR experiment. M is the nuclear spin magnetization of the sample. B_0 is the bias magnetic field, and $B_1 \sin \omega_1 t$ is the "pseudo-magnetic" field due to interaction with ultralight dark matter. The spin-1/2 level diagram indicates spin polarization as larger population in the ground spin sublevel, and spin coherence induced by the excitation field B_1, if it is resonant with the spin Larmor frequency

---

[1] Notable exceptions include electrically detected NMR [21] and beta-NMR [22, 23].

nucleus with non-zero spin is given by

$$\boldsymbol{\mu} = \frac{g_I \mu_N \boldsymbol{I}}{\hbar},$$

(6.12)

where g_I is the g-factor of the nuclear spin $\boldsymbol{I}$. In NMR it is typically convenient to rewrite the nuclear magnetic moment in terms of a nucleus' *gyromagnetic ratio*,

$$\gamma = \frac{g_I \mu_N}{\hbar},$$

(6.13)

where g_I is the nuclear g-factor and μ_N is the nuclear magneton, such that the magnetic moment may be written in condensed form as

$$\boldsymbol{\mu} = \gamma \boldsymbol{I}.$$

(6.14)

The total magnetization of an ensemble of spins can then be written as

$$M = N\hbar\gamma P_0 \boldsymbol{I},$$

(6.15)

where N is the number density of nuclear spins, and P_0 is the ratio of the spin state population difference to the total population, generally referred to as the *spin polarization*. Explicitly, for spin-1/2 nuclei,

$$P_0 = \frac{n_\uparrow - n_\downarrow}{n_\uparrow + n_\downarrow},$$

(6.16)

where $n_\uparrow$ represents spins with $m_I = +1/2$, and $n_\downarrow$ represents spins with $m_I = -1/2$. For nuclear spin polarization at thermal equilibrium, the Boltzmann distribution gives

$$n_\uparrow = e^{-E_+/(k_B T)}$$

(6.17)

$$n_\downarrow = e^{-E_-/(k_B T)},$$

(6.18)

where $E_\pm$ is the energy of the state with $m_I = \pm 1/2$, k_B is the Boltzmann constant, and T is the temperature of the system. In a large magnetic field, the dominant energy term is the Zeeman interaction,

$$\mathcal{H}_Z = -\hbar\gamma \boldsymbol{B} \cdot \boldsymbol{I},$$

(6.19)

where $\boldsymbol{B}$ is the applied magnetic field. For a magnetic field $\boldsymbol{B_0}$ in the $\hat{z}$ direction, the energy is

$$E = -\hbar\gamma B_0 m_I,$$

(6.20)

so Eq. (6.16) may be written as

$$P_0 = \frac{e^{\hbar\gamma B_0/(2k_B T)} - e^{-\hbar\gamma B_0/(2k_B T)}}{e^{\hbar\gamma B_0/(2k_B T)} + e^{-\hbar\gamma B_0/(2k_B T)}} = \tanh\left(\frac{\hbar\gamma B_0}{2k_B T}\right). \tag{6.21}$$

Under practically achievable conditions, the so-called high-temperature approximation is valid, so we keep only the leading term of the Taylor expansion of the hyperbolic tangent,

$$P_0 \approx \frac{\hbar\gamma B_0}{2k_B T}, \tag{6.22}$$

such that

$$\boldsymbol{M} \approx \frac{N\hbar^2\gamma^2 B_0}{2k_B T}\boldsymbol{I}. \tag{6.23}$$

From this, we can see that the magnitude of the observable NMR signal depends on the spin density, gyromagnetic ratio, and (assuming thermal spin polarization) the magnetic field strength.

? Problem 6.1 Magnetic Field Produced by a Spherical Sample

(a) Calculate the nuclear magnetization of liquid Xe in a field of 1 T at 170 K. Assume a density of 3.1 g/mL, atomic weight 131.3 g/mol, a 26.4% abundance of ^{129}Xe, and a xenon gyromagnetic ratio of -7.441×10^7 rad T^{-1} s^{-1}.

(b) Consider a spherical sample of ^{129}Xe with diameter 1 cm. What is the magnetic field produced by the sample (at 1 T and 170 K) at a distance of 1 cm from the center of the sample along the magnetization axis? What is the magnetic flux through a circular coil of radius 1 cm located 1 cm from the center of the sample along the magnetization axis?

(c) ^{129}Xe polarizations *far* above those achievable at thermal equilibrium can be achieved using a technique called Spin-Exchange Optical Pumping (SEOP) [24, 25]. If a nuclear spin polarization of 10% is achieved, what is the magnetic flux that would arise in the coil arrangement above?

Solution on page 334.

Because the energy difference between nuclear spin states is generally small compared to thermal energy, $\hbar\gamma B_0 \ll k_B T$, spin polarization at thermal equilibrium is many orders of magnitude smaller than unity. Considering that the NMR signal is proportional to polarization, sensitivity (either in dark matter searches or in chemical analysis) can be greatly enhanced through the use of so-called *hyperpolarization* techniques, which generate non-equilibrium spin states with polarization approaching 100%. Examples of such techniques include dynamic nuclear polarization (DNP) [26, 27], spin-exchange optical pumping (SEOP) [24], metastability exchange optical pumping (MEOP) [28], and parahydrogen-induced polarization (PHIP) [29–31].

6.2.2 Nuclear Spin Dynamics

A phenomenological description of the evolution an ensemble of nuclear spins in a magnetic field is given by the Bloch equation:

$$\frac{d\boldsymbol{M}}{dt} = \gamma \boldsymbol{M} \times \boldsymbol{B} - \frac{M_x \hat{\boldsymbol{x}} + M_y \hat{\boldsymbol{y}}}{T_2} - \frac{M_z - M_0}{T_1} \hat{\boldsymbol{z}} \,, \tag{6.24}$$

where γ is the nuclear gyromagnetic ratio, $\boldsymbol{M}$ is the nuclear magnetization vector, M_0 is the magnitude of the equilibrium magnetization, as derived in Eq. (6.23), and $\boldsymbol{B}$ is the magnetic field, oriented along the z-axis. We have also introduced two characteristic relaxation times: the longitudinal relaxation time T_1 and the transverse relaxation time T_2. The magnetization dynamics are a combination of relaxation and precession about the magnetic field at the Larmor frequency

$$\omega_L = \gamma B \,. \tag{6.25}$$

The longitudinal relaxation time may be interpreted as the characteristic time it takes for the spin system to reach equilibrium. For example, if an unpolarized sample is placed into a magnetic field, the magnetization will build up as

$$M(t) = M_0 \left(1 - e^{-t/T_1}\right). \tag{6.26}$$

For hyperpolarized spin systems, the magnetization generally decays to its equilibrium value with characteristic time T_1 as well.

The transverse relaxation time may be thought of as the nuclear spin coherence lifetime, corresponding to the exponential decay of precessing magnetization in the xy plane. As an example, we consider the case where we have prepared a spin state where our sample is magnetized along $\hat{\boldsymbol{x}}$ (by applying, for example, a $-\pi/2$ pulse along $\hat{\boldsymbol{y}}$ to rotate spins initially oriented along $\hat{\boldsymbol{z}}$ to along $\hat{\boldsymbol{x}}$). Then the magnetization along $\hat{\boldsymbol{x}}$ will be

$$M_x(t) = M_0 \cos{(\gamma B t)} e^{-t/T_2}. \tag{6.27}$$

The Fourier transform of this signal yields a Lorentzian peak at the Larmor frequency with full-width at half maximum (FWHM)

$$w_{1/2} = \frac{1}{\pi T_2}. \tag{6.28}$$

More specifically, T_2 refers to the "intrinsic" dephasing time that would occur in a perfectly homogeneous magnetic field. The transverse relaxation time in a real magnetic field (i.e., possessing some inhomogeneity) is T_2^*.

To understand how nuclear spins respond to oscillating magnetic (or axion) fields, it is useful to transform into a rotating reference frame. Consider a driving field along the x-axis: $B_x(t) = 2B_1 \cos \omega t$, which can be decomposed into two counter-rotating components. In the reference frame rotating around the z-axis at the frequency ω the magnetization components are $\tilde{M}_x$, $\tilde{M}_y$, and the M_z component is unchanged. The connection between the laboratory-frame and the rotating-frame magnetization components is:

$$M_x = \tilde{M}_x \cos \omega t - \tilde{M}_y \sin \omega t \tag{6.29}$$

$$M_y = \tilde{M}_x \sin \omega t + \tilde{M}_y \cos \omega t. \tag{6.30}$$

The steady-state solution of the Bloch equations in the rotating frame is:

$$\tilde{M}_x = \frac{\Delta \omega \gamma B_1 T_2^2}{1 + (T_2 \Delta \omega)^2 + \gamma^2 B_1^2 T_1 T_2} M_0 \tag{6.31}$$

$$\tilde{M}_y = \frac{\gamma B_1 T_2}{1 + (T_2 \Delta \omega)^2 + \gamma^2 B_1^2 T_1 T_2} M_0 \tag{6.32}$$

$$M_z = \frac{1 + (\Delta \omega T_2)^2}{1 + (T_2 \Delta \omega)^2 + \gamma^2 B_1^2 T_1 T_2} M_0, \tag{6.33}$$

where $\Delta \omega = \omega - \omega_L$ is the drive detuning.

For a resonant ($\Delta \omega = 0$) driving field that is far from saturation ($\gamma B_1 \ll 1/\sqrt{T_1 T_2}$), these equations become substantially simpler. We see that $\tilde{M}_x = 0$, $\tilde{M}_z \approx M_0$, and

$$M_y \approx \gamma B_1 T_2 M_0 \cos \omega t. \tag{6.34}$$

Together with Eqs. (6.10) and (6.11), this result allows us to convert the strength of the axionlike dark matter EDM or gradient interaction to an experimental observable: the transverse nuclear magnetization.

6.2.3 Nuclear Spin Interactions

We will now consider some basic features of NMR spectra that arise due to various spin interactions. An understanding of NMR spectra is needed for design and calibration of NMR-based dark matter searches.

6.2.3.1 Chemical Shielding

Because most experiments are not conducted with bare nuclei, it is necessary to generalize the Zeeman interaction of Eq. (6.19) to include local susceptibility effects:

$$\mathcal{H}_{CS} = \hbar \gamma \boldsymbol{I} \cdot (1 - \boldsymbol{\sigma}_{\mathrm{cs}}) \cdot \boldsymbol{B}_0, \qquad (6.35)$$

where the chemical shielding tensor $\boldsymbol{\sigma}_{\mathrm{cs}}$ describes the effect of electrons producing counteracting magnetic fields that "shield" the nucleus from the external field. This *chemical shielding* interaction is particularly useful for chemists: different chemical environments within a given molecule, labeled by subscript j, give rise to different shielding $\sigma_{\mathrm{cs},j}$, and therefore different peak shifts, which can be interpreted in terms of electron density. For practical reasons, it is often convenient to refer to the *chemical shift* relative to some reference, defined as $\delta = \sigma_{\mathrm{ref}} - \sigma_{\mathrm{cs}}$. The shift is typically measured in units of "parts per million" (ppm). For example, $\delta = 10^{-6}$ corresponds to a 1 ppm shift to higher frequency.[2] From the perspective of dark matter searches, these shifts can be problematic, as maximum sensitivity is achieved if all spins in a sample have the same Larmor frequency. For example, consider the case of ethanol, CH_3CH_2OH, which contains 102.8 moles of ^1H per liter: 1/2 of the hydrogens are in the CH_3 environment, 1/3 are in the CH_2 environment, and 1/6 of all hydrogens are in the OH environment. For comparison, methanol, CH_3OH, contains 98.9 moles of ^1H per liter, and 3/4 of them are in the CH_3 environment. So on a per-peak basis, methanol gives roughly 50% more signal for the peaks associated with the CH_3 environment, as illustrated in Fig. 6.2b.

The tensor nature of the chemical shielding is also of note, as its principal axis system is defined in the molecular frame. In liquid samples, rapid molecular tumbling averages out all components except for the isotropic part of the tensor, σ_{iso}. In the case of solid samples, however, the anisotropy of the chemical shielding tensor persists, and different molecular orientations yield different NMR frequencies. In single-crystal samples, NMR spectra consist of a countable number of peaks, which shift depending on crystal orientation relative to the magnetic field. Polycrystalline or powder samples, however, are composed of a large number of randomly oriented

---

[2] In CW-NMR operating at a constant frequency, the chemical shift refers to a change in the magnetic field, so $\delta = 10^{-6}$ corresponds to a 1 ppm shift to lower field. This is the historical reason for the somewhat confusing NMR tradition of plotting spectra with an inverted x axis.

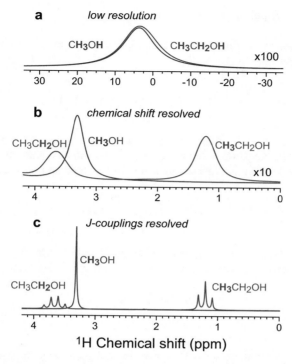

Fig. 6.2 ^{1}H NMR spectra of ethanol (blue) and methanol (red), acquired at 1.4 T (60 MHz ^{1}H Larmor frequency), with the chemical-shift scale referenced to tetramethylsilane (TMS). For simplicity, the hydroxyl (-OH) signals at 4.8 ppm are not shown in parts (**b**) and (**c**). (**a**) In the low-resolution case—corresponding here to a line width of 10 ppm—ethanol yields slightly more signal than methanol due the larger concentration of ^{1}H. (**b**) For line widths on the order of 1 ppm (the line width shown is 15 Hz, or 0.25 ppm), chemical shifts can be resolved. Because the (non-hydroxyl) hydrogens in ethanol are in two different chemical environments, the signal is separated into two peaks—the signal at 1.2 ppm corresponds to the CH$_3$ hydrogens and the signal at 3.7 ppm corresponds to the CH$_2$ hydrogens—each of which is smaller than the methanol signal. (**c**) At higher resolution (the line width shown here is approximately 1.8 Hz, or 30 ppb), further structure due to J-couplings can be seen in ethanol: the CH$_3$ signal is split into a triplet by the CH$_2$ hydrogens, and the CH$_2$ signal is split into a quartet by the CH$_3$ hydrogens. Note that the hydroxyl protons do not induce splittings due to rapid chemical exchange

crystallites, which leads to a distribution of NMR resonances. Such distributions are often hundreds of ppm wide, which reduces sensitivity. It is worth pointing out that the so-called powder broadening is an example of *inhomogeneous* broadening—the signal is broad not because of T_2 relaxation, but because of the distribution of orientations.

6.2.3.2 Direct Dipole-Dipole Coupling

The Hamiltonian describing the direct dipole-dipole coupling of two nuclear spins I_1 and I_2 is given by

$$\mathcal{H}_{12} = \frac{\mu_0}{4\pi} \frac{\hbar^2 \gamma_1 \gamma_2}{r_{12}^3} \left[I_1 \cdot I_2 - 3 \left(I_1 \cdot \hat{r}_{12} \right) \left(I_2 \cdot \hat{r}_{12} \right) \right], \qquad (6.36)$$

where γ_1, γ_2 are the gyromagnetic ratios of the two spins and r_{12} is the vector between them. In magnetic fields much larger than the characteristic dipolar field, the part of this Hamiltonian that commutes with the Zeeman interaction is:

$$\mathcal{H}_{12} \approx \frac{\mu_0}{4\pi} \frac{\hbar^2 \gamma_1 \gamma_2}{r_{12}^3} \left(1 - 3\cos^2 \alpha \right) \left[\hat{I}_{1z} \hat{I}_{2z} - \frac{1}{4} \left(\hat{I}_{1+} \hat{I}_{2-} + \hat{I}_{1-} \hat{I}_{2+} \right) \right], \qquad (6.37)$$

where α is the angle between r_{12} and the magnetic field vector, $\hat{I}_{1\pm}$ and $\hat{I}_{2\pm}$ are the raising and lowering operators for the respective nuclear spins 1 and 2, where

$$\hat{I}_+ = \hat{I}_x + i\hat{I}_y, \qquad (6.38)$$

$$\hat{I}_- = \hat{I}_x - i\hat{I}_y. \qquad (6.39)$$

In NMR literature the dipolar Hamiltonian acting on the spin ensemble is often written as

$$\mathcal{H}_D = \hbar \sum_{N;N'>N} I_N \cdot D_{NN'} \cdot I_{N'}, \qquad (6.40)$$

where $D_{NN'}$ is a rank-2 symmetric tensor, and the sum is over all spin pairs in the ensemble. Because the dipole-dipole coupling tensor is traceless, it is fully averaged out by molecular tumbling in isotropic liquids. In "dilute" powdered solids composed of isolated spin pairs, it is possible to resolve a double-peaked feature called "Pake doublet" that arises due to the distribution of angles α for different spin pairs. Solids with high nuclear spin density, which are of greater interest for dark matter searches, cannot be considered in terms of isolated spin pairs, so one must also consider the distribution of distances r_{IS}. This generally gives rise to broad "blobby" NMR resonances (a more detailed analysis of solid-state NMR lineshapes is given in Ref. [20]).

In some experiments dipolar broadening can be greatly reduced using a technique called *magic angle spinning* (MAS) [32]. By rotating a sample rapidly[3] about an axis, tensor components transverse to the spinning axis are averaged to zero.

---

[3] Rapidly here means that the rotation frequency should significantly exceed the largest dipolar-coupling frequency.

By setting the angle between the spinning axis and the magnetic field such that $1 - 3\cos^2\alpha = 0$, the effective time-averaged dipole-dipole coupling is zero. This technique can also be used to average out powder broadening due to chemical-shift anisotropy. While the added complexity of MAS may seem daunting for dark matter experiments, the improved signal intensity may prove useful.

6.2.3.3   Indirect Spin-Spin Coupling

The indirect dipole-dipole coupling, known as J-coupling, may be thought of as a second-order hyperfine effect where one nucleus N affects the electronic state of a molecule through hyperfine couplings to the molecular electron density, and this perturbation is then transmitted from the molecular electronic state to a second nucleus N' through its hyperfine interaction with the molecular electron density. The Hamiltonian may be written in the form

$$\mathcal{H}_J = \hbar \sum_{N; N' \neq N} \boldsymbol{I}_N \cdot \boldsymbol{J}_{NN'} \cdot \boldsymbol{I}_{N'} \,, \tag{6.41}$$

where $\boldsymbol{I}_N$ and $\boldsymbol{I}_{N'}$ are the spins of nuclei N and N', and $\boldsymbol{J}_{NN'}$ is the second-rank J-coupling tensor. $\boldsymbol{J}_{NN'}$ may in general be represented as a sum of irreducible spherical tensors

$$\boldsymbol{J}_{NN'} = \boldsymbol{J}_{NN'}^{(0)} + \boldsymbol{J}_{NN'}^{(1)} + \boldsymbol{J}_{NN'}^{(2)} \,, \tag{6.42}$$

where the isotropic component $\boldsymbol{J}_{NN'}^{(0)}$ transforms as a scalar, the antisymmetric component $\boldsymbol{J}_{NN'}^{(1)}$ transforms as a pseudovector, and the symmetric component $\boldsymbol{J}_{NN'}^{(2)}$ transforms as a symmetric rank-2 spherical tensor. The J-coupling is often referred to as the "scalar" coupling because typical high-resolution NMR experiments utilize isotropic liquid samples where rapid molecular tumbling averages higher-order tensor components to zero. In solid-state experiments, $\boldsymbol{J}_{NN'}^{(2)}$ adds to the dipole-dipole coupling and transforms the same way. The rank-1 component does not commute with the high-field Zeeman Hamiltonian and has never been conclusively measured.

In terms of dark matter searches, the J-coupling is most important for experiments with liquid samples [33, 34], where it causes peak splitting. Consider again the cases of ethanol (CH_3CH_2OH) and methanol (CH_3OH). Assuming the presence of even a small amount of water, the OH hydrogens undergo rapid exchange and thus their couplings to the other spins are averaged to zero. The three CH_3 hydrogens in methanol are magnetically equivalent, so there are no observable J-couplings and the methanol spectrum consists of a single resonance. In ethanol, there are two sets of equivalent spins, the three CH_3 hydrogens and the two CH_2 hydrogens. The two CH_2 hydrogens split the CH_3 resonance into a triplet with relative amplitudes 1:2:1, and the three CH_3 hydrogens split the CH_2 resonance into a quartet with relative amplitudes 1:3:3:1. As a result, the largest peak in ethanol is about three times smaller than the methanol peak, as shown in Fig. 6.2c.

6.2.3.4 Quadrupolar Coupling

The quadrupolar coupling Hamiltonian has the form

$$\mathcal{H}_Q(\Theta) = \frac{eQ}{2I(2I-1)} \boldsymbol{I} \cdot \boldsymbol{V}(\Theta) \cdot \boldsymbol{I}, \tag{6.43}$$

where e is the electric charge, Q is the nuclear quadrupole moment, I is the nuclear spin quantum number, and $\boldsymbol{V}(\Theta)$ is the electric field gradient tensor for an arbitrary molecular orientation Θ [19]. As a rank-2 interaction, the quadrupolar coupling is non-zero only for nuclei with spin $I \geq 1$. This is one of the notable advantages of studying spin-1/2 nuclei, because the coupling of quadrupolar nuclei to the electric field gradient is a major source of relaxation, leading to short coherence times and thus to broad resonance lines.

6.2.4 Zero-to-Ultralow-Field NMR

While the vast majority of NMR experiments are performed in a large applied magnetic field, an alternative method, zero-to-ultralow-field (ZULF) NMR [35], also exists and has found use in dark matter searches [33, 34]. In ZULF NMR, the magnetic field is small enough that the Zeeman interaction may be treated as a perturbation on other spin couplings. As of the time of this writing, all ZULF NMR dark matter searches have relied on the electron-mediated J-coupling as the primary interaction, perturbed by a small bias magnetic field [33, 34].

6.3   Detecting Spin Evolution due to Axionlike Dark Matter

The most sensitive scheme for NMR detection depends on the frequency range being explored. In general, the goal is to search for oscillating magnetic fields originating from the evolution of spins due to interaction with ALP dark matter. Spins act as transducers for the cosmic signal, converting the oscillating ALP field to an oscillating magnetic field at the same frequency. The optimal detection modality is the most sensitive magnetometer for the frequency corresponding to the ALP mass range under investigation. Additionally, the working environment of the device has to be considered. For example, the most sensitive vapor cell magnetometers require that the ambient background magnetic field is below 100 nT in order to operate. This requirement might be at odds with the conditions needed for the spin sample

Table 6.1 Example of different NMR detection modalities with their frequency range and sensitivities to oscillating magnetic fields

Frequency range	Sensor	Field sensitivity	Ref.
dc–100 Hz	Alkali SERF	160 aT/$\sqrt{\text{Hz}}$	[38]
0.4–300 kHz	RF alkali vapors	1000 aT/$\sqrt{\text{Hz}}$	[39, 40]
10 Hz–2.5 MHz	dc SQUID	150 aT/$\sqrt{\text{Hz}}$	[41]
>2.5 MHz	Inductive coil	2000 aT/$\sqrt{\text{Hz}}$	[12]

in an ALP search: for instance, in the case of ALP Compton frequencies ω_a in the ~100 MHz range, applied magnetic fields of several tesla are required in order for the NMR resonant frequency to match ω_a. Superconducting quantum interference devices (SQUIDs) require cryogenic temperatures which demands either a stand-off distance for thermal insulation or a cold nuclear sample. The latter might be problematic due the inherently broader magnetic resonance linewidth of solids.

This section gives an overview of the most sensitive detection modalities used in NMR spectroscopy at various frequencies. The frequency range to be considered is as open as the mass range of the potential dark matter candidates. However, NMR measurements at the highest possible frequencies require the largest possible magnetic field and the field range of commercially available magnets is limited. Commercially available high-field magnet systems feature proton frequencies up to 1.2 GHz corresponding to a field of 28.2 T in a 54 mm diameter bore. The maximum demonstrated dc magnetic field in a research facility at the time of writing was 45.5 T [36] corresponding to a proton Larmor frequency of 1.93 GHz and therefore an ALP mass of 8 μeV. An overview of even higher magnetic fields that can be produced for short times can be found in Ref. [37]. Here we begin with general considerations of the axion-induced NMR signals and then review the working principles of a selection of sensitive devices used to search for NMR signals, namely optical atomic magnetometers in the spin-exchange-relaxation free (SERF) regime for the lowest bandwidth and highest magnetic field sensitivity, RF vapor cell magnetometers in an intermediate regime, followed by SQUIDs, and finally inductive pick-up coils. The section closes with a brief discussion of magnetic noise suppression techniques. Table 6.1 summarizes the different detection modalities mentioned in this chapter and their basic properties.

6.3.1 Axion-Induced NMR Signals

Figure 6.3 illustrates how to detect oscillating magnetic signals of unknown origin and frequency using nuclear magnetic resonance. As discussed in Sect. 6.1.1, the interaction of nuclear spins with UBDM appears as an effective magnetic interaction Hamiltonian [Eqs. (6.1) and (6.10) for the EDM coupling and Eqs. (6.2) and (6.11) for the gradient coupling] modulated at the ALP field's Compton frequency. This interaction can be detected as an oscillating torque on the magnetization in a frame

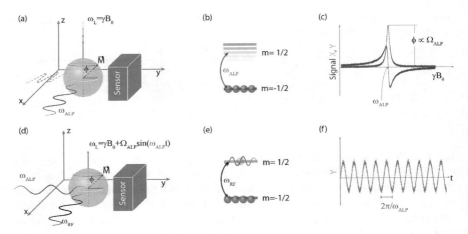

Fig. 6.3 Two ways to detect ALPs with nuclear magnetic resonance using a spherical sample of polarized nuclei and a sensor. (**a–c**) Continuous wave nuclear magnetic resonance (CW-NMR) with the cosmic ALP field driving spin population. The energy levels are scanned with the applied magnetic field and a resonance occurs if the Larmor frequency of the nuclear sample corresponds to the Compton frequency of the ALP. This detection mechanism is relevant for dipole moment and gradient coupling. In gradient coupling searches this detection is sensitive to two directions of the signal (x and y) for leading B_0 along z. In oscillating electric dipole moment searches an additional electric field needs to be applied along the magnetic background field. (**d–f**) Modulation of the magnetic field. Note that for gradient coupling searches the properties of the ALP pseudo-magnetic field are given by the field itself, while for EDM coupling searches an additional electric field needs to be applied. And here different directions with respect to the leading magnetic field can be chosen, so that it is possible to choose between situation (**a**) and (**d**)

co-rotating at the Larmor frequency of the nuclear sample, Fig. 6.3a. Depending on the experimental setup an ALP field can also be detected as a periodic modulation of the magnetic resonance frequency itself, see Fig. 6.3b. This can be searched for by observing the magnetic resonance frequency, for example, by probing it with another oscillating magnetic field at frequency ω_{RF} and observing the out-of-phase quadrature (Y) component with lock-in detection demodulated at ω_{RF}. The direction of the ALP pseudo-magnetic field depends on the coupling that is being investigated. In case of the gradient coupling, the oscillating pseudo-magnetic field is a property of the axion field itself. EDM coupling searches require an additional applied electric field which determines the direction of the pseudo-magnetic field.

Generally speaking, in order to obtain the biggest possible NMR signal from the ALP interaction one aims to have the largest number of nuclear spins subjected to the same magnetic field, such that they precess at the same Larmor frequency ω_L. The signal for the measurement configuration displayed in Fig. 6.3a is proportional to the magnitude of the oscillating component of the transverse magnetization M_x. The overall magnetization magnitude M, Eq. (6.23), results from N particles with magnetic moment μ, Eq. (6.12). The fundamental limitations on NMR sensitivity were already pointed out by Felix Bloch in his 1946 paper [42]: assuming uncorrelated, randomly oriented spin-1/2 particles in a volume V: there will be a statistically incomplete cancellation of the magnetization in any direction of order

$$M_{\text{SPN}} \approx \mu \frac{\sqrt{N}}{V} . \tag{6.44}$$

So even without any ALP signal, there will be a fluctuating signal at the detector proportional to this magnetization: this is the spin-projection noise (SPN). The maximum signal-to-noise therefore scales as

$$\frac{M}{M_{\text{SPN}}} \approx \sqrt{N}. \tag{6.45}$$

This means the larger the volume and the density of nuclear spins in a homogeneous magnetic field the more sensitive is the ALP search. The average power spectrum of the spin-noise signal will feature a Lorentzian lineshape with a width given by the transverse relaxation time T_2^* of the spin ensemble and a center frequency given by the Larmor frequency,

$$\omega_L = \gamma B_0 , \tag{6.46}$$

which depends on the applied background field B_0 and the gyromagnetic ratio γ of the nuclear species. Protons feature the highest known nuclear gyromagnetic ratio (of all stable nuclei) with $\gamma_p/(2\pi) = 42.6\,\text{MHz/T}$. In a magnetic resonance search, the accessible Larmor frequencies ω_L determine the sensitive mass range of the ALP-search experiment. Searches following Fig. 6.3b schematic, i.e., measurements of an oscillating center frequency of a magnetic resonance, result in a slightly different limit that depends on the linewidth of the magnetic resonance as well.

? Problem 6.2 Spin Noise

Consider the 10% hyperpolarized sample of liquid ^{129}Xe from Problem 6.1, with nuclear magnetic moment $|\mu_{\text{Xe}}| = 3.9\times10^{-27}$ J/T and liquid Xenon number density of $n_{\text{Xe}} = 1.4 \times 10^{28}\,\text{m}^{-3}$. Calculate the volume of a spherical liquid xenon sample needed to perform a spin noise-limited dark matter search. Assume that the magnetic field detector is optimally coupled to the spherical sample and has the sensitivity $B_{\text{det}} = 200$ aT, which corresponds to approximately 1 s of integration time for the most sensitive detectors listed in Table 6.1.

Solution on page 336.

6.3.2 Inductive Coil Detection

The most basic approach to detect an NMR signal is to use a pick-up coil coupled to the spin ensemble. The transverse magnetization precesses around the leading field at the Larmor frequency, creating an oscillating magnetic flux, which induces a Faraday voltage across the coil. A resonant circuit is often used to couple this voltage to a sensitive amplifier. There are many configurations analyzed in the NMR literature, see, for example, Ref. [43]. One example of such circuit is shown in Fig. 6.4. In this series capacitor-matched circuit the values of capacitors C_1 and C_2 are chosen so that the circuit resonance ω_0 is near the spin Larmor frequency, and probe impedance is matched to the transmission line and the amplifier input impedance. The resistance R includes dissipation due to the circuit elements, as well as the spin ensemble itself. The circuit dissipation can be quantified by the quality factor $Q = \omega_0 L / R$.

Inductive pick-up coils have been engineered since the invention of nuclear magnetic resonance and are highly sophisticated devices, commercially available in a broad range of frequencies (up to several GHz), Fig. 6.5. The voltage induced in the pick-up coil by the precessing spin magnetization is shown as a voltage source V_s in Fig. 6.4. This voltage is proportional to the Larmor frequency, due to Faraday's law. On resonance, and in the limit of small circuit dissipation ($Q \gg 1$), we can write a simple expression for the resulting voltage that appears at the input of the amplifier:

$$V_s' = \frac{V_s}{2} \sqrt{\frac{Q R_a}{\omega_0 L}}, \tag{6.47}$$

where R_a is the amplifier input impedance, usually matched to the 50 Ω transmission line impedance.

Consider the signal-to-noise ratio that can be achieved with this detection method. One of the noise sources is the thermal Nyquist noise due to the dissipation

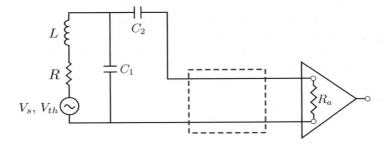

Fig. 6.4 A schematic of an NMR series capacitor-matched detection circuit. The blue dashed box denotes a transmission line (such as a coaxial cable) that couples the resonant detection probe on the left to the amplifier on the right

in the circuit, which appears as a noise voltage source in series with the signal source V_s, and with power spectral density $\tilde{V}_{th}^2(\omega) = 2Rk_BT/\pi$, where k_B is the Boltzmann constant and T is the circuit temperature. Another noise source is the amplifier noise, which usually referred to the amplifier input. The relative importance of these, and other noise sources, depends on the details of the NMR sample and of the probe circuit. Many of the modern NMR machines are limited by the thermal Nyquist noise, and therefore probes are sometimes cooled to reduce this noise.

6.3.3   Superconducting Quantum Interference Devices

Superconducting quantum interference devices (SQUIDs) can be used as current sensors to detect magnetic resonance signals. Since SQUIDs are usually low-frequency devices, optimization of the coupling circuit is, in general, different from the high-frequency inductive detection shown in Fig. 6.5. For example, in the case of non-resonant coupling the pick-up coil inductance should be roughly matched to the SQUID input coil self-inductance [45].

The basic building block of a dc SQUID is a loop of superconducting material interrupted via one or two Josephson junctions, i.e., non-conducting barriers in the superconducting loop. This loop is inductively coupled to a sensing coil, which is often superconducting itself, as seen in Fig. 6.6. The SQUID is a complicated non-linear superconducting device, whose operation is treated in Refs. [47, 48], which also present a comprehensive introduction to SQUID technology and applications. However, the SQUID characteristics can be linearized by operating it in feedback mode, where a feedback loop supplies a signal that cancels the flux from the sensing coil.

Fig. 6.5 Different coil geometries for inductive detection. Image and caption taken with permission from Ref. [44]. Common NMR coil geometries: (**a**) solenoid, (**b**) saddle coil, (**c**) inductively coupled high-temperature superconducting coils (drawings by Jason Kitchen of the National High Magnetic Field Laboratory)

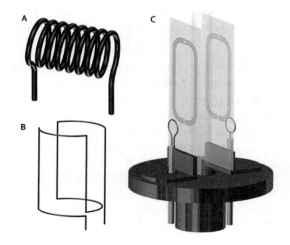

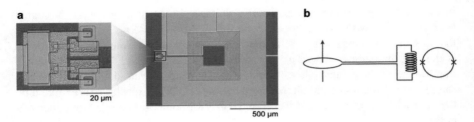

Fig. 6.6 Superconducting quantum interference device (SQUID) magnetometer. Image and caption taken with permission from Ref. [46]. (**a**) Photograph of a thin-film SQUID fabricated at Berkeley (right) and a close up of the Josephson junction area (left). (**b**) Configuration of a flux transformer coupled to a SQUID to form a magnetometer

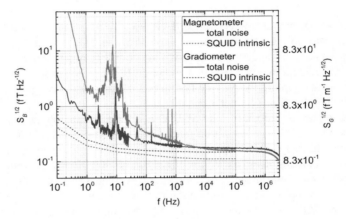

Fig. 6.7 Amplitude noise spectrum of a SQUID magnetometer and a SQUID gradiometer. Image and caption taken with permission from Ref. [41]. Measured magnetic flux density noise $S_{B,m}^{1/2}$ for the two setups with 45 mm diameter pick-up coils: magnetometer (solid green curve) and gradiometer (solid blue curve). The calculated intrinsic SQUID noise levels $S_{B,i}^{1/2}$ are given by the dotted curves. For the gradiometer, the noise is referred to the bottom pick-up loop, and the gradient noise is shown on the right

SQUIDs have been used to measure NMR since the 1990s [49–51] in a wide variety of fields. SQUID sensitivity can be characterized in terms of magnetic field at the location of the pick-up coil. A wideband ultra-sensitive magnetometer [41] from 2017 demonstrates a noise floor of $150\,\mathrm{aT}/\sqrt{\mathrm{Hz}}$ in a frequency range between 20 kHz and 2.5 MHz (Fig. 6.7). Similar performance has been achieved in the Search for Halo Axions with Ferromagnetic Toroids (SHAFT) experiment that searches for electromagnetic coupling of ultralight axionlike dark matter [45]. We note that the superconducting nature of SQUIDs limit their use to low-temperature environments, usually near 4 K (although there are high-T_c SQUIDs, whose performance is not quite as good). The SQUID sensors should also be carefully shielded from external magnetic fields. There are also other superconducting devices in development whose performance may offer significant improvements compared to the SQUID characteristics [52].

6.3.4 Atomic Vapor Sensors

Optical atomic magnetometry [54] is based on the manipulation of atomic spins with laser light and the subsequent observation of their evolution under the influence of a magnetic field. Overall the principles of optical atomic magnetometry are very similar to those of NMR discussed in earlier sections, however, in atoms the nuclear spin and electron angular momentum (orbital and spin) are coupled and so the dynamics involve the total atomic angular momentum. For alkali atoms, the ground state magnetic moment is dominated by the electron spin of the valence electron. Atomic magnetometers have been around since the 1970s [55]. Excellent review articles and books have been written on this topic [54, 56, 57]. Admitting numerous variations, Fig. 6.8 illustrates the most common ingredients of an optical atomic magnetometer. Figure 6.8a shows a vapor cell filled with a dilute vapor of alkali atoms. Figure 6.8b shows an experimental configuration using crossed probe and pump beams orthogonal to an applied background magnetic field B_0, and an oscillating field close to the Larmor frequency ω_L to excite the magnetic resonance. A (truncated) atomic level scheme and optical transitions for ^{87}Rb can be seen in Fig. 6.8c. Applying an on-resonant circular polarized light beam along the direction of the magnetic field optically pumps the atoms, through consecutive absorption and spontaneous emission cycles, into the highest magnetic sublevel of the ground state. The atomic spins in the vapor are thus oriented and can be measured as a macroscopic, collective (electron spin) magnetization. The evolution of these spins due to the magnetic field occurs at the driving frequency ω_{RF} and a magnetic resonance centered around the Larmor frequency ω_L is observable using demodulation with a lock-in amplifier. The width and amplitude of this complex Lorentzian are the key quantities to optimize for sensitive magnetometry. The center frequency of the Lorentzian, i.e., the Larmor frequency, is a measure of the magnetic field. For small magnetic fields (in the regime of the linear Zeeman effect) it can be written as:

$$\hbar\omega_L = \mu_B g_F |\boldsymbol{B}|, \qquad (6.48)$$

where $\mu_B/\hbar = 2\pi \times 14\,\text{GHz/T}$ is the Bohr magneton, g_F the Landé factor, and $|\boldsymbol{B}|$ the magnitude of the background magnetic field. This oscillation is then read out, for example, with an off-resonant laser beam via modulation of its polarization due to the Faraday effect [58]. The cells can be evacuated (i.e., contain a low density of single species alkali atoms), include various wall coatings or buffer gases to reduce relaxation rates and contain different species or combination of species of alkali atoms. The fundamental atomic shot-noise limited sensitivity δB_{SNL} of such a magnetometer is dominated by two quantities: the spin-relaxation rate Γ_{rel} and

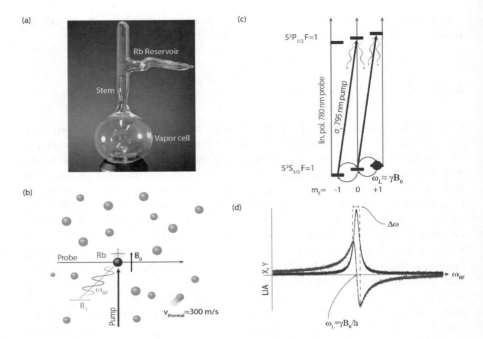

Fig. 6.8 Optical magnetometry with alkali vapor cells (**a**) An example of a glass vapor cell used for magnetometry. It is two centimeters in diameter, has a reservoir for Rb and a stem to separate the sensing volume and the reservoir. In this particular case, the inner walls of the cell are coated with an alkene film [53] enabling coherence times of up to 77 s. (**b**) In the cell atoms fly ballistically with a large velocity and are interrogated and manipulated with laser beams. (**c**) The atomic energy level of ^{87}Rb atoms with interactions. (**d**) The resulting magnetic resonance is often demodulated with a lock-in and is well described by a complex Lorentzian centered around the Larmor frequency ω_L

the number of spins N that are measured simultaneously. For measurement times $\tau \gg 1/\Gamma_{\text{rel}}$ [59] it is given by:

$$\delta B_{\text{SNL}} \approx \frac{1}{\gamma}\sqrt{\frac{\Gamma_{\text{rel}}}{N\tau}}. \tag{6.49}$$

Similar to NMR-based ALP searches, the fundamental sensitivity improves with the number of atoms and with a reduction in the relaxation rate. The longer the more spins can be observed precessing the better the magnetic field resolution. Unfortunately, N and Γ_{rel} are often correlated. For example, increasing the vapor pressure of the alkali atoms by heating the cell (and therefore the number of atoms to be interrogated) also increases spin-exchange- and spin-destruction-collision rates, which in turn increase the relaxation rate.

Radiofrequency vapor cell magnetometry has been used to measure NMR at 60 kHz [40]. A complication is that, due to different gyromagnetic ratios, the nuclear spins and the alkali atoms cannot be subjected to the same magnetic field. This problem can be solved by placing the nuclear sample in a magnetic solenoid coil penetrating the magnetic shield of the vapor cell magnetometer [60].

6.3.4.1 Spin-Exchange-Collision-Free (SERF) Magnetometry

Increasing the density of the alkali vapor and therefore the spin-exchange collision rate leads to an interesting new regime. First, magnetic resonances broaden as a function of spin-exchange rate, and then begin to get narrower. This is the so-called spin-exchange relaxation free regime (SERF). If the spin-exchange collision rate is much higher than the Larmor precession frequency, this decoherence mechanism effectively averages out. This behavior was first reported in Ref. [61], explained by the same group [62], and has been used for magnetometry since 2002 [63]. In 2010, the Romalis group at Princeton demonstrated a record-breaking sensitivity of $160\,\text{aT}/\sqrt{\text{Hz}}$ in a gradiometric configuration [38]. SERF magnetometers can be used for ZULF NMR [35, 64, 65], see Sect. 6.2.4. Due to the fact that SERF magnetometers employ a magnetic resonance centered around a near-zero magnetic field, the accessible frequency range depends on the linewidth of the resonance, normally below 1 kHz. Searches for UBDM using SERF magnetometers include the Cosmic Axion Spin Precession Experiment (CASPEr) ZULF sidebands and CASPEr comagnetometer experiments, as shown in Fig. 6.9.

6.3.5 Magnetic Noise Suppression

As discussed in Sects. 6.1.1.1 and 6.1.1.2, a UBDM field that couples to atomic spins mimics an oscillating magnetic field, therefore real oscillating magnetic fields are one of the most important sources of systematic errors. And, of course, magnetic fields are everywhere: the Earth's magnetic field itself is $\sim 10^{10}$ times larger than the smallest field that can be detected by commercially available atomic magnetometers (averaged over 1 s), radio waves over a wide range of frequencies travel through space, and electronic currents generate associated magnetic fields. Thus a conventional laboratory environment is teeming with complicated patterns of oscillating magnetic fields and magnetic field gradients of many orders at many frequencies, especially at the power line frequency and its harmonics. Most sensitive magnetometers require, therefore, a sophisticated shielded environment to avoid being saturated by magnetic noise. The effort to invest in magnetic shielding depends on the UBDM candidate mass range that is being searched. While static and slowly varying magnetic fields require complicated, multilayered magnetic shields constructed from materials with a high magnetic permeability to guide the magnetic flux around the sample or possibly superconducting shields, higher frequency magnetic noise can be effectively shielded by conductive enclosures. For sub-kHz magnetic noise, most vapor cell magnetometer are still limited by intrinsic magnetic field noise of the shield. Note that, in many cases (such as if the coupling

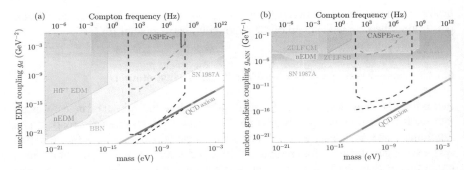

Fig. 6.9 Existing bounds and sensitivity projections for the: (**a**) EDM and (**b**) gradient coupling of axionlike dark matter taken from [12] with permission. The purple line shows the QCD axion coupling band in (**a**) and (**b**). The darker purple color shows the mass range motivated by theory [2]. The blue regions mark the mass ranges where the ADMX and HAYSTAC experiments (see Chap. 4) have probed the QCD axion-photon coupling [75, 76]. The green region is excluded by analysis of cooling in supernova SN1987A (see Chap. 3), with color gradient indicating theoretical uncertainty [2].The region shaded in red is the exclusion at 95% confidence level placed by CASPEr-E in [12]. The dashed green line marks the projected 5σ sensitivity of the CASPEr-E search with a 4.6 mm sample, as used in [12]. The dashed blue line marks the projected 5σ sensitivity of the CASPEr-E search with an 80 cm sample, operating at 100 mK temperature. The black dashed line marks the sensitivity limited by quantum spin-projection noise [77]. This would be sufficient to detect the EDM coupling of the QCD axion across a 6-decade mass range from ≈ 0.3 peV to ≈ 500 neV. The other bounds are as follows. (**a**) The pink region is excluded by the neutron EDM (nEDM) experiment [78]. The blue region is excluded by the HfF$^+$ EDM experiment [79]. The yellow region is excluded by analysis of Big Bang nucleosynthesis (BBN) [80]. (**b**) The pink region is excluded by the neutron EDM (nEDM) experiment [78]. The blue region is excluded by the zero-to-ultralow field comagnetometer (ZULF CM) experiment [33]. The gray region is excluded by the zero-to-ultralow field sideband (ZULF SB) experiment [34]. The yellow region is excluded by the new-force search with K-^{3}He comagnetometer [81]. The bounds are shown as published, although corrections should be made to some of the low-mass limits, due to stochastic fluctuations of the axionlike dark matter field [82]

of the UBDM is primarily to nuclei), signals from UBDM fields are unaffected by the magnetic shielding [66], while in some cases the magnetic shielding can in fact have significant effects (e.g., for hidden photons as discussed in Chap. 7). Detailed discussion of magnetic shielding can be found in Chapter 12 of Ref. [54].

To further improve the sensitivity of UBDM searches relying on magnetic resonance, other mechanisms have to be deployed to extend the discovery reach of the apparatus. One method of reducing sensitivity to local magnetic field changes is comagnetometry [67], i.e., where the magnetic field (and the UBDM signal) is simultaneously measured in two (or more) different ways in the same volume such that the magnetic responses and UBDM responses of the two methods are distinct. As elucidated in Sect. 6.1, the overall coupling strength of an ALP field to an atomic nucleus depends on the nuclear composition and the electronic state of the sample [68]. Consider the case where the sample consists of two ALP-sensitive species with opposing signs for the ALP interaction and identical signs for the magnetic interaction. When signals from both species are measured simultaneously and independently, subtracting the resulting signals reduces the effects of magnetic fields while enhancing the measurable effects of an ALP-spin interaction. This method has been successfully deployed in the experiments described in Refs. [3, 69, 70].

As a last thought for this section, if comagnetometer measurements are able to achieve sufficient insensitivity to magnetic fields, what will be the next most important systematic? The answer could be given by a common application of nuclear comagnetometry: gyroscopy [71, 72]. In fact, systematic errors due to rotations are major impediments to recent comagnetometer experiments searching for Lorentz invariance [73, 74].

6.4   Experimental Searches

Finally, we present a selection of magnetic-resonance-based experiments constraining ALP parameter space and place them in context with respect to astrophysical limits (see Chap. 3). The Cosmic Axion Spin Precession experiment (CASPEr) is a multi-pronged approach proposed in 2014 [77] to search for ALP dark matter over a wide range of ALP masses using NMR and undertaken in Boston, USA, and Mainz, Germany. The nEDM experiment searches for a permanent electric dipole moment of the neutron (nEDM) at the Paul Scherrer Institut in Switzerland. Reanalyzing their accumulated years of nEDM data (including a data set collected at the Institut Laue-Langevin in Grenoble, France between 1998 and 2002) for oscillating signals allowed the team to place tight constraints on both the EDM and the gradient coupling for low mass ALPs. The HfF$^+$ EDM experiment at the Joint Institute for Laboratory Astrophysics (JILA) in Boulder, USA, by the group of Eric Cornell searches for a permanent electron EDM. It is a precision experiment measuring electron spin precession with trapped molecular ions. Data collected in 2016 and 2017 were reanalyzed for oscillating signals and used to place constraints on the EDM coupling. Figure 6.9 summarizes the constraints on EDM and gradient ALP couplings. Other closely related magnetic resonance experiments include the QUAX experiment [13, 83, 84] that searches for electron-ALP interactions as well as a proposal to use antiferromagnetically doped topological insulators [85] to search for high mass axions in the 0.7–3.5 meV range.

References

1. M. Pospelov, A. Ritz, Nucl. Phys. B **573**, 177 (2000)
2. P.W. Graham, S. Rajendran, Phys. Rev. D **88**, 035023 (2013)
3. C. Abel, N.J. Ayres, G. Ban, G. Bison, K. Bodek, V. Bondar, M. Daum, M. Fairbairn, V.V. Flambaum, P. Geltenbort et al., Phys. Rev. X **7**, 041034 (2017)
4. S.P. Chang, S. Hacıömeroğlu, O. Kim, S. Lee, S. Park, Y.K. Semertzidis, Phys. Rev. D **99**, 083002 (2019)
5. L. Schiff, Phys. Rev. **132**, 2194 (1963)

6. I.B. Khriplovich, S.K. Lamoreaux, *CP Violation Without Strangeness* (Springer, Berlin, 1997)
7. O.P. Sushkov, V.V. Flambaum, I.B. Khriplovich, Sov. Phys. JETP **60**, 873 (1984)
8. V. Flambaum, D. DeMille, M. Kozlov, Phys. Rev. Lett. **113**, 103003 (2014)
9. V. Spevak, N. Auerbach, V. Flambaum, Phys. Rev. C **56**, 1357 (1997)
10. T. Mukhamedjanov, O. Sushkov, Phys. Rev. A **72**, 034501 (2005)
11. J. Ludlow, O. Sushkov, J. Phys. B **46**, 085001 (2013)
12. D. Aybas, J. Adam, E. Blumenthal, A.V. Gramolin, D. Johnson, A. Kleyheeg, S. Afach, J.W. Blanchard, G.P. Centers, A. Garcon et al., Phys. Rev. Lett. **126**, 141802 (2021)
13. N. Crescini, D. Alesini, C. Braggio, G. Carugno, D. D'Agostino, D. Di Gioacchino, P. Falferi, U. Gambardella, C. Gatti, G. Iannone et al., Phys. Rev. Lett. **124**, 171801 (2020)
14. R.R. Ernst, G. Bodenhausen, A. Wokaun et al., *Principles of Nuclear Magnetic Resonance in One and Two Dimensions*, vol. 14 (Clarendon Press, Oxford, 1987)
15. E. Fukushima, *Experimental Pulse NMR: A Nuts and Bolts Approach* (CRC Press, Boca Raton, 2018)
16. C. Abel, S. Afach, N.J. Ayres, C.A. Baker, G. Ban, G. Bison, K. Bodek, V. Bondar, M. Burghoff, E. Chanel et al., Phys. Rev. Lett. **124**, 081803 (2020)
17. P.W. Graham, D.E. Kaplan, J. Mardon, S. Rajendran, W.A. Terrano, L. Trahms, T. Wilkason, Phys. Rev. D **97**, 055006 (2018)
18. A. Abragam, *The Principles of Nuclear Magnetism* (Oxford University Press, Oxford, 1961)
19. M.H. Levitt, *Spin Dynamics: Basics of Nuclear Magnetic Resonance* (Wiley, Chichester, 2013)
20. C.P. Slichter, *Principles of Magnetic Resonance*, vol. 1 (Springer, Heidelberg, 1990)
21. M. Dobers, K. von Klitzing, J. Schneider, G. Weimann, K. Ploog, Phys. Rev. Lett. **61**, 1650 (1988)
22. R.D. Harding, S. Pallada, J. Croese, A. Antušek, M. Baranowski, M. Bissell, L. Cerato, K. Dziubinska-Kühn, W. Gins, F. Gustafsson et al., Phys. Rev. X **10**, 041061 (2020)
23. Y.G. Abov, A. Gulko, F. Dzheparov et al., Phys. At. Nucl. **69**, 1701 (2006)
24. T.G. Walker, W. Happer, Rev. Mod. Phys. **69**, 629 (1997)
25. P. Nikolaou, A.M. Coffey, L.L. Walkup, B.M. Gust, N. Whiting, H. Newton, S. Barcus, I. Muradyan, M. Dabaghyan, G.D. Moroz, M.S. Rosen, S. Patz, M.J. Barlow, E.Y. Chekmenev, B.M. Goodson, **110**, 14150 (2013)
26. A. Abragam, M. Goldman, Rep. Prog. Phys. **41**, 395 (1978)
27. W.T. Wenckebach, *Essentials of Dynamic Nuclear Polarization* (Spindrift Publications, Burgh-Haamstede, 2016)
28. F.D. Colegrove, L.D. Schearer, G.K. Walters, Phys. Rev. **132**, 2561 (1963)
29. C.R. Bowers, D.P. Weitekamp, J. Am. Chem. Soc. **109**, 5541 (1987)
30. C.R. Bowers, *Sensitivity Enhancement Utilizing Parahydrogen* (Wiley, Chichester, 2007)
31. R.A. Green, R.W. Adams, S.B. Duckett, R.E. Mewis, D.C. Williamson, G.G. Green, Prog. NMR Spec. **67**, 1 (2012)
32. E.R. Andrew, A. Bradbury, R.G. Eades, Nature **183**, 1802 (1959)
33. T. Wu, J.W. Blanchard, G.P. Centers, N.L. Figueroa, A. Garcon, P.W. Graham, D.F. Jackson Kimball, S. Rajendran, Y.V. Stadnik, A.O. Sushkov, A. Wickenbrock, D. Budker, Phys. Rev. Lett. **122**, 191302 (2019)
34. A. Garcon, J.W. Blanchard, G.P. Centers, N.L. Figueroa, P.W. Graham, D.F. Jackson Kimball, S. Rajendran, A.O. Sushkov, Y.V. Stadnik, A. Wickenbrock et al., Sci. Adv. **5**, eaax4539 (2019)
35. J.W. Blanchard, D. Budker, eMagRes **5**, 1395 (2016)
36. S. Hahn, K. Kim, K. Kim, X. Hu, T. Painter, I. Dixon, S. Kim, K.R. Bhattarai, S. Noguchi, J. Jaroszynski, D.C. Larbalestier, Nature **570**, 496 (2019)
37. R. Battesti, J. Beard, S. Böser, N. Bruyant, D. Budker, S.A. Crooker, E.J. Daw, V.V. Flambaum, T. Inada, I.G. Irastorza et al., Phys. Rep. **765**, 1 (2018)
38. H.B. Dang, A.C. Maloof, M.V. Romalis, Appl. Phys. Lett. **97**, 151110 (2010)
39. I.M. Savukov, S.J. Seltzer, M.V. Romalis, K.L. Sauer, Phys. Rev. Lett. **95**, 063004 (2005)
40. I. Savukov, S. Seltzer, M. Romalis, J. Magn. Reson. **185**, 214 (2007)
41. J.H. Storm, P. Hömmen, D. Drung, R. Körber, Appl. Phys. Lett. **110**, 072603 (2017)
42. F. Bloch, Phys. Rev. **70**, 460 (1946)

43. M.A. McCoy, R.R. Ernst, Chem. Phys. Lett. **159**, 587 (1989)
44. J.H. Ardenkjaer-Larsen, G.S. Boebinger, A. Comment, S. Duckett, A.S. Edison, F. Engelke, C. Griesinger, R.G. Griffin, C. Hilty, H. Maeda et al., Angew. Chem. Int. Ed. **54**, 9162 (2015)
45. A.V. Gramolin, D. Aybas, D. Johnson, J. Adam, A.O. Sushkov, Nature Phys. **17**, 79 (2021)
46. J. Clarke, M. Hatridge, M. Mößle, Annu. Rev. Biomed. Eng. **9**, 389 (2007)
47. A.I.B.E. J. Clarke, *The SQUID Handbook. Vol. I Fundamentals and Technology of SQUIDs and SQUID Systems* (Wiley, Weinheim, 2004)
48. A.I.B.E. J. Clarke, *The SQUID Handbook. Vol. II: Applications of SQUIDs and SQUID Systems* (Wiley, Weinheim, 2006)
49. N.Q. Fan, J. Clarke, Rev. Sci. Inst. **62**, 1453 (1991)
50. D.M. TonThat, J. Clarke, Rev. Sci. Inst. **67**, 2890 (1996)
51. Chem. Phys. Lett. **272**, 245 (1997)
52. T. Yamamoto, K. Inomata, M. Watanabe, K. Matsuba, T. Miyazaki, W.D. Oliver, Y. Nakamura, J. Tsai, Appl. Phys. Lett. **93**(4), 042510 (2008)
53. M.V. Balabas, T. Karaulanov, M.P. Ledbetter, D. Budker, Phys. Rev. Lett. **105**, 070801 (2010)
54. D. Budker, D.F. Jackson Kimball, *Optical Magnetometry* (Cambridge University Press, Cambridge, 2013)
55. J. Dupont-Roc, S. Haroche, C. Cohen-Tannoudji, Phys. Lett. A **28**, 638 (1969)
56. D. Budker, M. Romalis, Nature Phys. **3**, 227 (2007)
57. A. Auzinsh, D. Budker, S.M. Rochester, *Optically Polarized Atoms* (Oxford University Press, Oxford, 2010)
58. D. Budker, W. Gawlik, D. Kimball, S. Rochester, V. Yashchuk, A. Weis, Rev. Mod. Phys. **74**, 1153 (2002)
59. M. Auzinsh, D. Budker, D.F. Kimball, S.M. Rochester, J.E. Stalnaker, A.O. Sushkov, V.V. Yashchuk, Phys. Rev. Lett. **93**, 173002 (2004)
60. V. Yashchuk, J. Granwehr, D. F. Kimball, S. Rochester, A. Trabesinger, J. Urban, D. Budker, A. Pines, Phys. Rev. Lett. **93**, 160801 (2004)
61. W. Happer, H. Tang, Phys. Rev. Lett. **31**, 273 (1973)
62. W. Happer, A.C. Tam, Phys. Rev. A **16**, 1877 (1977)
63. J.C. Allred, R.N. Lyman, T.W. Kornack, M.V. Romalis, Phys. Rev. Lett. **89**, 130801 (2002)
64. M.P. Ledbetter, I.M. Savukov, V.M. Acosta, D. Budker, M.V. Romalis, Phys. Rev. A **77**, 033408 (2008)
65. M.P. Ledbetter, T. Theis, J.W. Blanchard, H. Ring, P. Ganssle, S. Appelt, B. Blümich, A. Pines, D. Budker, Phys. Rev. Lett. **107**, 107601 (2011)
66. D.F. Jackson Kimball, J. Dudley, Y. Li, S. Thulasi, S. Pustelny, D. Budker, M. Zolotorev, Phys. Rev. D **94**, 082005 (2016)
67. S.K. Lamoreaux, Nucl. Inst. Meth. Phys. Res. A **284**(1), 43 (1989)
68. D.F. Jackson Kimball, New J. Phys. **17**, 073008 (2015)
69. T. Wu, J.W. Blanchard, D.F. Jackson Kimball, M. Jiang, D. Budker, Phys. Rev. Lett. **121**, 023202 (2018)
70. I.M. Bloch, Y. Hochberg, E. Kuflik, T. Volansky, J. High Energy Phys. **2020**, 167 (2020)
71. T. Kornack, R. Ghosh, M. Romalis, Phys. Rev. Lett. **95**, 230801 (2005)
72. T.G. Walker, M.S. Larsen, Adv. At. Mol. Opt. Phys. **65**, 373 (2016)
73. J. Brown, S. Smullin, T. Kornack, M. Romalis, Phys. Rev. Lett. **105**, 151604 (2010)
74. M. Smiciklas, J. Brown, L. Cheuk, S. Smullin, M.V. Romalis, Phys. Rev. Lett. **107**, 171604 (2011)
75. N. Du, N. Force, R. Khatiwada, E. Lentz, R. Ottens, L. Rosenberg, G. Rybka, G. Carosi, N. Woollett, D. Bowring et al., Phys. Rev. Lett. **120**, 151301 (2018)
76. B. Brubaker, L. Zhong, Y. Gurevich, S. Cahn, S. Lamoreaux, M. Simanovskaia, J. Root, S. Lewis, S. Al Kenany, K. Backes et al., Phys. Rev. Lett. **118**(6), 061302 (2017)
77. D. Budker, P.W. Graham, M. Ledbetter, S. Rajendran, A.O. Sushkov, Phys. Rev. X **4**, 021030 (2014)
78. C. Abel, N.J. Ayres, G. Ban, G. Bison, K. Bodek, V. Bondar, M. Daum, M. Fairbairn, V.V. Flambaum, P. Geltenbort et al., Phys. Rev. X **7**, 041034 (2017)

79. T.S. Roussy, D.A. Palken, W.B. Cairncross, B.M. Brubaker, D.N. Gresh, M. Grau, K.C. Cossel, K.B. Ng, Y. Shagam, Y. Zhou et al., Phys. Rev. Lett. **126**, 171301 (2021)
80. K. Blum, R.T. D'Agnolo, M. Lisanti, B.R. Safdi, Phys. Lett. B **737**, 30 (2014)
81. G. Vasilakis, J.M. Brown, T.W. Kornack, M.V. Romalis, Phys. Rev. Lett. **103**, 261801 (2009)
82. G.P. Centers, J.W. Blanchard, J. Conrad, N.L. Figueroa, A. Garcon, A.V. Gramolin, D.F. Jackson Kimball, M. Lawson, B. Pelssers, J.A. Smiga et al., Nat. Commun. **12**, 7321 (2021)
83. G. Ruoso, A. Lombardi, A. Ortolan, R. Pengo, C. Braggio, G. Carugno, C.S. Gallo, C.C. Speake, J. Phys. Conf. Ser. **718**, 042051 (2016)
84. N. Crescini, C. Braggio, G. Carugno, P. Falferi, A. Ortolan, G. Ruoso, Nucl. Instrum. Methods. Phys. Res. B **842**, 109 (2017)
85. D.J.E. Marsh, K.C. Fong, E.W. Lentz, L. Šmejkal, M.N. Ali, Phys. Rev. Lett. **123**, 121601 (2019)

Chapter 7
Dark Matter Radios

Derek F. Jackson Kimball and Arran Phipps

Abstract Many theories predict that ultralight bosonic dark matter (UBDM) can couple to photons and thus generate electromagnetic signals. In such scenarios, UBDM can be searched for using a radio: an antenna connected to a tunable LC circuit that is in turn connected to an amplifier. Such "dark matter radios" are particularly useful tools to search the broad range of UBDM wavelengths where resonant cavity dimensions are too large to be practical. In this chapter, we discuss how dark matter radios can be used to search for UBDM, focusing on the case of hidden photons.

7.1 Hidden Photons

Chapters 4 and 5 of this text have described searches for UBDM in the form of axions or axionlike particles (ALPs) that utilize techniques based on their coupling to electromagnetic fields. In this chapter, we describe another technique that can be used to look for electromagnetic couplings of UBDM (including axions and ALPs), but instead focus on a different class of ultralight bosons known as hidden photons [1].

Hidden sectors described by extra $\mathbb{U}(1)$ symmetries[1] are a common feature in theories going beyond the Standard Model such as string theory [2]. Even though such hidden sectors may be quite complicated [3, 4], their observable effects can be

---

[1] The $\mathbb{U}(1)$ gauge symmetry is related to the fact that particle wave functions can have an overall complex phase which is unobservable in experiments. Thus, all observables must be invariant with respect to this phase—this is the $\mathbb{U}(1)$ symmetry. $\mathbb{U}(1)$ gauge invariance directly implies conservation of charge, just as translational symmetry gives conservation of momentum, rotational symmetry gives conservation of angular momentum, and symmetry with respect to time translation gives energy conservation.

D. F. Jackson Kimball · A. Phipps (✉)
Department of Physics, California State University, East Bay, Hayward, CA, USA
e-mail: derek.jacksonkimball@csueastbay.edu; arran.phipps@csueastbay.edu

© The Author(s) 2023
D. F. Jackson Kimball, K. van Bibber (eds.), *The Search for Ultralight Bosonic Dark Matter*, https://doi.org/10.1007/978-3-030-95852-7_7

parameterized by the effective operators coupling them to Standard Model particles and fields [5]. One possibility is direct couplings: Standard Model particles and fields can be "charged" with respect to the hidden sector [6]. It turns out that the only other generic possibility is a kinetic mixing between the new $\mathbb{U}(1)$ symmetry and electromagnetism [1]: this is the origin of the hidden photon.

The effects of this hidden photon can be understood in particularly simple terms: it behaves exactly like a photon except that

- It has a nonzero mass $m_{\gamma'}$.
- It interacts with charged particles primarily through its mixing into a "real" photon field, parameterized by a kinetic mixing parameter κ.

The Lagrangian describing photons and hidden photons can be written in what is known as the "mass basis" as follows:

$$\mathcal{L} = -\frac{1}{16\pi}\left(F_{\mu\nu}F^{\mu\nu} + \mathcal{F}_{\mu\nu}\mathcal{F}^{\mu\nu}\right) + \frac{m_{\gamma'}^2 c^2}{2\hbar^2}X_\mu X^\mu - \frac{1}{c}J^\mu\left(A_\mu + \kappa X_\mu\right), \qquad (7.1)$$

where A_μ and $F_{\mu\nu}$ are the gauge potential and field strength tensor of electromagnetism (the ordinary photon field), X_μ and $\mathcal{F}_{\mu\nu}$ are the gauge potential and field strength of the hidden photon field, J^μ is the ordinary electromagnetic current, and here and throughout this chapter we use Gaussian (cgs) units. Notice that in the limit where $m_{\gamma'} \to 0$, because of the symmetry between the photon and hidden photon fields, one can redefine a linear combination $A_\mu + \kappa X_\mu$ that couples to the electromagnetic current J^μ and a sterile component $X_\mu - \kappa A_\mu$ that does not interact at all electromagnetically. Essentially, this means that all hidden photon interactions are suppressed by powers of $m_{\gamma'}^2$ in the small mass limit. This is the argument that significantly reduces many astrophysical bounds on hidden photons [7].

In vacuum, the hidden photon field obeys the wave equation

$$\left(\frac{1}{c^2}\frac{\partial^2}{\partial t^2} - \nabla^2 + \frac{m_{\gamma'}^2 c^2}{\hbar^2}\right)X_\mu = 0 \,, \qquad (7.2)$$

and the constraint $\partial_\mu X^\mu = 0$ is assumed (equivalent to the Lorenz gauge condition).

The two key features of hidden photons, that they have nonzero mass and only weakly interact with SM particles via the kinetic mixing with photons, have three important consequences: (1) their nonzero mass means that hidden photons can have the right characteristics to behave as cold dark matter [8], (2) their kinetic mixing with photons means that hidden photons can weakly excite electromagnetic systems, and (3) their weak coupling with SM particles and macroscopic Compton wavelength allow hidden photons to have a long penetration depth in conductors and superconductors.

7.2 Hidden Photon Electrodynamics

A useful way to understand the effect of the hidden photon on electrodynamics is to rewrite the Lagrangian given in Eq. (7.1) in the "interaction basis" by making the substitutions:

$$\bar{A}_\mu = A_\mu + \kappa X_\mu ,\tag{7.3}$$

$$\bar{X}_\mu = X_\mu - \kappa A_\mu ,\tag{7.4}$$

which yields for the Lagrangian

$$\mathcal{L} = -\frac{1}{16\pi}\left(\bar{F}_{\mu\nu}\bar{F}^{\mu\nu} + \bar{\mathcal{F}}_{\mu\nu}\bar{\mathcal{F}}^{\mu\nu}\right) + \frac{m_{\gamma'}^2 c^2}{2\hbar^2}\bar{X}_\mu\bar{X}^\mu$$

$$- \frac{1}{c}J^\mu\bar{A}_\mu + \kappa\frac{m_{\gamma'}^2 c^2}{\hbar^2}\bar{X}^\mu\bar{A}_\mu .\tag{7.5}$$

From Eq. (7.5), we can see that the effect of hidden photons on Standard Model particles can be derived from the existence of an effective current density:

$$\bar{\mathcal{J}}^\mu = -\kappa\frac{m_{\gamma'}^2 c^3}{\hbar^2}\bar{X}^\mu ,\tag{7.6}$$

since the last term in Eq. (7.5) describes the coupling to SM particles.

This gives us

$$\mathcal{L} = -\frac{1}{16\pi}\left(\bar{F}_{\mu\nu}\bar{F}^{\mu\nu} + \bar{\mathcal{F}}_{\mu\nu}\bar{\mathcal{F}}^{\mu\nu}\right) + \frac{m_{\gamma'}^2 c^2}{2\hbar^2}\bar{X}_\mu\bar{X}^\mu - \frac{1}{c}\left(J^\mu + \bar{\mathcal{J}}^\mu\right)\bar{A}_\mu .\tag{7.7}$$

? Problem 7.1 Interaction Basis

Derive Eq. (7.5) from Eqs. (7.1), (7.3), and (7.4).

Solution on page 336.

Note that the timelike component of the four-potential, $\bar{X}^0$, is suppressed compared to the spacelike component, i.e., the vector potential X. The relationship between $\bar{X}^0$ and X is derived from the equivalent of the Lorenz gauge condition,

$$\frac{1}{c}\frac{\partial\bar{X}^0}{\partial t} = \nabla \cdot X .\tag{7.8}$$

Assuming a plane wave solution for $\bar{X}^\mu \propto e^{i(\mathbf{k}\cdot\mathbf{r}-\omega t)}$, Eq. (7.8) yields

$$\bar{X}^0 = -\frac{c}{\omega}\mathbf{k}\cdot\mathcal{X}. \tag{7.9}$$

Using the relationship between the wavevector $\mathbf{k}$ and the hidden photon's velocity,

$$\mathbf{k} = m_{\gamma'}\mathbf{v}/\hbar, \tag{7.10}$$

and the fact that $\hbar\omega \approx m_{\gamma'}c^2$, we find that

$$\bar{X}^0 \approx -\frac{\mathbf{v}}{c}\cdot\mathcal{X}. \tag{7.11}$$

Therefore, the timelike component of the hidden photon four-potential is suppressed by a factor $\approx v/c \approx 10^{-3}$ as compared to the spacelike component. For the same reason, the spacelike component of the hidden photon four-current (i.e., the effective charge density) is suppressed by $\approx v/c$ compared to the hidden photon current density, $\mathcal{J}$, see Eq. (7.6).

? Problem 7.2 Oscillation Frequency of Hidden Electromagnetic Fields

Show that a gas of nonrelativistic hidden photons with high mode occupation number manifests as a field oscillating at approximately the hidden photon Compton frequency.

Solution on page 337.

7.3 Hidden Electric and Magnetic Fields as Dark Matter

If hidden photons comprise the majority of dark matter, they must be nonrelativistic and the energy of the hidden photon field is stored primarily in the hidden-electric field $\mathcal{E}'$. This can be understood by analogy with the relativistic properties of ordinary electric and magnetic fields. If an observer is in the rest frame of a static charge distribution, they will measure a static electric field $\mathbf{E}$ sourced by the charges but no magnetic field. If the observer moves at a constant velocity v with respect to the charge distribution, they will measure a motional magnetic field of magnitude $B \sim (v/c)E$. In the case where $v \ll c$, the observer measures that most of the energy is stored in the electric field. This contrasts with the case of electromagnetic waves propagating in vacuum, where there is equal energy stored in both $\mathbf{E}$ and $\mathbf{B}$.

Because hidden photons have a nonzero rest mass, there exists a hidden photon rest frame in which the hidden photon energy is stored entirely in the oscillating hidden electric field $\mathcal{E}'$—analogous to the rest frame of a static charge distribution. If the hidden photons are the dark matter, they must be nonrelativistic (see Chaps. 1 and 3), and so the vast majority of the hidden photon energy density is associated with $\mathcal{E}'$.

The dark matter energy density in our local region of the Milky Way galaxy, $\rho_{\rm dm} \approx 0.4$ GeV/cm^3, can thus be estimated using the hidden electric field analog of the usual formula from electromagnetism

$$\rho_{\rm dm} = \frac{1}{8\pi}(\mathcal{E}')^2 \,, \tag{7.12}$$

from which we find that

$$\mathcal{E}' \approx 40 \text{ V/cm} \,. \tag{7.13}$$

The associated hidden magnetic field is

$$\mathcal{B}' \approx \frac{v}{c}\mathcal{E}' \approx 10^{-4} \text{ G} \,. \tag{7.14}$$

The hidden fields $\mathcal{E}'$ and $\mathcal{B}'$ result from the interference of a large number of virialized hidden photons having high mode occupation number, and thus the hidden field properties have a stochastic nature [9, 10]. The stochastic behavior of virialized UBDM fields is analogous to that of thermal (chaotic) light [11]. The finite coherence time $\tau_{\rm coh}$ and coherence length $\lambda_{\rm coh}$ of $\mathcal{E}'$ and $\mathcal{B}'$ result from the velocity spread of the hidden photons ($\delta v \approx v \approx 10^{-3}c$, see Chap. 1):

$$\tau_{\rm coh} \approx 4000 \text{ s} \times \frac{10^{-12} \text{ eV}}{m_{\gamma'}c^2} \,, \tag{7.15}$$

and

$$\lambda_{\rm coh} \approx 10^6 \text{ km} \times \frac{10^{-12} \text{ eV}}{m_{\gamma'}c^2} \,. \tag{7.16}$$

The amplitude, phase, and polarization of $\mathcal{E}'$ and $\mathcal{B}'$ remain roughly constant over $\tau_{\rm coh}$ and $\lambda_{\rm coh}$.

Also of note is the fact that due to the nonzero rest mass of hidden photons, hidden photon fields can possess longitudinal modes unlike photon fields in vacuum [12]. The existence of longitudinal modes of the hidden fields affects both astrophysical constraints and experimental strategies [12].

7.4   Dark Matter Radio Experimental Scheme

Because the coupling of the hidden photon field to Standard Model particles is entirely through kinetic mixing into real electromagnetic fields, it can be difficult to distinguish hidden photon dark matter signals from electromagnetic noise. Therefore, in order to achieve the highest possible sensitivity to hidden photon dark matter, it is generally advantageous to enclose the detector within electromagnetic shielding (for example, a superconducting shield). The hidden photon dark matter will penetrate the shield and can produce a signal in the detector. As noted above, however, the hidden photon field also affects charges within the shield and generates a real electromagnetic field inside the shield that can interfere with the signal from the hidden photons. Therefore, careful consideration of the "hidden photon electrodynamics" is needed to understand the measurable signal inside the shield.

Consider a cylindrical superconducting shield of radius R and length ℓ with axis along z. Working in the interaction basis, let us consider a single mode of the hidden photon field described by the vector potential $\mathcal{X}$ which we assume points along $\hat{\mathbf{z}}$, parallel to the shield axis. The effective current density can be described by

$$\mathcal{J}(\mathbf{r}, t) = -\kappa \frac{m_{\gamma'}^2 c^3}{\hbar^2} X_0 e^{i(\mathbf{k}\cdot\mathbf{r} - \omega t)} \hat{\mathbf{z}} , \qquad (7.17)$$

where X_0 is the amplitude of the vector potential. As discussed above, the spacelike components of the effective four-current (the charge density) and four-potential (the scalar potential) are suppressed by $\approx v/c$ and can be neglected in the following considerations.

7.4.1   Electric Field Due to Hidden Photons Within Shields

To solve for the electric field $\mathbf{E}$ inside the cylindrical shield, we start from Maxwell's equations for $\mathbf{E}$, namely Gauss's Law and Faraday's Law:

$$\nabla \cdot \mathbf{E} = 4\pi\rho \approx 0 , \qquad (7.18)$$

$$\nabla \times \mathbf{E} = -\frac{1}{c}\frac{\partial \mathbf{B}}{\partial t} . \qquad (7.19)$$

Taking the curl of Eq. (7.19), we find

$$\nabla \times (\nabla \times \mathbf{E}) = -\frac{1}{c}\nabla \times \frac{\partial \mathbf{B}}{\partial t} = -\frac{1}{c}\frac{\partial}{\partial t}(\nabla \times \mathbf{B}) . \qquad (7.20)$$

The left-hand side of Eq. (7.20) can be simplified by using the identity

$$\nabla \times (\nabla \times \mathbf{E}) = \nabla(\nabla \cdot \mathbf{E}) - \nabla^2 \mathbf{E} \approx -\nabla^2 \mathbf{E} , \qquad (7.21)$$

where the right-hand side follows from Eq. (7.18). The right-hand side of Eq. (7.20) can be rewritten in terms of $\mathbf{E}$ and $\mathcal{J}$ using another of Maxwell's equations, Ampère's Law,

$$\nabla \times \mathbf{B} = \frac{1}{c}\left(4\pi\mathcal{J} + \frac{\partial\mathbf{E}}{\partial t}\right) , \tag{7.22}$$

yielding

$$-\frac{1}{c}\frac{\partial}{\partial t}(\nabla \times \mathbf{B}) = -\frac{4\pi}{c^2}\frac{\partial\mathcal{J}}{\partial t} - \frac{1}{c^2}\frac{\partial^2\mathbf{E}}{\partial t^2} . \tag{7.23}$$

Combining Eqs. (7.21) and (7.23), we obtain

$$\left(\frac{1}{c^2}\frac{\partial^2}{\partial t^2} - \nabla^2\right)\mathbf{E} = -\frac{4\pi}{c^2}\frac{\partial\mathcal{J}}{\partial t} . \tag{7.24}$$

So far the analysis has been quite general. Now, let us assume the z-polarized current density of Eq. (7.17), giving us a more specific form of the wave equation for $\mathbf{E}$, namely

$$\left(\frac{1}{c^2}\frac{\partial^2}{\partial t^2} - \nabla^2\right)\mathbf{E} = 4\pi i\frac{\omega}{c^2}\mathcal{J}(\mathbf{r}, t) .$$

$$= -4\pi i\kappa\frac{m_{\gamma'}^2\omega c}{\hbar^2}X_0 e^{i(\mathbf{k}\cdot\mathbf{r}-\omega t)}\hat{\mathbf{z}} . \tag{7.25}$$

It can be seen from Eq. (7.25) that $\mathbf{E}$ is aligned with $\mathcal{J}$. Because the electric field arising from the hidden photon field generates forces on charges within the cylindrical superconducting shields, the charges will rearrange themselves so as to cancel the field parallel to the surface of the shield ($r = R$), resulting in the boundary condition $\mathbf{E}(R) \cdot \hat{\mathbf{z}} = 0$. Furthermore, based on Eq. (7.25), the time dependence of $\mathbf{E}$ must match the time dependence of $\mathcal{J}$, and so the time derivative in Eq. (7.25) can be resolved:

$$\left(-\frac{\omega^2}{c^2} - \nabla^2\right)\mathbf{E} = -4\pi i\kappa\frac{m_{\gamma'}^2\omega c}{\hbar^2}X_0 e^{i(\mathbf{k}\cdot\mathbf{r}-\omega t)}\hat{\mathbf{z}} . \tag{7.26}$$

Next, let us take the limit where $kR \ll 1$; in other words, the de Broglie wavelength (approximately equal to the coherence length for virialized dark matter) of the hidden photon field is far larger than the dimensions of the shield. In practical terms, this is the scenario for which dark matter radios have an advantage compared to resonant cavities. The $kR \ll 1$ limit allows us to assume $e^{i\mathbf{k}\cdot\mathbf{r}} \approx 1$, further simplifying our equation for the spatial dependence of $\mathbf{E}$:

$$\left(\frac{\omega^2}{c^2} + \nabla^2\right)\mathbf{E} = 4\pi i\kappa \frac{m_{\gamma'}^2\omega c}{\hbar^2} X_0 e^{-i\omega t}\hat{\mathbf{z}} \,. \tag{7.27}$$

Since we ignore edge effects based on taking the $\ell \gg R$ limit for the shield and have cylindrical symmetry, $\mathbf{E}$ can be assumed to be independent of z and the angular variable ϕ. Equation (7.27) is solved by assuming a spatial dependence for the electric field of

$$E(r) = E_0\left(1 - \frac{J_0(\omega r/c)}{J_0(\omega R/c)}\right), \tag{7.28}$$

where $J_n(x)$ is the nth order Bessel function of the first kind [13, 14]. Note that $E(R) = 0$, thus satisfying the boundary condition imposed by the shield. Substituting Eq. (7.28) into Eq. (7.27) enables us to solve for E_0, and we find that

$$E_0 = 4\pi i\kappa \frac{m_{\gamma'}^2 c^3}{\hbar^2\omega} X_0 \,. \tag{7.29}$$

Since $\omega \approx m_{\gamma'}c^2/\hbar$,

$$E_0 = 4\pi i\kappa \frac{m_{\gamma'}c}{\hbar} X_0 \,. \tag{7.30}$$

Thus, the full description of the electric field generated inside the shield by the hidden photon field is

$$\mathbf{E}(\mathbf{r}, t) = 4\pi i\kappa \frac{m_{\gamma'}c}{\hbar} X_0\left(1 - \frac{J_0(\omega r/c)}{J_0(\omega R/c)}\right) e^{-i\omega t}\hat{\mathbf{z}} \,. \tag{7.31}$$

For $\omega R/c \ll 1$, we can Taylor expand the Bessel functions and find that

$$\mathbf{E}(\mathbf{r}, t) \approx 4\pi i\kappa \frac{m_{\gamma'}c}{\hbar} X_0 \frac{\omega^2}{c^2}\left(R^2 - r^2\right) e^{-i\omega t}\hat{\mathbf{z}} \,,$$

$$\approx 4\pi i\kappa \frac{X_0}{\lambda_{\gamma'}^3}\left(R^2 - r^2\right) e^{-i\omega t}\hat{\mathbf{z}} \,, \tag{7.32}$$

which shows that the electric field at the center of the shield is suppressed by a factor $\approx R^2/\lambda_{\gamma'}^2$ in comparison to the field in the absence of a shield, where $\lambda_{\gamma'} = \hbar/(m_{\gamma'}c)$ is the hidden photon Compton wavelength.

7.4.2 *Magnetic Field Due to Hidden Photons Within Shields*

The magnetic field induced within the shield by the hidden photons can be derived from Faraday's Law [Eq. (7.19)],

$$\nabla \times \mathbf{E} = \frac{i\omega}{c}\mathbf{B} , \tag{7.33}$$

where we have used the fact that the time dependence of $\mathbf{B}$ is described by $e^{-i\omega t}$. Taking the curl of the $\mathbf{E}$ described by Eq. (7.31), we obtain

$$\mathbf{B}(\mathbf{r}, t) = 4\pi\kappa\frac{m_{\gamma'}c}{\hbar}X_0\left(\frac{J_1(\omega r/c)}{J_0(\omega R/c)}\right)e^{-i\omega t}\hat{\boldsymbol{\phi}} . \tag{7.34}$$

Taylor expansion of the Bessel functions for $\omega R/c \ll 1$ yields

$$\mathbf{B}(\mathbf{r}, t) \approx 4\pi\kappa\frac{m_{\gamma'}c}{\hbar}X_0\left(\frac{\omega r}{c}\right)e^{-i\omega t}\hat{\boldsymbol{\phi}} ,$$

$$\approx 4\pi\kappa\frac{X_0}{\lambda_{\gamma'}^2}re^{-i\omega t}\hat{\boldsymbol{\phi}} . \tag{7.35}$$

The relative amplitudes of the electric and magnetic fields can be compared by taking the ratio:

$$\frac{|\mathbf{B}|}{|\mathbf{E}|} \approx \frac{r\lambda_{\gamma'}}{R^2 - r^2} , \tag{7.36}$$

and therefore the measurable magnetic field is $\sim \lambda_{\gamma'}/R$ larger than the electric field. For this reason, dark matter radios are designed to detect $\mathbf{B}$.

? Problem 7.3 DM Energy Density and the Magnetic Field Within Shields

Due to the effects of the shielding, the electric and magnetic fields measurable with a DM Radio scheme are different from those in a vacuum. Relate the approximate value of the magnetic field within the shield given by Eq. (7.35) to the local DM density, $\rho_{DM} \approx 0.4\,\mathrm{GeV/cm^3}$, obtaining the result:

$$\mathbf{B}(\mathbf{r}, t) = 4\pi\kappa\sqrt{8\pi\rho_{\mathrm{dm}}}\frac{r}{\lambda_{\gamma'}}e^{-i\omega t}\hat{\boldsymbol{\phi}} . \tag{7.37}$$

Estimate the amplitude of this field in G.

Solution on page 338.

7.4.3 DM Radio Inside a Cylindrical Shield

Essentially, a dark matter radio is an antenna readout by a tunable LC circuit connected to an amplifier. This methodology is complementary to microwave cavity searches as discussed in Chap. 4. A crucial difference between a microwave cavity and a dark matter radio is the frequency range that can be probed. The resonant frequency of a microwave cavity is inversely proportional to its size, which creates a practical limit on the lowest Compton frequencies that can be probed. On the other hand, the resonant frequency of an LC circuit is $\omega_0 = 1/\sqrt{LC}$, and thus large inductors and capacitors enable searches for hidden photons with much lower Compton frequencies.

Let us consider a schematic design of a dark matter radio as proposed in Ref. [15]. Figure 7.1 shows the schematic diagram of the DM Radio experiment. A hollow cylindrical superconducting sheath is housed within a superconducting shield. External electromagnetic fields are screened by the superconductors, but the hidden photon field penetrates inside. This gives rise to the circumferential magnetic field as described by Eqs. (7.34), (7.35), and (7.37), as shown in Fig. 7.2.

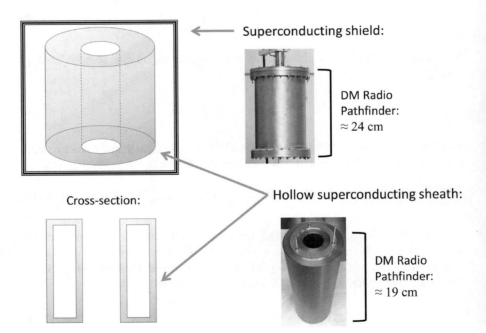

Fig. 7.1 Top left: schematic diagram of the DM Radio setup discussed in Ref. [15], showing the hollow superconducting sheath housed within a superconducting shield. Bottom left: cross-section of the hollow superconducting sheath. Top right: outer superconducting shield of the first generation "DM Radio Pathfinder" experiment. Bottom right: hollow superconducting sheath for DM Radio Pathfinder

Fig. 7.2 The oscillating effective current due to the hidden photon field $\mathcal{J}(\mathbf{r}, t)$ (pale violet arrows) induces a circumferential magnetic field $\mathbf{B}(\mathbf{r}, t)$ (green arrows) inside the hollow superconducting sheath. The screening currents (yellow arrows) induced in the superconductor cancel $\mathbf{B}(\mathbf{r}, t)$ within the bulk of the superconductor. A concentric circular slit is cut in the bottom of the superconducting sheath. The two sides of the slit are connected through an inductive loop, and the current through the loop is measured by a SQUID

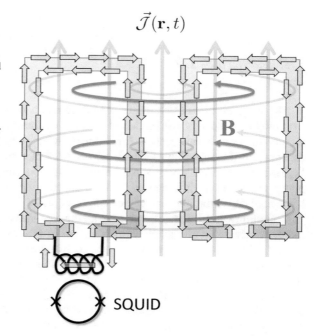

The next step is to measure the field $\mathbf{B}$ generated within the superconducting sheath using a resonant LC circuit. The DM Radio approach is to use the superconducting sheath as the "pick-up loop" for detection, and thus the sheath acts as the inductor in the circuit. The current through the inductor is read out by cutting a concentric circular slit in the bottom of the sheath and connecting the two sides of the slit through an inductive loop that siphons off the screening current. The field from the inductive loop can be measured with a SQUID.

A slitted sheath of inner radius r_1, outer radius r_2, and height h has an approximate inductance of

$$L \approx \frac{2}{c^2} h \ln\left(\frac{r_2}{r_1}\right). \tag{7.38}$$

? Problem 7.4 Inductance

Derive Eq. (7.38). If one models the concentric cylindrical slitted sheath as an N-turn toroidal solenoid with equivalent inductance, show that the corresponding number of turns would be

$$N^2 = \frac{r_1 + r_2}{r_1 - r_2} \ln\left(\frac{r_2}{r_1}\right).$$

Solution on page 339.

The resonant circuit also needs a capacitor. Since the mass and, therefore, the Compton frequency of the hidden photon are unknown, the resonant frequency of the circuit needs to be scanned, for example, with a tunable capacitor. The tunable capacitor in the approach of DM Radio is a set of concentric hexagonal niobium "cylinders" between which sapphire plates are inserted (Fig. 7.3). The sapphire plates are a dielectric material, and by adjusting how far they extend into the hexagonal capacitor, the capacitance can be adjusted.

Combining these elements, we have all the essential components of a dark matter radio. A schematic circuit diagram is shown in Fig. 7.4; note that the LC circuit naturally has nonzero resistance R due to loss mechanisms. (For the DM Radio Pathfinder experiment, $R \approx 5 \times 10^{-3}\ \Omega$.) The induced electromotive force (EMF), $\mathcal{V}_{\gamma'}$, in the dark matter radio is due to the changing magnetic flux through the inductor induced by the hidden photon field. For conceptual simplicity, here let us model the concentric cylindrical slitted sheath as an N-turn toroidal solenoid with equivalent inductance, as discussed in Problem 7.4 (in fact, for the DM Radio Pathfinder experiment, such a solenoid is used as the resonator). The field is given by Eq. (7.37) from which we derive

Fig. 7.3 Tunable concentric hexagonal capacitor used in the DM Radio experiment

Fig. 7.4 Schematic diagram
of a dark matter radio circuit

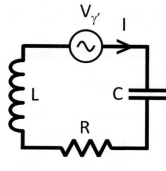

$$\mathcal{V}_{\gamma'} \approx 4i\kappa\sqrt{2\pi\rho_{\mathrm{dm}}}\frac{NV}{\lambda_{\gamma'}^2}e^{-i\omega t} . \tag{7.39}$$

? Problem 7.5 DM Radio EMF

Derive Eq. (7.39).

Solution on page 340.

The EMF $\mathcal{V}_{\gamma'}$ induced by the hidden photon field drives a current through the resonant RLC circuit. The advantage of the resonant RLC circuit is that when the resonance frequency is tuned near the hidden photon Compton frequency, the circuit will "ring up" and enhance the measurable signal. This enhancement is described by the Q-factor of the circuit, given by the ratio of the energy stored in the inductor to the energy lost per cycle due to dissipation (equivalent to the ratio of the resonant frequency to the linewidth)

$$Q = \frac{\omega_0}{\Delta\omega} = \frac{\omega_0 L}{R} . \tag{7.40}$$

The magnetic flux Φ through a cross-sectional area of the sheath, i.e., the flux through one loop of the equivalent N-turn toroidal solenoid, is given by (see solution to Problem 7.5):

$$\Phi = 4\kappa\sqrt{2\pi\rho_{\mathrm{dm}}}\frac{V}{\lambda_{\gamma'}}e^{-i\omega t} , \tag{7.41}$$

and the flux through the solenoid due to the hidden photon field is $N\Phi$. Because the RLC circuit rings up, on resonance this translates to a total flux of $QN\Phi$ due to the current accumulated in the inductor due to the hidden photon-induced EMF $\mathcal{V}_{\gamma'}$.

? Problem 7.6 DM Radio Q-factor

Show that on resonance, the flux through the inductor in the circuit shown in Fig. 7.4 is given by $\Phi_L = QN\Phi$. Thus, the resonant RLC circuit enhances the measurable flux by the Q-factor.

Solution on page 341.

The magnetic flux in the inductor can be measured with a SQUID. The flux through the SQUID is scaled by its area (A_s) relative to the cross-sectional area of the inductor $A = h(r_2 - r_1)$,

$$\Phi_{\text{SQUID}} \approx \left(\frac{A_s}{A}\right) QN\Phi, \tag{7.42}$$

and SQUIDs can have areas of $A_s \approx 1$ cm^2. A typical commercial DC SQUID has a noise floor of $\delta\Phi \approx 10^{-6}\Phi_0/\sqrt{\text{Hz}}$, where $\Phi_0 \approx 2 \times 10^{-7}$ G $\cdot$ cm^2 is the magnetic flux quantum and, as argued in Ref. [15], is an efficient detector for frequencies below about 100 MHz. To evaluate the sensitivity of a DM Radio experiment, the SQUID sensitivity must be compared to other sources of noise. A main source of noise comes from the circuit itself: the Johnson noise δV_{th} from the resistance. This thermal noise can be estimated by considering the current noise δI_{th} generated in the inductor:

$$\delta I_{\text{th}} = \frac{\delta V_{\text{th}}}{R} = \sqrt{\frac{2k_B T \Delta\omega}{\pi R}} = \sqrt{\frac{2k_B T Q \Delta\omega}{\pi \omega_0 L}} = \sqrt{\frac{2k_B T}{\pi L}}, \tag{7.43}$$

where T is the temperature of the circuit, k_B is Boltzmann's constant, and we assume for the bandwidth of the measurement $\Delta\omega = \omega_0/Q$. This thermal current generates noise in the magnetic flux measured by the SQUID:

$$\delta\Phi_{\text{th}} \approx cL\delta I_{\text{th}} \approx c\sqrt{\frac{2k_B T L}{\pi}}. \tag{7.44}$$

It turns out that under typical conditions, even for a cryogenic system with $T \ll 1$ K, $\delta\Phi_{\text{th}}$ is much greater than the SQUID noise floor.

From these properties and Eq. (7.41), we can estimate the sensitivity (on resonance) of a prototypical DM Radio experiment to κ:

$$\delta\kappa \approx \frac{1}{m_{\gamma'}c^2} \frac{\delta\Phi_{\text{th}}}{QNV} \frac{A}{A_s} \frac{\hbar c}{4\sqrt{2\pi\rho_{\text{dm}}}}, \tag{7.45}$$

and integrating for a time τ would improve the sensitivity by a factor of $\approx \sqrt{1/(\tau\Delta\omega)} \approx \sqrt{Q/(\omega_0\tau)}$. Choosing $V \approx 3 \times 10^6$ cm^3, $A \approx 10^4$ cm^2, $N \approx 2$,

$Q \approx 10^6$, and $T = 1$ K, we obtain the relationship:

$$\delta\kappa \approx \left(5 \times 10^{-8}\right) \times \left(\frac{10^{-12} \text{ eV}}{m_{\gamma'} c^2}\right). \tag{7.46}$$

7.5   Out-of-Band Sensitivity

Consider a DM Radio-style resonator operating at a fixed frequency. It is unlikely that the resonant frequency is exactly matched to the oscillation frequency of the hidden photons. The farther detuned the dark matter frequency is from resonance, the weaker the resonant enhancement of the signal. A large mismatch will result in very little signal power, but for small detunings the degradation is mild. With this in mind, over what frequency range does the resonator have sensitivity to a dark matter signal? A reasonable choice might be to assert that the sensitivity should be limited to the standard half-power bandwidth of the resonator: $\Delta f = f_0/Q$.

Note, however, that the ability to detect a dark matter signal actually depends on the signal-to-noise ratio. A relatively weaker signal due to detuning can be detected just as easily if the noise power has also decreased by the same amount. It was previously shown that the dark matter signal manifests as an effective voltage source in series with the resonator. The degradation of signal power due to detuning follows the Lorentzian line shape of the resonator. The thermal noise due to loss mechanisms (the resistor R in the circuit model) *also* appears as a series voltage source, and its power *also* follows the Lorentzian line shape at detuned frequencies. Thus, the ratio of signal power to thermal noise power remains unchanged even away from the LC resonance.

Since the total noise power is the sum of the frequency-dependent thermal noise power and the frequency-independent white noise of the amplifier, there is still some degradation in detection ability with detuning, but the effective *sensitivity bandwidth* can be much greater than the intrinsic resonator bandwidth and depends on the noise properties of the readout amplifier. This concept is shown in Fig. 7.5.

The out-of-band sensitivity offers several advantages. First, the total time required to scan a range of frequencies to a particular level of dark matter coupling is reduced as each frequency step of the resonator covers a greater bandwidth than the resonator bandwidth alone. Second, the larger frequency steps of the resonator during a scan relax the engineering requirements of the detector tuning system. Finally, it makes the use of quantum-limited amplifiers for dark matter radios desirable in order to maximize the sensitivity bandwidth. A detailed analysis of the quantum limits on dark matter radios, including amplifier back-action, can be found in Ref. [16]. While this same analysis applies to microwave cavity detectors, their total noise tends to be dominated by the readout amplifier and the sensitivity bandwidth is about the same as the resonator bandwidth. The out-of-band sensitivity is a consequence of the lower operating frequency, resulting in a higher thermal occupation of the resonator.

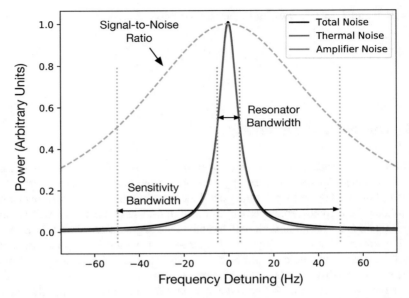

Fig. 7.5 Total noise, thermal noise, and amplifier noise as a function of detuning from the *LC* oscillator resonant frequency. The signal-to-noise ratio is only slightly degraded away from resonance, resulting in a much greater sensitivity bandwidth compared to the intrinsic resonator bandwidth

7.6 Sensitivity of Dark Matter Radio Experiments

There are a number of DM Radio experiments proposed, planned, or underway aimed at exploring unconstrained parameter space for UBDM [15, 17–19], and several have reported initial results [20–22]. These experiments employ a variety of tools and techniques going beyond the basic scheme described in Sect. 7.4 and target axion and ALPs as well as hidden photons.

The sensitivity of DM Radio experiments can be compared to astrophysical constraints (see Chap. 3). In contrast to axions and ALPs, there are no strong constraints on hidden photons from supernovae for any parameters [7]. Constraints from star-cooling fall off rapidly for smaller values of $m_{\gamma'}$, and so the dominant constraint on κ comes from measurements of the cosmic microwave background (CMB). The idea of the CMB limits is that there can be resonant conversion of CMB photons into hidden photons when the "effective mass" of the photon due to interactions with the primordial plasma matches $m_{\gamma'}$ [23]. (This resonant conversion effect is similar to the Mikheyev–Smirnov–Wolfenstein (MSW) effect for neutrinos [24, 25].) Therefore, the constraints from the CMB have a sharp cutoff at around $m_{\gamma'}c^2 \approx 10^{-14}$ eV [23, 26]. Constraints below $m_{\gamma'}c^2 \approx 10^{-14}$ eV come from limits on heating of the ionized interstellar medium by hidden photon dark matter [27].

A wide view of the hidden photon parameter space is shown in Fig. 7.6. The green areas show regions of parameter space excluded by a reinterpretation of published

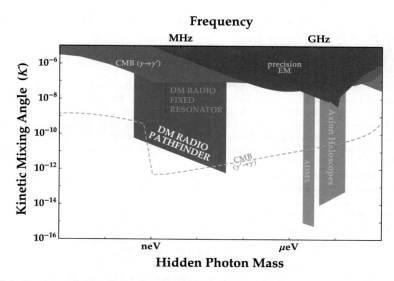

Fig. 7.6 Current exclusion limits for the hidden photon parameter space. Plot adapted from Ref. [22]

ADMX and precursor experiment axion limits applied to hidden photons [28]. The thin red line shows the first exclusion limit produced by a dark matter radio—the DM Radio Fixed Resonator [22]. Shown in blue is the expected exclusion limit for a 1-year scan of the DM Radio Pathfinder, a larger detector currently operating at Stanford University. These prototype dark matter detectors have served as stepping-stones for two future axion-focused detectors being constructed by the DM Radio collaboration: DM-Radio-50L and DM-Radio-m^3. The DM-Radio-50L detector will use a 50 liter, $\sim$1T toroidal magnet to probe axionlike particles between 5 kHz and 5 MHz, corresponding to the 20 peV–20 neV axion mass range. The DM-Radio-m^3 detector will use a cubic meter, 4T magnet to probe the QCD axion between 5 MHz and 300 MHz ($\sim$20 neV–800 neV) with sensitivity to the DFSZ QCD axion above 30 MHz. Both of these detectors may also be modified to search for hidden photons over the same span of frequencies. Interestingly, it has recently been realized that the Earth itself can be used as a transducer for a DM-Radio-like experiment at much lower frequencies [29].

References

1. B. Holdom, Phys. Lett. B **166**, 196 (1986)
2. M. Cvetic, P. Langacker, Phys. Rev. D **54**, 3570 (1996)
3. N. Arkani-Hamed, D.P. Finkbeiner, T.R. Slatyer, N. Weiner, Phys. Rev. D **79**, 015014 (2009)
4. L. Ackerman, M.R. Buckley, S.M. Carroll, M. Kamionkowski, Phys. Rev. D **79**, 023519 (2009)

5. M.S. Safronova, D. Budker, D. DeMille, D.F. Jackson Kimball, A. Derevianko, C.W. Clark, Rev. Mod. Phys. **90**, 025008 (2018)
6. P. Langacker, Rev. Mod. Phys. **81**, 1199 (2009)
7. M. Pospelov, A. Ritz, M. Voloshin, Phys. Rev. D **78**, 115012 (2008)
8. A.E. Nelson, J. Scholtz, Phys. Rev. D **84**, 103501 (2011)
9. G.P. Centers, J.W. Blanchard, J. Conrad, N.L. Figueroa, A. Garcon, A.V. Gramolin, D.F. Jackson Kimball, M. Lawson, B. Pelssers, J.A. Smiga, et al., Nat. Commun. **12**, 7321 (2021)
10. A. Derevianko, Phys. Rev. A **97**, 042506 (2018)
11. R. Loudon, *The Quantum Theory of Light* (Oxford University Press, Oxford, 2000)
12. P.W. Graham, J. Mardon, S. Rajendran, Y. Zhao, Phys. Rev. D **90**, 075017 (2014)
13. G.B. Arfken, H.J. Weber, *Mathematical Methods for Physicists* (Miami University, Oxford, 1999)
14. M.L. Boas, *Mathematical Methods in the Physical Sciences* (Wiley, New York, 2006)
15. S. Chaudhuri, P.W. Graham, K. Irwin, J. Mardon, S. Rajendran, Y. Zhao, Phys. Rev. D **92**, 075012 (2015)
16. S. Chaudhuri, K.D. Irwin, P.W. Graham, J. Mardon, arXiv:1904.05806 (2019).
17. P. Sikivie, N. Sullivan, D.B. Tanner, Phy. Rev. Lett. **112**(13), 131301 (2014)
18. P. Arias, A. Arza, B. Döbrich, J. Gamboa, F. Méndez, Eur. Phys. J. C **75**, 310 (2015)
19. Y. Kahn, B.R. Safdi, J. Thaler, Phy. Rev. Lett. **117**, 141801 (2016)
20. J.L. Ouellet, C.P. Salemi, J.W. Foster, R. Henning, Z. Bogorad, J.M. Conrad, J.A. Formaggio, Y. Kahn, J. Minervini, A. Radovinsky, et al., Phys. Rev. Lett. **122**(12), 121802 (2019)
21. A.V. Gramolin, D. Aybas, D. Johnson, J. Adam, A.O. Sushkov, Nature Phys. **17**, 79 (2021)
22. A. Phipps, S. Kuenstner, S. Chaudhuri, C. Dawson, B. Young, C. FitzGerald, H. Froland, K. Wells, D. Li, H. Cho, et al., in *Microwave Cavities and Detectors for Axion Research* (Springer, New York, 2020), p. 139
23. A. Mirizzi, J. Redondo, G. Sigl, J. Cosmol. Astropart. Phys. **2009**, 026 (2009)
24. L. Wolfenstein, Phys. Rev. D **17**, 2369 (1978)
25. S.P. Mikheyev, A.Y. Smirnov, Yad. Fiz **42**, 1441 (1985)
26. P. Arias, D. Cadamuro, M. Goodsell, J. Jaeckel, J. Redondo, A. Ringwald, J. Cosmol. Astropart. Phys. **2012**, 013 (2012)
27. S. Dubovsky, G. Hernández-Chifflet, J. Cosmol. Astropart. Phys. **2015**, 054 (2015)
28. P. Arias, D. Cadamuro, M. Goodsell, J. Jaeckel, J. Redondo, A. Ringwald, J. Cosm. Astropart. Phys. **2012**, 013 (2012)
29. M. A. Fedderke, P. W. Graham, D. F. Jackson Kimball, S. Kalia, Phys. Rev. D **104**, 075023 (2021)

Chapter 8
Laboratory Searches for Exotic Spin-Dependent Interactions

Andrew A. Geraci and Yun Chang Shin

Abstract The possible existence of exotic spin-dependent interactions with ranges from the subatomic scale to astrophysical scales has been of great theoretical interest for the last few decades. Typically, these exotic interactions are mediated by ultralight bosons with very weak coupling strength. If they indeed exist, such long-range interactions would indicate new physics beyond the Standard Model. A wide variety of experimental tests have been made to search for novel long-range spin-dependent interactions. Most experimental searches have focused on monopole-dipole or dipole-dipole interactions that could be induced by the exchange of ultralight bosons such as axions or axionlike particles. These ultralight bosons could also provide an answer to some of the most challenging problems in modern particle physics and astronomy: for example, the strong-CP problem in quantum chromodynamics (QCD), where C represents the charge conjugate symmetry and P represents the parity symmetry, and the explanation of dark matter and dark energy. In this chapter, we discuss the theoretical motivations as well as experimental searches for exotic spin-dependent interactions mediated by ultralight bosons in recent decades. We also introduce ongoing experimental efforts, such as Axion Resonant InterAction DetectioN Experiment (ARIADNE) and the QUest for AXion (QUAX)-$g_s g_p$ experiment. The high sensitivities of these tests will allow vast expansion of the discovery potential for exotic spin-dependent interactions.

A. A. Geraci (✉)
Center for Fundamental Physics, Department of Physics and Astronomy, Northwestern University, Evanston, IL, USA
e-mail: andrew.geraci@northwestern.edu

Y. C. Shin
Center for Axion and Precision Physics Research, IBS, Daejeon, Republic of Korea
e-mail: yunshin@ibs.re.kr

© The Author(s) 2023
D. F. Jackson Kimball, K. van Bibber (eds.), *The Search for Ultralight Bosonic Dark Matter*, https://doi.org/10.1007/978-3-030-95852-7_8

8.1   Introduction

In nature, there are four different fundamental forces that explain interactions of objects: electromagnetic, strong, weak, and gravitational forces. The gravitational and electromagnetic forces produce long-range interactions, and the strong and weak forces produce interactions at subatomic scales. Understanding the fundamental forces of nature has long been a profound goal in physics research. Each of the fundamental forces is characterized by its strength, effective range, and the nature of the particles that mediate the interaction. The Standard Model of particle physics is a theory describing three of the four known fundamental interactions and all elementary particles; the quantum description of the role of gravity in the Standard Model is incomplete. Although the Standard Model has had great success in describing most interactions and particles, there is a considerable piece of the puzzle missing: the nature of the dark matter. Recently, new theories have postulated the existence of new "exotic" interactions that may explain various anomalous phenomena that remain unexplained by the Standard Model [1]. In these theories, exotic interactions with very weak strength have ranges from sub-mm scales all the way to cosmological scales. The search for such exotic interactions has been motivated in recent decades by the cosmological dilemmas of dark matter and dark energy [2, 3].

8.1.1   Dark Matter and New Spin-Dependent Interactions

Over the last few decades, astrophysical observations have indicated that most of mass-energy density of our universe is in the form of non-baryonic components belonging to what is known as the "dark sector." Recent measurements show that ordinary baryonic matter contributes only about 4% of the energy content in the universe. The remaining 96% of the energy content belongs to the dark sector with two sub-components called "dark energy" and "dark matter." The dark energy is believed to accelerate the expansion of the universe and contributes about 68% of the energy content in our universe. Although the accelerating expansion of our universe has been verified by astronomical observations, the existence of dark energy is still under debate [4, 5].

On the other hand, the reality of dark matter is becoming more evident based on the observation of the abundance of light elements and the measurement of the cosmic microwave background (CMB), as well as from the measured galactic rotation curves, galactic cluster dynamics, and gravitational lensing studies (see Chap. 1). Dark matter is considered to be responsible for approximately 85% of the matter density and about 25% of total energy density in the universe.

The fact that dark matter cannot be explained in the framework of the Standard Model of particle physics creates the need for a new theory that extends beyond the Standard Model. There are two main categories of beyond-Standard-Model theories to explain dark matter: one class of theories suggest that dark matter is composed of neutralinos or other weakly interacting massive particles (WIMPs) and the other suggests that dark matter is made of axions or other weakly interacting sub-eV particles (WISPs), another term for the ultralight bosons described throughout this text. WIMPs have been a promising solution for the dark matter problem since they were first introduced [6]. However, non-observation of WIMPs at the Large Hadron Collider (LHC) and in other direct-detection dark matter experiments over past few decades gives rise to strong motivation to look for WISPs. A wide variety of theories beyond the Standard Model have predicted the existence of such weakly coupled scalar, pseudoscalar, vector, and axial-vector bosons with very light mass as dark matter candidates [2].

These ultralight bosons arise from spontaneous symmetry breaking of global $\mathbb{U}(1)$ symmetries at a scale of f and their effective couplings to standard model particles are suppressed by a factor on the scale f [7–10], as discussed in Chap. 2. Examples include majorons which result from breaking of the global $\mathbb{U}(1)$ B $-$ L symmetry where B is the baryon number and L is the lepton number. Majorons could explain the small neutrino mass $m_\nu \sim m_l^2/f_L$ where m_l is the mass of the associated charged lepton and f_L is the symmetry-breaking scale for lepton number [11, 12]. Familons are WISPs that arise from breaking of family symmetry which normally refers to various discrete, global, or local symmetries between quark-lepton families or generations. Familons couple to a divergence of current changing flavor quantum number and therefore can be emitted in flavor changing decays [13].

These ultralight bosons could be an answer not only for the dark matter problem, but also for many other fundamental questions in physics, such as the CP problem in QCD [7, 14, 15]. One well-known example of such an ultralight boson is the axion. The axion remains the most prominent solution of the strong CP problem many decades after its prediction and is also a very well-motivated dark matter candidate, as discussed in Chaps. 1 and 2 and elsewhere throughout this book. Now the question is if these ultralight bosons could also be linked to new interactions that are yet to be observed? If such an exotic long-range interaction mediated by an ultralight boson is discovered, it would have a profound impact on our understanding of nature.

8.1.2 New Spin-Dependent Interactions

Weakly coupled, long-range interactions are a generic consequence of a spontaneously broken continuous symmetry as shown by Goldstone's theorem. Goldstone's theorem states that if a system that is invariant under a continuous, global symmetry has this symmetry broken so that the ground (vacuum) state is not invariant with respect to the global symmetry (referred to as a spontaneously broken symmetry—see Chap. 2), there must be a state in the spectrum of excitations of the system with zero energy (which can be created from the ground state by performing a spacetime-dependent symmetry transformation). This state is called a Goldstone mode or Goldstone boson. Since the Goldstone mode is "gapless" so that its energy vanishes when its momentum vanishes, i.e., as $\omega \to 0$ then $k \to 0$ (see Fig. 2.1 and surrounding discussion), the Goldstone mode oscillates with infinite wavelength ($\lambda \to \infty$) to minimize the energy of the wave by reducing the momentum. Thus the Goldstone mode corresponds to massless particles traveling with the speed of light. Therefore, when the underlying symmetry is exact up to an energy scale f, the process always produces a massless Goldstone boson [16].

However, when the symmetry is explicitly broken, meaning that the symmetry becomes an approximate symmetry instead of an exact symmetry, the Goldstone boson acquires a very small mass rather than being exactly massless. Such particles are referred as to "pseudo-Goldstone bosons" (again, see Chap. 2). The pseudo-Goldstone bosons acquire a small mass of order $m_b \sim \Lambda^2/f$ that depends on the symmetry-breaking scale Λ of the continuous Lagrangian [17]. If such bosons have sufficiently small mass they have a macroscopic Compton wavelength and can mediate macroscopic interactions [18].

The ultralight pseudo-Goldstone bosons can couple to fundamental fermions through scalar and pseudoscalar vertices for spin-0 bosons, vector, axial-vector, tensor, and pseudotensor vertices for spin-1 bosons. The interactions can be classified in terms of a multipole expansion: well-known examples of possible allowed interactions are monopole-monopole, monopole-dipole, and dipole-dipole. Monopole-dipole interactions include a scalar coupling, g_s, and a pseudoscalar coupling, g_p, thereby violating P and T symmetry. This can be seen in the nonrelativistic limit, where this interaction is proportional to $g_s g_p \boldsymbol{\sigma} \cdot \boldsymbol{r}$ where $\boldsymbol{\sigma}$ is the spin of one particle, and $\boldsymbol{r}$ is the distance between two particles: the spin $\boldsymbol{\sigma}$ is P-even and T-odd, whereas the position vector $\mathbf{r}$ is P-odd and T-even, making their scalar product, $\boldsymbol{\sigma} \cdot \boldsymbol{r}$, P- and T-odd. Dipole-dipole interactions are dependent on the spins of two particles, $\boldsymbol{\sigma}_1$ and $\boldsymbol{\sigma}_2$, and can be either velocity-independent or velocity-dependent. Spin-dependent but velocity-independent dipole-dipole interactions are proportional to $g_p^2 \boldsymbol{\sigma}_1 \cdot \boldsymbol{\sigma}_2$ [19–21]. Depending on the model, the monopole and dipole couplings can occur for electrons and/or nuclei.

An interesting point is that, although the couplings can be extremely feeble for the interaction between single particles, a macroscopic object composed of many particles, e.g., on the order of $\sim 10^{22}$–10^{23}, would produce a coherent field, thus enhancing the signal as compared to a single particle and potentially making it detectable with a sufficiently sensitive laboratory experiment [18]. Many of the experimental tests of such interactions have been done with polarized gases [22]. Torsion balance experiments also have recently set new limits on both monopole-dipole interactions and dipole-dipole interactions. But laboratory constraints on possible new interactions in the mesoscopic range, which is roughly defined in a scale of between $\sim \mu$m to $\sim$ mm, have not yet been well developed [23]. In many cases, laboratory measurements combined with astrophysical data have produced the most stringent constraints on the products of the coupling constants g_s and g_p. These coupling strengths are constrained by experiments or astrophysical observations. For the QCD axion, g_s and g_p are related to the axion mass as they are fixed by the axion decay constant f_a (see discussion in Chap. 2):

$$6 \times 10^{-27} \left(\frac{10^9 \text{GeV}}{f_a} \right) < g_s < 10^{-21} \left(\frac{10^9 \text{GeV}}{f_a} \right) , \tag{8.1}$$

and

$$g_p = \frac{C_f m_f}{f_a} = C_f \times 10^{-9} \left(\frac{m_f}{1 \text{GeV}} \right) \left(\frac{10^9 \text{GeV}}{f_a} \right) , \tag{8.2}$$

where C_f is a dimensionless coupling constant for the particular fermion considered.

8.2   Spin-Dependent Interactions Mediated by Light Bosons: Classification

Spin-dependent interactions mediated by ultralight bosons were first described by Moody and Wilczek along with some suggestions for experimental tests in Ref. [18]. They proposed experimental tests to detect axions via the macroscopic forces mediated by axion exchange. A phenomenological theory was also developed by Dobrescu and Mocioiu [1]. They listed all possible spin-dependent interactions satisfying rotational invariance and standard assumptions of quantum field theory. Fadeev et al. later revisited and derived nonrelativistic potentials mediated by spin-0 and spin-1 bosons [24]. They have updated the Dobrescu and Mocioiu's work with more details and several corrections, for example, including contact-terms in the coordinate-space nonrelativistic potentials, which can affect atomic- and molecular-scale experiments, in particular [25].

Exotic spin-dependent interactions that are mediated by spin-0 and spin-1 bosons between two fermions with masses m_1 and m_2, and spins σ_1 and σ_2 can be derived from the elastic scattering of two fermions in the nonrelativistic limit as shown in Fig. 8.1. The scattering is mediated by a boson of mass m_b and the four-momentum q is transferred from fermion 2 to fermion 1.

Fig. 8.1 The Feynman diagram of elastic scattering between two fermions with masses m_1 and m_2 mediated by ultralight bosons with mass m_b with four-momentum transferred from vertex 2 to vertex 1

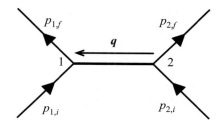

The Lagrangian describing the interaction between fermions ψ mediated by a spin-0 boson ϕ is given as:

$$\mathcal{L}_\psi = \phi \sum_\psi \bar{\psi} \left(g_{s,\psi} + i\gamma_5 g_{p,\psi}\right) \psi , \tag{8.3}$$

where $\gamma_5 = -i\gamma_0\gamma_1\gamma_2\gamma_3$ are Dirac matrices. The interaction constants $g_{s,\psi}$ and $g_{p,\psi}$ parametrize the scalar and pseudoscalar interaction strengths, respectively.

The Lagrangian describing the interaction mediated by exchange of a massive spin-1 boson (denoted Z') can be separated from that describing the interaction mediated by a massless spin-1 boson (denoted γ') due to the presence of a longitudinal polarization appearing for Z' bosons. In the case of a massive spin-1 Z' boson, the Lagrangian becomes

$$\mathcal{L}_{Z'} = Z'_\mu \sum_\psi \bar{\psi} \left(g_{V,\psi} + \gamma_5 g_{A,\psi}\right) \psi , \tag{8.4}$$

where $g_{V,\psi}$ and $g_{A,\psi}$ are the vector and axial-vector interaction strengths, respectively. The interaction Lagrangian describing the exchange of a massless spin-1 γ' boson is:

$$\mathcal{L}_{\gamma'} = \frac{v_h}{\Lambda^2} P_{\mu\nu} \sum_\psi \bar{\psi}\sigma^{\mu\nu} \left(\text{Re}(C_\psi) + i\gamma_5\text{Im}(C_\psi)\right) \psi , \tag{8.5}$$

where ψ denotes the fermion field, v_h is the Higgs vacuum expectation value, Λ is the ultraviolet energy cut-off scale of the Lagrangian in Eq. (8.5),

$$P_{\mu\nu} = \partial_\mu A_\nu - \partial_\nu A_\mu \tag{8.6}$$

is the field strength tensor of the massless spin-1 boson γ',

$$\sigma^{\mu\nu} = \frac{i}{2}[\gamma^\mu, \gamma^\nu] , \tag{8.7}$$

and $\text{Re}(C_\psi)$ and $\text{Im}(C_\psi)$ denote the respective interaction strengths. The details of these cases can be found in Ref. [24].

The nonrelativistic momentum-space potential $V(q)$ can be estimated from the scattering matrix $M(q)$ in the leading order as

$$V(q) \approx -\frac{M(q)}{4m_1m_2} . \tag{8.8}$$

The nonrelativistic coordinate-space potential can be obtained by applying the three-dimensional Fourier transformation to Eq. (8.8):

$$V(r) = \int \frac{d^3 q}{(2\pi)^3} V(q) e^{i q \cdot r} = \int \frac{d^3 q}{(2\pi)^3} \frac{-M(q)}{4 m_1 m_2} e^{i q \cdot r}. \tag{8.9}$$

8.2.1 Interactions Mediated by Massive Spin-0 Bosons

Interactions mediated by a massive spin-0 boson can be derived from the Lagrangian in Eq. (8.3). They can be classified by three different type of interactions: scalar-scalar, pseudoscalar-scalar, and pseudoscalar-pseudoscalar interactions.

8.2.1.1 Scalar-Scalar Interaction

The scattering matrix of the interaction between two fermions via the exchange of a spin-0 boson at the tree level can be calculated from the Feynman rules in momentum space,

$$i M(q) = [i \bar{u}(p_{f,1}) g_{s,1} u(p_{i,1})][i \bar{u}(p_{f,2}) g_{s,2} u(p_{i,2})] \frac{i}{q^2 - m_b^2}, \tag{8.10}$$

where m_b is the mass of the boson which mediates the interaction and $u(p)$ is the spinor. The four-momentum of the mediating boson is defined as

$$q = p_{1,f} - p_{1,i} = p_{2,i} - p_{2,f}, \tag{8.11}$$

and the average momenta of each fermion are defined as

$$p_1 = \frac{p_{1,i} + p_{1,f}}{2}, \tag{8.12}$$

$$p_2 = \frac{p_{2,i} + p_{2,f}}{2}, \tag{8.13}$$

where the labels 1, 2 denote the fermions and i, f denote the initial and final state of the fermions as defined from Fig. 8.1. In the nonrelativistic limit ($|p| \ll m$), the momentum-space Dirac spinor $u(p)$ can be expanded up to the first order in p as

$$u(p) \approx \sqrt{m} \begin{pmatrix} \left(1 - \frac{\sigma \cdot p}{2m}\right) \xi \\ \left(1 + \frac{\sigma \cdot p}{2m}\right) \xi \end{pmatrix}, \tag{8.14}$$

where ξ is a 2×1 matrix with normalization $\xi^\dagger \xi = 1$ (ξ is often also called a spinor in the literature).

In the nonrelativistic limit, by using Eqs. (8.10) and (8.14) in the expression for the coordinate-space potential (8.9), the potential of the scalar-scalar interaction becomes

$$V_{ss}(r) = -g_{s,1}g_{s,2}\mathcal{V}_{ss} = -g_{s,1}g_{s,2} \int \frac{d^3q}{(2\pi)^3} \frac{1}{|\mathbf{q}|^2 + m_b^2} e^{i\mathbf{q}\cdot\mathbf{r}}, \tag{8.15}$$

where we note that in the nonrelativistic limit the spacelike component of the momentum dominates, $q^2 = q_0^2 - |\mathbf{q}|^2 \approx -|\mathbf{q}|^2$. The integration in spherical coordinates results in the potential $V_{ss}(r)$:

$$V_{ss}(r) = -\frac{g_{s,1}g_{s,2}}{4\pi} \frac{e^{-m_b r}}{r}, \tag{8.16}$$

which includes the well-known Yukawa-type factor.

? Problem 8.1 Yukawa Potential in the Monopole-Monopole Interaction

Derive the Yukawa-type potential in Eq. (8.16) from Eq. (8.15) by computing the integral in spherical coordinates.

Solution on page 342.

8.2.1.2   Pseudoscalar-Scalar Interaction

In the case of a pseudoscalar-scalar interaction between two fermions, the scattering matrix can be obtained from the Feynman diagram as follows:

$$iM(q) = [i^2\bar{u}(p_{f,1})g_{p,1}\gamma_5 u(p_{i,1})][i\bar{u}(p_{f,2})g_{s,2}u(p_{i,2})]\frac{i}{q^2 - m_b^2}, \tag{8.17}$$

where we assumed a pseudoscalar coupling on vertex 1 ($g_{p,1}$) and a scalar coupling on vertex 2 ($g_{s,2}$), respectively.

In the nonrelativistic limit, the spinor products in the scattering amplitude become

$$\bar{u}(p_f)u(p_i) \approx 2m,$$
$$\bar{u}(p_f)\gamma^5 u(p_i) \approx \mp\boldsymbol{\sigma}\cdot\mathbf{q}, \tag{8.18}$$

where the sign of the pseudoscalar vertex depends on the direction of momentum transfer.

This scalar-pseudoscalar interaction is normally called a monopole-dipole interaction in the sense of the multipole expansion. From Eqs. (8.18) and (8.17), the potential describing the monopole (g_s) and dipole ($i g_p \boldsymbol{\sigma} \cdot \boldsymbol{q}/2m$) interaction between two fermions can be expressed based on Eq. (8.9) as,

$$
V_{ps}(r) = -\frac{g_{p,1} g_{s,2}}{2 i m_1} \mathcal{V}_{ps}(r) = -\frac{g_{p,1} g_{s,2}}{2 i m_1} \int \frac{d^3 q}{(2\pi)^3} \frac{\boldsymbol{\sigma}_1 \cdot \boldsymbol{q}}{|q|^2 + m_b^2} e^{i q \cdot r}, \qquad (8.19)
$$

where the integration function $\mathcal{V}_{ps}(r)$ can be obtained by taking the inner product between the spin vector and the gradient of $\mathcal{V}_{ss}(r)$:

$$
\mathcal{V}_{ps}(r) = -i \boldsymbol{\sigma}_1 \cdot \nabla \mathcal{V}_{ss}(r). \qquad (8.20)
$$

With the calculation of $\mathcal{V}_{ps}(r)$, the interaction potential becomes

$$
V_{ps}(r) = -\frac{g_{p,1} g_{s,2}}{8 \pi m_1} \left(\boldsymbol{\sigma}_1 \cdot \hat{\boldsymbol{r}} \right) \left(\frac{m_b}{r} + \frac{1}{r^2} \right) e^{-m_b r}. \qquad (8.21)
$$

? Problem 8.2 Spin-Dependent Interaction via Spin-0 Boson Exchange: Monopole-Dipole Interaction

Derive the monopole-dipole interaction potential (8.21) by completing the calculation in Eq. (8.20). Show that the strength of the interaction can be expressed as an "effective" magnetic field or pseudo-magnetic field.

Solution on page 343.

8.2.1.3 Pseudoscalar-Pseudoscalar Interaction

The scattering matrix describing a pseudoscalar-pseudoscalar interaction is given by

$$
i M(q) = [i^2 \bar{u}(p_{f,1}) g_{p,1} \gamma_5 u(p_{i,1})][i^2 \bar{u}(p_{f,2}) g_{p,2} \gamma_5 u(p_{i,2})] \frac{i}{q^2 - m_b^2}. \qquad (8.22)
$$

In this matrix, the momentum is transferred from vertex two to vertex one. From Eq. (8.18), the scattering matrix in the nonrelativistic limit becomes,

$$
i M(q) = -(\boldsymbol{\sigma}_1 \cdot \boldsymbol{q})(\boldsymbol{\sigma}_2 \cdot \boldsymbol{q}) \frac{-i}{|q|^2 + m_b^2}. \qquad (8.23)
$$

From Eq. (8.9), the potential for the pseudoscalar-pseudoscalar interaction between two fermions is

$$V_{pp}(r) = \frac{g_{p,1}g_{p,2}}{4m_1m_2}\mathcal{V}_{pp}(r) = \frac{g_{p,1}g_{p,2}}{4m_1m_2}\int\frac{d^3q}{(2\pi)^3}\frac{(\sigma_1\cdot q)(\sigma_2\cdot q)}{|q|^2+m_b^2}e^{iq\cdot r}, \quad (8.24)$$

where the labels 1 and 2 indicate each fermion with mass m_1 and m_2, respectively. The spin product in the integral part in Eq. (8.24) can be expressed via the summation of each spin state $a, b = 1, 2, 3$:

$$\mathcal{V}_{pp}(r) = \sum_a\sum_b\int\frac{d^3q}{(2\pi)^3}\frac{\sigma_{1,a}\sigma_{2,b}q_aq_b}{|q|^2+m_b^2}e^{iq\cdot r}. \quad (8.25)$$

Analogously to the method used in Eq. (8.20), $\mathcal{V}_{pp}(r)$ can be derived from $\mathcal{V}_{ss}(r)$ in the following way:

$$\mathcal{V}_{pp}(r) = -\sum_a\sum_b\sigma_{1,a}\sigma_{2,b}\partial_a\partial_b\mathcal{V}_{ss}. \quad (8.26)$$

The evaluation of partial derivatives in Eq. (8.26) yields

$$4\pi\,\partial_a\partial_b\mathcal{V}_{ss} = \left(\partial_a\partial_b e^{-m_br}\right)\frac{1}{r}+2\left(\partial_a e^{-m_br}\right)\left(\partial_b\frac{1}{r}\right)+e^{-m_br}\left(\partial_a\partial_b\frac{1}{r}\right). \quad (8.27)$$

By completing the calculation of Eq. (8.27), the potential for the pseudoscalar-pseudoscalar interaction becomes

$$V_{pp}(r) = -\frac{g_{p,1}g_{p,2}}{16\pi m_1m_2}e^{-m_br}$$
$$\times\left[\sigma_1\cdot\sigma_2\left(\frac{1}{r^3}+\frac{m_b}{r^2}+\frac{4\pi}{3}\delta(r)\right)\right. \quad (8.28)$$
$$\left.-(\sigma_1\cdot\hat{r})(\sigma_2\cdot\hat{r})\left(\frac{m_b^2}{r}+\frac{3m_b}{r^2}+\frac{3}{r^3}\right)\right].$$

? Problem 8.3 Spin-Dependent Interaction via Spin-0 Boson Exchange: Dipole-Dipole Interaction

Derive the dipole-dipole interaction potential (8.28) by using Eq. (8.27).

Solution on page 344.

8.2.2 Interactions Mediated by Massive Spin-1 Bosons

Interactions between two fermions mediated by a massive spin-1 boson (Z') can be described using the Lagrangian in Eq. (8.4). In this case, the terms describing the possible vertex interactions are given by

$$
\begin{aligned}
g_V \bar{u}(p_f)\gamma^\mu u(p_i)\,, \\
g_A \bar{u}(p_f)\gamma^\mu \gamma^5 u(p_i)\,.
\end{aligned}
\tag{8.29}
$$

Using approximate nonrelativistic solutions, the spinor products for each vertex can be simplified as

$$
\begin{aligned}
\bar{u}(p_f)\gamma^0 u(p_i) &\approx 2m, \\
\bar{u}(p_f)\boldsymbol{\gamma} u(p_i) &\approx 2\boldsymbol{p} \pm i\boldsymbol{q} \times \boldsymbol{\sigma}, \\
\bar{u}(p_f)\gamma^0 \gamma^5 u(p_i) &\approx 2\boldsymbol{\sigma} \cdot \boldsymbol{p}, \\
\bar{u}(p_f)\boldsymbol{\gamma}\gamma^5 u(p_i) &\approx 2m\boldsymbol{\sigma},
\end{aligned}
\tag{8.30}
$$

where $\pm$ in the vector component is the direction of vector vertex.

8.2.2.1 Vector-Vector Interaction

The scattering matrix element of the vector-vector interaction is given by

$$
iM(q) = g_{V,1}g_{V,2}[i\bar{u}(p_{f,1})\gamma^\mu u(p_{i,1})][i\bar{u}(p_{f,2})\gamma^\nu u(p_{i,2})]i\frac{g_{\mu\nu} - q_\mu q_\nu/m_b^2}{m_b^2 - q^2}.
\tag{8.31}
$$

Since the vector current is conserved in the interaction, the $q_\mu q_\nu$ term in the propagator can be neglected. Therefore the scattering matrix can be simplified in the nonrelativistic limit as

$$
iM(q) \approx -ig_{V,1}g_{V,2}\frac{4m_1 m_2 - (2\boldsymbol{p}_1 - i\boldsymbol{q} \times \boldsymbol{\sigma}_1) \cdot (2\boldsymbol{p}_2 + i\boldsymbol{q} \times \boldsymbol{\sigma}_2)}{|\boldsymbol{q}|^2 + m_b^2}.
\tag{8.32}
$$

The potential describing the vector-vector interaction in momentum space is approximately

$$
V(q) \approx \frac{g_{V,1}g_{V,2}}{m_b^2 + |\boldsymbol{q}|^2}
\tag{8.33}
$$

$$
\times \left(1 - \frac{(\boldsymbol{q} \times \boldsymbol{\sigma}_1) \cdot (\boldsymbol{q} \times \boldsymbol{\sigma}_2)}{4m_1 m_2} - \frac{\boldsymbol{p}_1 \cdot \boldsymbol{p}_2}{m_1 m_2} - i\frac{\boldsymbol{p}_1 \cdot (\boldsymbol{q} \times \boldsymbol{\sigma}_2) - \boldsymbol{p}_2 \cdot (\boldsymbol{q} \times \boldsymbol{\sigma}_1)}{2m_1 m_2}\right).
$$

In the nonrelativistic limit, the above expression can be simplified as

$$V(q) \approx \frac{g_{V,1}g_{V,2}}{m_b^2 + |q|^2} \left(1 - \frac{(q \times \sigma_1) \cdot (q \times \sigma_2)}{4m_1 m_2}\right). \tag{8.34}$$

The dot product of the cross products between spin and momentum of each fermion can be decomposed into

$$(q \times \sigma_1) \cdot (q \times \sigma_2) = |q|^2(\sigma_1 \cdot \sigma_2) - (q \cdot \sigma_1)(q \cdot \sigma_2). \tag{8.35}$$

The coordinate-space potential can be obtained as

$$V_{VV}(r) = g_{V,1}g_{V,2} \int \frac{d^3q}{(2\pi)^3} \frac{e^{iq \cdot r}}{m_b^2 + |q|^2} \left(1 - \frac{|q|^2(\sigma_1 \cdot \sigma_2) - (q \cdot \sigma_1)(q \cdot \sigma_2)}{4m_1 m_2}\right). \tag{8.36}$$

The first term of Eq. (8.36) is simply the Yukawa potential which we already studied in the case of scalar-scalar interaction, Eq. (8.16).

$$\int \frac{d^3q}{(2\pi)^3} \frac{e^{iq \cdot r}}{m_b^2 + |q|^2} = \frac{e^{-m_b r}}{4\pi r}. \tag{8.37}$$

The second and third terms of Eq. (8.36) can be evaluated by using the results from our analysis of the pseudoscalar-pseudoscalar interaction,

$$\int \frac{d^3q}{(2\pi)^3} \frac{q^2 e^{iq \cdot r}}{m_b^2 + |q|^2} = -\nabla^2 \left(\frac{e^{-m_b r}}{4\pi r}\right) = \left(\delta(r) - \frac{m_b^2}{4\pi r}\right) e^{-m_b r}, \tag{8.38}$$

and

$$\int \frac{d^3q}{(2\pi)^3} \frac{(q \cdot \sigma_1)(q \cdot \sigma_2) e^{iq \cdot r}}{m_b^2 + |q|^2} = \frac{\sigma_1 \cdot \sigma_2}{4\pi} \left(\frac{1}{r^3} + \frac{m_b}{r^2} + \frac{4\pi}{3}\delta(r)\right) e^{-m_b r}$$
$$- \frac{(\sigma_1 \cdot \hat{r})(\sigma_2 \cdot \hat{r})}{4\pi} \left(\frac{3}{r^3} + \frac{3m_b}{r^2} + \frac{m_b^2}{r}\right). \tag{8.39}$$

From these relationships, the interaction potential of the vector-vector interaction becomes

$$V_{VV}(r) = g_{V,1}g_{V,2}\frac{e^{-m_b r}}{4\pi r}$$

$$+ \frac{g_{V,1}g_{V,2}}{16\pi m_1 m_2}(\boldsymbol{\sigma}_1 \cdot \boldsymbol{\sigma}_2)\left(\frac{1}{r^3} + \frac{m_b}{r^2} + \frac{m_b^2}{r} - \frac{8\pi}{3}\delta(r)\right)e^{-m_b r} \qquad (8.40)$$

$$- \frac{g_{V,1}g_{V,2}}{16\pi m_1 m_2}(\boldsymbol{\sigma}_1 \cdot \hat{\boldsymbol{r}})(\boldsymbol{\sigma}_2 \cdot \hat{\boldsymbol{r}})\left(\frac{3}{r^3} + \frac{3m_b}{r^2} + \frac{m_b^2}{r}\right)e^{-m_b r} .$$

8.2.2.2 Axial-Vector-Vector Interaction

The potential describing the axial-vector-vector interaction can be calculated by assuming an axial-vector coupling in the first vertex and a vector coupling in the second vertex. In this case, the scattering matrix is given by

$$M(q) = -g_{A,1}g_{V,2}i[i\bar{u}(p_{f,1})\gamma^\mu\gamma_5 u(p_{i,1})][i\bar{u}(p_{f,2})\gamma^\nu u(p_{i,2})]i\frac{g_{\mu\nu} - q_\mu q_\nu/m_b^2}{m_b^2 - q^2} .$$
$$(8.41)$$

Since the vector current is conserved, $q_\mu J^\mu = 0$ imposes the condition that the $q_\mu q_\nu$ term vanishes. In the nonrelativistic limit, the momentum-space potential is

$$V(q) \approx \frac{g_{A,1}g_{V,2}}{4m_1 m_2(m_b^2 + |\boldsymbol{q}|^2)}\left[4m_2\boldsymbol{\sigma}_1 \cdot \boldsymbol{p}_1 - 2m_1\boldsymbol{\sigma}_1 \cdot (2\boldsymbol{p}_2 + i\boldsymbol{q} \times \boldsymbol{\sigma}_2)\right] .$$
$$(8.42)$$

It can be simplified as

$$V(q) \approx \frac{g_{A,1}g_{V,2}}{m_b^2 + |\boldsymbol{q}|^2}\left[\boldsymbol{\sigma}_1 \cdot \left(\frac{\boldsymbol{p}}{m_1} - \frac{\boldsymbol{p}_2}{m_2}\right) + i\frac{\boldsymbol{q} \cdot (\boldsymbol{\sigma}_1 \times \boldsymbol{\sigma}_2)}{2m_2}\right]. \qquad (8.43)$$

The coordinate-space potential can thus be obtained by applying a Fourier transform:

$$V_{AV}(r) = \frac{g_{A,1}g_{V,2}}{4\pi}\boldsymbol{\sigma}_1 \cdot \left(\frac{\boldsymbol{p}}{m_1} - \frac{\boldsymbol{p}_2}{m_2}\right)\frac{e^{-m_b r}}{r}$$

$$- \frac{g_{A,1}g_{V,2}}{8\pi m_2}(\boldsymbol{\sigma}_1 \times \boldsymbol{\sigma}_2) \cdot \hat{\boldsymbol{r}}\left(\frac{m_b}{r} + \frac{1}{r^2}\right)\frac{e^{-m_b r}}{r} . \qquad (8.44)$$

8.2.2.3 Axial-Vector-Axial-Vector Interaction

The main difference between an axial-vector-axial-vector interaction and a vector-vector interaction is that the axial current of each fermion is not conserved whereas the vector current is conserved. Therefore, in the case of the axial-vector-axial-vector interaction, it is necessary to consider the $q_\mu q_\nu$ term in the propagator. The scattering matrix becomes

$$M(q) = -g_{A,1}g_{A,2}i[i\bar{u}(p_{f,1})\gamma^\mu\gamma_5 u(p_{i,1})][i\bar{u}(p_{f,2})\gamma^\nu\gamma_5 u(p_{i,2})]i\frac{g_{\mu\nu} - q_\mu q_\nu/m_b^2}{m_b^2 - q^2}, \tag{8.45}$$

and so consequently the momentum-space potential can be written as

$$V(q) = \frac{g_{A,1}g_{A,2}}{m_b^2 + |q|^2}\left(\frac{\sigma_1 \cdot p_1}{m_1}\frac{\sigma_2 \cdot p_2}{m_2} - \sigma_1 \cdot \sigma_2 + \frac{(q \cdot \sigma_1)(q \cdot \sigma_2)}{m_b^2}\right). \tag{8.46}$$

The first term in Eq. (8.46) can be neglected because $p_i/m_i \ll 1$ in the nonrelativistic limit. The coordinate-space potential then becomes

$$\begin{aligned}
V_{AA}(r) = &-\frac{g_{A,1}g_{A,2}}{4\pi}(\sigma_1 \cdot \sigma_2)\frac{e^{-m_b r}}{r} + \frac{g_{A,1}g_{A,2}}{4\pi m_b^2}e^{-m_b r} \\
&\times \left[(\sigma_1 \cdot \sigma_2)\left(\frac{1}{r^3} + \frac{m_b}{r^2} + \frac{4\pi}{3}\delta(r)\right)\right. \\
&\left. -(\sigma_1 \cdot \hat{r})(\sigma_2 \cdot \hat{r})\left(\frac{3}{r^3} + \frac{3m_b}{r^2} + c\right)\right].
\end{aligned} \tag{8.47}$$

8.2.3   Interactions Mediated by Massless Spin-1 Bosons

For massless spin-1 bosons, the interaction Lagrangian is described as

$$\mathcal{L}_{\gamma'} = \frac{v_h}{\Lambda^2}P_{\mu\nu}\bar{\psi}\sigma^{\mu\nu}\left[\text{Re}(C_\psi) + i\gamma_5\text{Im}(C_\psi)\right]\psi, \tag{8.48}$$

where the $P_{\mu\nu}$ is the field strength tensor of the massless gauge boson γ', and $\sigma^{\mu\nu} = \frac{i}{2}[\gamma^\mu, \gamma^\nu]$. Using anti-symmetric properties of the tensors, the Lagrangian can be written as

$$\mathcal{L}_{\gamma'} = 2i\frac{v_h}{\Lambda^2}\partial_\mu A_\nu\bar{\psi}\gamma^\mu\gamma^\nu\left[\text{Re}(C_\psi) + i\gamma_5\text{Im}(C_\psi)\right]\psi. \tag{8.49}$$

Using $\partial_\mu = iq_\mu$ for the operator, then

$$\mathcal{L}_{\gamma'} = -2\frac{v_h}{\Lambda^2}q_\mu A_\nu\bar{\psi}\gamma^\mu\gamma^\nu\left[\text{Re}(C_\psi) + i\gamma_5\text{Im}(C_\psi)\right]\psi. \tag{8.50}$$

The matrix elements $\bar{u}(p_f)\gamma^\mu\gamma^\nu u(p_i)$ and $\bar{u}(p_f)\gamma^\mu\gamma^\nu\gamma_5 u(p_i)$ in the scattering matrix need to be estimated. In the nonrelativistic limit, the q_μ can be approximated to only have a spacelike component. Therefore, the spinor products in the leading order become ($\pm$ corresponds to the particle 1 and 2 cases, respectively):

$$q_i \bar{u}(p_f) \gamma^i \gamma^0 u(p_i) = m q_i \left((1 + \tfrac{\sigma \cdot p}{2m}) \xi^\dagger, \ (1 - \tfrac{\sigma \cdot p}{2m}) \xi^\dagger \right)$$

$$\times \begin{pmatrix} 0 & \sigma^i \\ -\sigma^i & 0 \end{pmatrix} \begin{pmatrix} 0 & 1 \\ 1 & 0 \end{pmatrix} \begin{pmatrix} \left(1 - \dfrac{\sigma \cdot p}{2m}\right) \xi \\ \left(1 + \dfrac{\sigma \cdot p}{2m}\right) \xi \end{pmatrix} \tag{8.51}$$

$$= 2i \, p_{1,2} \cdot (q \times \sigma) \pm (\sigma \cdot q)^2,$$

and

$$q_i \bar{u}(p_f) \gamma^i \gamma^j u(p_i) = m q_i \left((\xi^\dagger, \ \xi^\dagger) \begin{pmatrix} 0 & \sigma^i \\ -\sigma^i & 0 \end{pmatrix} \begin{pmatrix} 0 & \sigma^j \\ -\sigma^j & 0 \end{pmatrix} \begin{pmatrix} \xi \\ \xi \end{pmatrix} \right) \tag{8.52}$$

$$= -2m(-i(q \times \sigma)_j).$$

In the case of $i \neq j$, the spinor products become

$$q_i \bar{u}(p_f) \gamma^i \gamma^0 \gamma_5 u(p_i) = m q_i \, (\xi^\dagger, \ \xi^\dagger) \begin{pmatrix} 0 & \sigma^i \\ -\sigma^i & 0 \end{pmatrix} \begin{pmatrix} 0 & 1 \\ 1 & 0 \end{pmatrix} \begin{pmatrix} -1 & 0 \\ 0 & 1 \end{pmatrix} \begin{pmatrix} \xi \\ \xi \end{pmatrix} \tag{8.53}$$

$$= -2m(\sigma \cdot q),$$

and

$$q_i \bar{u}(p_f) \gamma^i \gamma^j \gamma_5 u(p_i) = m q_i \left((1 + \tfrac{\sigma \cdot p}{2m}) \xi^\dagger, \ (1 - \tfrac{\sigma \cdot p}{2m}) \xi^\dagger \right)$$

$$\times \begin{pmatrix} -\sigma^i \sigma^j & 0 \\ 0 & -\sigma^i \sigma^j \end{pmatrix} \begin{pmatrix} -1 & 0 \\ 0 & 1 \end{pmatrix} \begin{pmatrix} \left(1 - \dfrac{\sigma \cdot p}{2m}\right) \xi \\ \left(1 + \dfrac{\sigma \cdot p}{2m}\right) \xi \end{pmatrix}$$

$$= -2(q \times (p \times \sigma))_j \ . \tag{8.54}$$

There are three different types of possible interactions mediated by massless spin-1 bosons: tensor-tensor, pseudotensor-tensor, and pseudotensor-pseudotensor. Details of concerning these interactions can be found in Ref. [24].

8.2.3.1 Tensor-Tensor Interaction

For the tensor-tensor interaction, the scattering matrix element is

$$iM(q) = \frac{4 v_h^2 \mathrm{Re}(C_1) \mathrm{Re}(C_2)}{\Lambda^4} [i q_i \bar{u}(p_{f,1}) \gamma^i \gamma^j u(p_{i,1})]$$

$$\times [iq_l \bar{u}(p_{f,2})\gamma^l \gamma^m u(p_{i,2})]\frac{-ig^{jm}}{q^2}. \qquad (8.55)$$

In the nonrelativistic limit, the scattering matrix becomes

$$iM(q) = \frac{4v_h^2 \mathrm{Re}(C_1)\mathrm{Re}(C_2)}{\Lambda^4}[-2m_1(\boldsymbol{q} \times \boldsymbol{\sigma}_1)]_j[-2m_2(\boldsymbol{q} \times \boldsymbol{\sigma}_2)]_m \frac{ig^{jm}}{|\boldsymbol{q}|^2}. \qquad (8.56)$$

Then the momentum-space potential is given by

$$V(q) = \frac{4v_h^2 \mathrm{Re}(C_1)\mathrm{Re}(C_2)}{\Lambda^4}\frac{e^{i\boldsymbol{q}\cdot\boldsymbol{r}}}{|\boldsymbol{q}|^2}(-(\boldsymbol{q} \times \boldsymbol{\sigma}_1) \cdot (\boldsymbol{q} \times \boldsymbol{\sigma}_2)). \qquad (8.57)$$

After the Fourier transform, the coordinate-space potential becomes

$$V_{TT}(r) = \frac{4v_h^2 \mathrm{Re}(C_1)\mathrm{Re}(C_2)}{4\pi \Lambda^4}\left[(\boldsymbol{\sigma}_1 \cdot \boldsymbol{\sigma}_2)\left(\frac{1}{r^3} - \frac{8\pi}{3}\delta(r)\right) - (\boldsymbol{\sigma}_1 \cdot \hat{\boldsymbol{r}})(\boldsymbol{\sigma}_2 \cdot \hat{\boldsymbol{r}})\frac{3}{r^3}\right]. \qquad (8.58)$$

8.2.3.2 Pseudotensor-Pseudotensor Interaction

For the pseudotensor-pseudotensor interaction, the leading order contribution comes from the timelike components. The corresponding scattering matrix is

$$iM(q) = \frac{4v_h^2 \mathrm{Im}(C_1)\mathrm{Im}(C_2)}{\Lambda^4}[i^2 q_i \bar{u}(p_{f,1})\gamma^i \gamma^0 \gamma_5 u(p_{i,1})]$$

$$\times [i^2 q_l \bar{u}(p_{f,2})\gamma^l \gamma^0 \gamma_5 u(p_{i,2})]\frac{-ig^{00}}{q^2}, \qquad (8.59)$$

In the nonrelativistic limit, it becomes

$$iM(q) = \frac{4v_h^2 \mathrm{Im}(C_1)\mathrm{Im}(C_2)}{\Lambda^4}[i^2(-2m_1)(\boldsymbol{\sigma}_1 \cdot \boldsymbol{q})][i^2(-2m_2)(\boldsymbol{\sigma}_2 \cdot \boldsymbol{q})]\frac{ig^{00}}{|\boldsymbol{q}|^2}. \qquad (8.60)$$

Then the momentum-space potential is given by

$$V(q) = \frac{4v_h^2 \mathrm{Im}(C_1)\mathrm{Im}(C_2)}{\Lambda^4}\frac{e^{i\boldsymbol{q}\cdot\boldsymbol{r}}}{|\boldsymbol{q}|^2}(\boldsymbol{q} \cdot \boldsymbol{\sigma}_1)(\boldsymbol{q} \cdot \boldsymbol{\sigma}_2), \qquad (8.61)$$

yielding the coordinate-space potential:

$$V_{\tilde{T}\tilde{T}}(r) = \frac{4v_h^2 \text{Im}(C_1)\text{Im}(C_2)}{4\pi\Lambda^4}\left[(\boldsymbol{\sigma}_1 \cdot \boldsymbol{\sigma}_2)\left(\frac{1}{r^3} + \frac{4\pi}{3}\delta(\boldsymbol{r})\right) - (\boldsymbol{\sigma}_1 \cdot \hat{\boldsymbol{r}})(\boldsymbol{\sigma}_2 \cdot \hat{\boldsymbol{r}})\frac{3}{r^3}\right].$$

$$(8.62)$$

8.2.3.3 Pseudotensor-Tensor Interaction

Unlike the tensor-tensor or pseudotensor-pseudotensor interaction, the leading order of the fermion current for the pseudotensor-tensor interaction depends on the fermion momenta. In the following calculation, we assume that the first vertex involves the pseudotensor coupling and the second vertex involves the tensor coupling. The pseudotensor term appears in leading order with the timelike component of the gamma matrix, while the tensor term appears in leading order with the spacelike component of the gamma matrix and is thus associated with momentum. Therefore, it is necessary to evaluate both the timelike and spacelike components in order to estimate the scattering amplitude.

Timelike Component: The timelike component of the scattering matrix is given by

$$M(q)_{00} = -i\frac{4v_h^2 \text{Im}(C_1)\text{Re}(C_2)}{\Lambda^4}[i^2 q_i \bar{u}(p_{f,1})\gamma^i \gamma^0 \gamma_5 u(p_{i,1})]$$

$$\times [iq_i \bar{u}(p_{f,2})\gamma^i \gamma^0 u(p_{i,2})]\frac{ig_{00}}{|q|^2} \qquad (8.63)$$

In the nonrelativistic limit, it becomes

$$M(q)_{00} \to -i\frac{4v_h^2 \text{Im}(C_1)\text{Re}(C_2)}{\Lambda^4}\frac{[-2m_1(\boldsymbol{\sigma}_1 \cdot \boldsymbol{q})][2i\,\boldsymbol{p}_2 \cdot (\boldsymbol{q} \times \boldsymbol{\sigma}_2) - (\boldsymbol{\sigma}_2 \cdot \boldsymbol{q})^2]}{|q|^2}.$$

$$(8.64)$$

Let us evaluate the above expression term by term. Note that the following calculations employ the path integral formalism, making use of the integration measure $\int \mathcal{D}q$ to carry out the functional integral over all possible trajectories (see, for example, discussion in Refs. [26, 27]). First,

$$\int \mathcal{D}q e^{iqr}\frac{1}{|q|^2}(\boldsymbol{\sigma}_1 \cdot \boldsymbol{q})\boldsymbol{p}_2 \cdot (\boldsymbol{q} \times \boldsymbol{\sigma}_2) = \sum_{i,l} p_2^i \epsilon^{ijk}\sigma_{2,k}\sigma_{1,l}\int \mathcal{D}q e^{iqr}\frac{q_j q_l}{|q|^2},$$

$$= \frac{1}{4\pi}\sum_{i,l} p_2^i \epsilon^{ijk}\sigma_{2,k}\sigma_{1,l}\left[\frac{\delta_{jl}}{r^3} - \frac{3r_j r_l}{r^5} + \frac{4\pi}{3}\delta_{jl}\delta(\boldsymbol{r})\right],$$

$$= \frac{1}{4\pi}\left[\frac{\boldsymbol{\sigma}_1 \cdot (\boldsymbol{\sigma}_2 \times \boldsymbol{p}_2)}{r^3} - 3\frac{\boldsymbol{p}_2 \cdot (\hat{\boldsymbol{r}} \times \boldsymbol{\sigma}_2)(\boldsymbol{\sigma}_1 \cdot \hat{\boldsymbol{r}})}{r^3} + \frac{4\pi}{3}\boldsymbol{\sigma}_1 \cdot (\boldsymbol{\sigma}_2 \times \boldsymbol{p}_2)\delta(\boldsymbol{r})\right],$$

$$= \frac{1}{4\pi}\left[\boldsymbol{p}_2 \cdot (\boldsymbol{\sigma}_1 \times \boldsymbol{\sigma}_2)\left(\frac{1}{r^3} + \frac{4\pi}{3}\delta(\boldsymbol{r})\right) + 3\frac{\boldsymbol{p}_2 \cdot (\boldsymbol{\sigma}_2 \times \hat{\boldsymbol{r}})(\boldsymbol{\sigma}_1 \cdot \hat{\boldsymbol{r}})}{r^3}\right].$$

$$(8.65)$$

Second,

$$
\int \mathcal{D}q e^{iqr} \frac{1}{|q|^2} (\boldsymbol{\sigma}_1 \cdot \boldsymbol{q})(\boldsymbol{\sigma}_2 \cdot \boldsymbol{q})^2 = \int \mathcal{D}q e^{iqr} \frac{1}{|q|^2} |q|^2 (\boldsymbol{\sigma}_1 \cdot \boldsymbol{q})
$$

$$
= -i \boldsymbol{\sigma}_1 \cdot \nabla \delta(\boldsymbol{r}).
$$

(8.66)

Inserting the expressions (8.65) and (8.66) into Eq. (8.64), we obtain

$$
\begin{aligned}
M(r)_{00} = &-\frac{4v_\hbar^2 \text{Im}(C_1)\text{Re}(C_2)m_1}{\pi \Lambda^4} \boldsymbol{p}_2 \cdot (\boldsymbol{\sigma}_1 \times \boldsymbol{\sigma}_2) \left(\frac{1}{r^3} + \frac{4\pi}{3}\delta(\boldsymbol{r}) \right) \\
&-\frac{4v_\hbar^2 \text{Im}(C_1)\text{Re}(C_2)m_1}{\pi \Lambda^4} \frac{3 \boldsymbol{p}_2 \cdot (\boldsymbol{\sigma}_2 \times \hat{\boldsymbol{r}})(\boldsymbol{\sigma}_1 \cdot \hat{\boldsymbol{r}})}{r^3} \\
&-\frac{8v_\hbar^2 \text{Im}(C_1)\text{Re}(C_2)m_1}{\Lambda^4} \boldsymbol{\sigma}_1 \cdot \nabla \delta(\boldsymbol{r}) .
\end{aligned}
$$

(8.67)

This timelike component of the pseudotensor-tensor potential can then be written as

$$
\begin{aligned}
V_{00}(r) = &\frac{4v_\hbar^2 \text{Im}(C_1)\text{Re}(C_2)}{8\pi \Lambda^4} (\boldsymbol{\sigma}_1 \times \boldsymbol{\sigma}_2) \left\{ -\frac{\boldsymbol{p}_2}{m_2}, \left(\frac{1}{r^3} + \frac{4\pi}{3}\delta(\boldsymbol{r}) \right) \right\} \\
&+\frac{4v_\hbar^2 \text{Im}(C_1)\text{Re}(C_2)}{8\pi \Lambda^4} \left\{ -\frac{\boldsymbol{p}_{2,i}}{m_2}, \frac{3(\boldsymbol{\sigma}_2 \times \hat{\boldsymbol{r}})_i (\boldsymbol{\sigma}_1 \cdot \hat{\boldsymbol{r}})}{r^3} \right\} \\
&-\frac{2v_\hbar^2 \text{Im}(C_1)\text{Re}(C_2)}{\Lambda^4} \frac{\boldsymbol{\sigma}_1 \cdot \nabla \delta(\boldsymbol{r})}{m_2} .
\end{aligned}
$$

(8.68)

Spacelike Component: The spacelike component of the scattering matrix is

$$
\begin{aligned}
M(q)_{lm} = &-i \frac{4v_\hbar^2 \text{Im}(C_1)\text{Re}(C_2)}{\Lambda^4} \\
&\times [i^2 q_i \bar{u}(p_{f,1}) \gamma^i \gamma^l \gamma_5 u(p_{i,1})][i q_i \bar{u}(p_{f,2}) \gamma^i \gamma^m u(p_{i,2})] \frac{i g_{lm}}{|q|^2} .
\end{aligned}
$$

In the nonrelativistic limit

$$
\begin{aligned}
M(q)_{jj} \rightarrow &-i \frac{4v_\hbar^2 \text{Im}(C_1)\text{Re}(C_2)}{\Lambda^4} \frac{[2m_2 i(\boldsymbol{q} \times \boldsymbol{\sigma}_2)_j][-2(\boldsymbol{q} \times (\boldsymbol{p}_1 \times \boldsymbol{\sigma}_1))_j]}{|q|^2} \\
&= -\frac{16v_\hbar^2 \text{Im}(C_1)\text{Re}(C_2)m_2}{\Lambda^4} \frac{[(\boldsymbol{q} \times \boldsymbol{\sigma}_2)_j][(\boldsymbol{q} \times (\boldsymbol{p}_1 \times \boldsymbol{\sigma}_1))_j]}{|q|^2} .
\end{aligned}
$$

(8.69)

Summing over the j index gives the inner product of the two vectors and yields

$$
(\boldsymbol{q} \times \boldsymbol{\sigma}_2) \cdot (\boldsymbol{q} \times (\boldsymbol{p}_1 \times \boldsymbol{\sigma}_1)) = (\boldsymbol{q} \cdot \boldsymbol{\sigma}_1)\boldsymbol{q} \cdot (\boldsymbol{\sigma}_2 \times \boldsymbol{p}_1) - (\boldsymbol{q} \cdot \boldsymbol{p}_1)(\boldsymbol{q} \cdot (\boldsymbol{\sigma}_2 \times \boldsymbol{\sigma}_1)) .
$$

(8.70)

Taking the Fourier transform with respect to the three-momentum $\boldsymbol{q}$, the first term gives

$$\boldsymbol{p}_1 \cdot (\boldsymbol{\sigma}_1 \times \boldsymbol{\sigma}_2) \left(\frac{1}{r^3} - \frac{4\pi}{3} \delta(\boldsymbol{r}) \right) + \frac{3(\boldsymbol{\sigma}_1 \cdot \hat{\boldsymbol{r}}) \boldsymbol{p}_1 \cdot (\boldsymbol{\sigma}_2 \times \hat{\boldsymbol{r}})}{r^3} . \tag{8.71}$$

From the conservation of energy-momentum, $\boldsymbol{q} \cdot \boldsymbol{p} = 0$, the second term in Eq. (8.70) vanishes. Thus combining the timelike and spacelike components together, the potential for the pseudotensor-tensor interaction is given by

$$
\begin{aligned}
V_{\tilde{T}T}(r) = &\frac{4v_\hbar^2 \text{Im}(C_1)\text{Re}(C_2)}{8\pi \Lambda^4} (\boldsymbol{\sigma}_1 \times \boldsymbol{\sigma}_2) \left\{ \frac{\boldsymbol{p}_1}{m_1} - \frac{\boldsymbol{p}_2}{m_2}, \left(\frac{1}{r^3} + \frac{4\pi}{3} \delta(\boldsymbol{r}) \right) \right\} \\
&+ \frac{4v_\hbar^2 \text{Im}(C_1)\text{Re}(C_2)}{8\pi \Lambda^4} \left\{ \frac{\boldsymbol{p}_1}{m_1} - \frac{\boldsymbol{p}_{2,i}}{m_2}, \frac{3(\boldsymbol{\sigma}_2 \times \hat{\boldsymbol{r}})_i (\boldsymbol{\sigma}_1 \cdot \hat{\boldsymbol{r}})}{r^3} \right\} \\
&- \frac{2v_\hbar^2 \text{Im}(C_1)\text{Re}(C_2)}{\Lambda^4} \frac{\boldsymbol{\sigma}_1 \cdot \nabla \delta(\boldsymbol{r})}{m_2} .
\end{aligned}
\tag{8.72}
$$

8.3 Searches for New Interactions Between Polarized Electrons and Unpolarized Nucleons

Several experiments have searched for the monopole-dipole interaction between polarized electrons and unpolarized nucleons mediated by axions or ALPs, described by the monopole-dipole interaction potential [Eq. (8.21)]

$$V_{\text{sp}}(\boldsymbol{r}) = \frac{\hbar^2 g_s^N g_p^e}{8\pi m_e} \left(\frac{1}{\lambda r} + \frac{1}{r^2} \right) e^{-r/\lambda} \boldsymbol{\sigma} \cdot \hat{\boldsymbol{r}} , \tag{8.73}$$

where $\boldsymbol{r}$ is the displacement vector between electron and nucleon, g_s^N and g_p^e are the scalar and pseudoscalar coupling constants of axions to the nucleon and to the electron, respectively, m_e is mass of electron, λ is the axion Compton wavelength, and $\boldsymbol{\sigma}$ is the spin unit vector. For example, Ni et al. tried to measure an induced magnetization in a paramagnetic salt located near a heavy copper mass [28]. Hammond et al. [29] observed the motion of three copper cylinders located between a source of polarized electrons consisting of two split toroidal electromagnets inside a magnetic shield system that enabled high spin polarization with negligible external magnetic field [30]. Youdin et al. searched for monopole-dipole couplings between a nearby lead mass and the spins of ^{133}Cs and ^{199}Hg atoms using co-located atomic magnetometers [31]. Torsion balances have been a successful method used for searches at the millimeter scale [32] as well as at Earth and solar system scales [33]. In this section we describe the working principles of three example experiments which have produced recent world-leading constraints: a torsion pendulum experiment, an electron-spin-resonance experiment, and a spectroscopy experiment using trapped ions [34].

8.3.1   Torsion Pendulum Experiments

The most stringent experimental constraint at distance scales below ≈ 1 mm on an axion-mediated interaction has been made with a torsion pendulum experiment by Hoedl et al. [32], which is a characteristic example of the sort of experimental method used to search for exotic interactions mediated by new bosons. The torsion pendulum apparatus consists of two parts: a split toroidal electromagnet and a planar torsion pendulum suspended between two magnet halves. The magnet halves are fixed in the apparatus (see Fig. 8.2). The pendulum is free to twist about the axis of the torsion fiber. The twist angle of the pendulum is optically monitored with an autocollimator [32]. The autocollimator measures small angles by comparing the position of a collimated laser beam that is reflected from the surface of the pendulum to the position of a reference laser beam. An axion-mediated force between the polarized electrons in the electromagnet and the unpolarized silicon atoms in the pendulum would generate a torque on the pendulum which is given by $g_s^N g_p^e G(x, \lambda_a)$, where $G(x, \lambda_a)$ is a geometrical factor and x is the distance

Fig. 8.2 Schematic diagram of a torsion pendulum experiment designed to search for a monopole-dipole coupling between electron spins and nucleons. Two toroidal electromagnet halves source polarized electrons. A planar torsion pendulum suspended between the two magnet halves sources unpolarized nucleons. The pendulum is freely suspended by a tungsten wire. Adapted from Ref. [32]

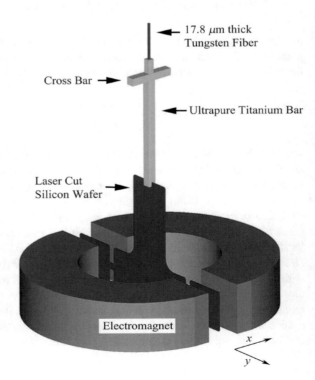

between the pendulum and the symmetry plane between the magnet halves. This interaction acts like an effective "magnetic field" (pseudo-magnetic field) to generate torque on the pendulum, thereby behaving as a torsion spring. When the effective magnetic field is switched from one direction to the other direction by changing the polarization of the electron spins in the electromagnet, if there is a new macroscopic interaction in the form of Eq. (8.73), the pendulum experiences a torque. Thus a spin-dependent interaction could be detected by measuring the change in the equilibrium twist angle of the torsion pendulum. The strength of the interaction depends on the distance between pole and pendulum. Figure 8.3 shows the exclusion limit based on this torsion balance experiment (red dashed line labeled Hoedl et al. [32]), as well as constraints from other experiments [35].

Although $g_s^N g_p^e$ would be very small for QCD axions due to the fact that $g_s^N \propto \theta_{QCD}$, where θ_{QCD} is the CP-violating phase appearing in the Lagrangian describing the strong interaction (see discussion in Chap. 2), this experiment has the advantage compared to axion haloscope searches (as discussed in, for example, Chaps. 4 and 6) that axions are sourced directly from the local object. Thus the effect would exist

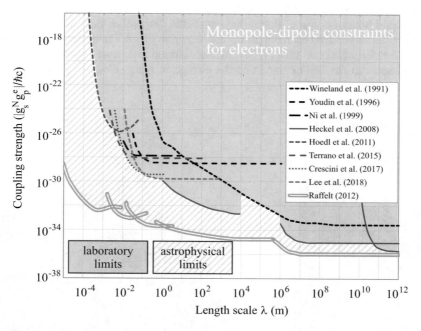

Fig. 8.3 Constraints on monopole-dipole couplings between nucleons and electrons $|g_s^N g_p^e|/(\hbar c)$ from laboratory experiments and astrophysical observations, adapted and updated from Ref. [35]. Constraints from experiments discussed in this chapter include those from Ref. [32], shown by the red dashed line labeled Hoedl et al. [32], Ref. [36], shown by the red dotted line labeled Crescini et al. (2017) [36], Ref. [34], shown by the black short-dashed line labeled Wineland et al. [19], and Ref. [37], shown by the dashed blue line labeled Lee et al. [37]. Further discussion of other constraints can be found in Ref. [35]

even if the axion is not the dominant component of dark matter, and means that the signal is not subject to the myriad uncertainties affecting the interpretation of limits from haloscope experiments due to the unknown local distribution of dark matter (see, for example, discussion in Chaps. 3 and 10). Because the axion-mediated interaction is locally sourced in this experiment, and, in fact, all the experiments considered in this chapter, the axion-induced signal can be purposefully modulated in a controlled fashion, making it potentially easier to distinguish from noise.

8.3.2 Electron-Spin Based Magnetometer Searches

Recent magnetometry experiments have provided the best constraints on monopole-dipole couplings of electron spins at distances of order 1–10 cm [36]. The QUAX-$g_s g_p$ experiment described in Ref. [36] is an adaptation of the QUAX experiment (QUest for AXions) to search for monopole-dipole interactions between an unpolarized source mass and the electron spins in a paramagnetic gadolinium oxy-orthosilicate $Gd_2 SiO_5$ crystal (GSO) crystal. The GSO crystal is cooled down to 4K in a liquid helium cryostat. Figure 8.4 shows the diagram of the setup; the derived constraints on electron-spin couplings are shown in Fig. 8.3 with the red dotted line labeled as Crescini et al. (2017) [36]. An unpolarized lead source mass is spun a few cm away from the GSO crystal and the change in magnetization is read out with

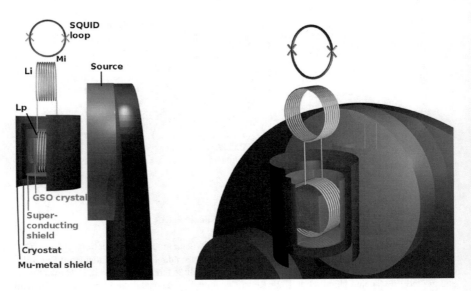

Fig. 8.4 Schematic diagram of the setup for the short-range spin-dependent force search of the QUAX-$g_s g_p$ experiment. An unpolarized lead source mass wheel is spun in proximity to a paramagnetic GSO crystal, and the resulting change in its polarization is read out with a SQUID magnetometer. Figure from Ref. [36]

a dc superconducting quantum interference device (SQUID) magnetometer. The setup includes concentric superconducting shields placed within an outermost μ-metal shield in order to reduce the flux trapped in the inner superconducting shields. The overall rejection factor of magnetic shield system is expected to be $\sim 10^{12}$, significantly reducing environmental magnetic disturbances. The distance between the center of mass of each lead source and the GSO crystal is modulated in time by mounting the masses on a rotating 70 cm diameter aluminum disk that rotates at a constant angular velocity. The minimum distance between each source and detector is 3.7 cm.

8.3.3 *Spectroscopic Constraints with Trapped Ions*

Constraints on novel electron monopole-dipole interactions have also been obtained using hyperfine spectroscopy of trapped and cooled ^{9}Be$^+$ ions in experiments performed by Wineland et al. [34]. Here by reversing the magnetic field along the direction of the Earth's gravitational field, novel spin-dependent frequency shifts can be constrained. Here the source of the interaction was assumed to be nucleons in the Earth and the resulting frequency shift between two Zeeman sublevels within the electronic ground state hyperfine manifold was measured. In the absence of novel spin-dependent interactions, the field reversal should result in no frequency shift. The frequency shift was determined to be less than $\sim 13.4\,\mu$Hz, resulting in the limits shown by the black short-dashed line in Fig. 8.3.

8.4 Monopole-Dipole Searches with Polarized Nuclear Spins and Unpolarized Nucleons

Several experimental techniques have been employed to search for novel spin-dependent monopole-dipole interactions V_{sp} between nucleons. Some experiments have used ultra-cold neutrons (UCNs) and ^{3}He to test spin-dependent interactions between polarized and unpolarized nucleons. Baessler et al. tried to find a deviation from the expected energy levels of UCNs in the Earth's gravitational field [38]. Serebrov et al. searched for a change in the UCN precession frequency due to a spin-dependent interaction [39]. Pethkhov et al. looked for a change in the spin relaxation time of ^{3}He induced by exotic spin-dependent interactions [40]. Fu et al. also set a constraint on the monopole-dipole interaction [41] by reanalyzing existing data on the spin relaxation times of polarized ^{3}He in the context of exotic spin-dependent interactions [42]. Here we describe in more detail a few examples based on magnetometry and nuclear magnetic resonance (NMR), including experiments under development.

8.4.1 Axion Searches with Comagnetometers

A powerful technique to search for axion-mediated interactions is to measure variations in the nuclear Larmor precession frequency as a source mass is brought near a polarized atomic vapor. There have been many successful approaches, for example, relying on comagnetometry with multiple nuclear spin species [43–46], nuclear spins along with alkali gases [37], and single-species liquid comagnetometers based on NMR measurements of different nuclei in identical molecules [47].

8.4.1.1 Noble Gas Comagnetometer

A method to search for non-magnetic, spin-dependent interactions is to use a sensitive low-field comagnetometer based on detection of free spin precession of gaseous, nuclear polarized samples [45]. The idea is to measure spin precession of two species, ^{3}He and ^{129}Xe gas, in the same volume. The Larmor frequencies of ^{3}He and ^{129}Xe in a guiding magnetic field B are given by $\omega_{L,\mathrm{He(Xe)}} = \gamma_{\mathrm{He(Xe)}} B$, with $\gamma_{\mathrm{He(Xe)}}$ being the gyromagnetic ratios of the respective gas species with $\gamma_{\mathrm{He}}/\gamma_{\mathrm{Xe}} = 2.75408159$. The goal of employing a comagnetometer is to separate out background magnetic fields and drifts from any anomalous spin-dependent interactions. One seeks to establish a signal which will vanish for ordinary magnetic fields but be sensitive to new physics. In practice perfect subtraction of ordinary backgrounds is challenging for several reasons, but the technique is quite powerful and has produced constraints on monopole-dipole interactions between nuclei at a variety of laboratory scale distances.

The influence of the ambient magnetic field and its temporal fluctuations cancels in the difference of measured Larmor frequencies of the co-located spin samples

$$\Delta\omega = \omega_{\mathrm{He}} - \frac{\gamma_{\mathrm{He}}}{\gamma_{\mathrm{Xe}}}\omega_{\mathrm{Xe}} \,. \tag{8.74}$$

This frequency shift in Eq. (8.74) can be separated into three parts [45]:

$$\Delta\omega(t) = \Delta\omega_{\mathrm{lin}} + \epsilon_{\mathrm{He}} A_{\mathrm{He}} e^{-t/T^*_{2,\mathrm{He}}} - \epsilon_{\mathrm{Xe}} A_{\mathrm{Xe}} e^{-t/T^*_{2,\mathrm{Xe}}} \,, \tag{8.75}$$

where $\epsilon_{\mathrm{He(Xe)}}$ are the respective geometry-dependent factors describing self-field effects, $T^*_{2,\mathrm{He(Xe)}}$ are the respective effective spin coherence relaxation times, and $A_{\mathrm{He(Xe)}}$ are the constants describing the amplitude of the spin polarization of the respective species. The first part is a constant frequency shift, $\Delta\omega_{\mathrm{lin}}$, due to Earth's rotation which is not compensated by comagnetometry. This effect is commonly referred to as the gyro-compass effect and is nicely demonstrated and well-described in Ref. [48]. The other two parts are related to the generalized Ramsey-Bloch-Siegert shift [49–51] arising from the self-fields caused by the precessing ^{3}He/^{129}Xe nuclear spins appearing in nonspherical vapor cells [52]. In the same manner,

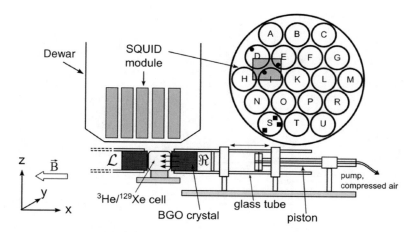

Fig. 8.5 Schematic diagram of the experimental setup from Ref. [45] used to search for anomalous monopole-dipole interactions. The ^{3}He/^{129}Xe cell is located at the center. The BGO crystal is placed on right (or left) side with respect to the sample. The crystal travels back and forth toward the cell with a certain frequency. The SQUID module located on the top of the cell monitors the precession frequency of each species. Adapted from Ref. [45]

the weighted accumulated phase difference acquired during precession, $\Delta\Phi(t) = \Phi_{He}(t) - (\gamma_{He}/\gamma_{Xe})\Phi_{Xe}$, can also be measured. Any anomalous frequency shift generated by non-magnetic spin interactions, such as the monopole-dipole interaction described in Eq. (8.21), could be analyzed by monitoring $\Delta\omega(t)$ and $\Delta\Phi(t)$, respectively (Fig. 8.5).

The experiment was done inside the magnetically shielded room (MSR) at the Physikalisch-Technische Bundesanstalt Berlin (PTB). A homogeneous magnetic guide field of ~350nT was provided in the MSR. The detection of spin precession was done with multi-channel low T_c dc SQUID device. The SQUID sensor detects a sinusoidal magnetic flux change due to the nuclear spin precession of the gas. The spin precession frequency shift due to any monopole-dipole interaction is induced by an unpolarized mass with high nucleon density. In this experiment, a cylindrical BGO crystal ($Bi_4Ge_3O_{12}$), which is non-conductive and non-magnetic ($\chi_{mag} \approx 0$), with nucleon density ($\rho = 7.13\,g/cm^3$) was used. The BGO crystal source mass is alternately moved to the left and right side of the ^{3}He/Xe cell. If a sufficiently strong monopole-dipole interaction exists, the movement of the BGO crystal would produce a frequency shift correlated with the motion of the source mass. In the case of a non-zero spin-dependent interaction, a shift $\Delta\omega_{sp}^w$ in the weighted frequency difference described by Eq. (8.74) can be extracted from respective frequency measurements in the "close" and "distant" BGO positions given by

$$\Delta\omega_{sp}^w = \frac{2V_\Sigma^c}{\hbar}\left(1 - \frac{\gamma_{He}}{\gamma_{Xe}}\right). \tag{8.76}$$

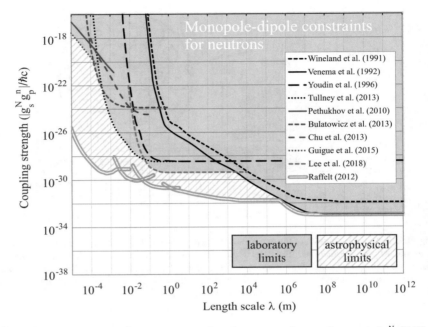

Fig. 8.6 Constraints on monopole-dipole couplings between nucleons and neutrons $|g_s^N g_p^n|/(\hbar c)$ from laboratory experiments and astrophysical observations, adapted and updated from Ref. [35]. Constraints from experiments discussed in this chapter include those from Ref. [45], shown by the black dotted line labeled Tullney et al. [45], Ref. [46], shown by the red dashed line labeled Bulatowicz et al. [46], Ref. [37], shown by the dashed blue line labeled Lee et al. [37], and Ref. [54], shown by the long-dashed red line labeled Chu et al. [54]. Further discussion of other constraints can be found in Ref. [35]

Here the average potential in the "close" BGO position, V_Σ^c, is obtained by integration of the monopole-dipole interaction potential V_{sp} over the volume of the massive unpolarized sample averaged over the volume of the polarized spin sample, each having a cylindrical shape [45]. The estimation of V_Σ^c assumes that the He and Xe nuclei can be described by the single-particle Schmidt model, see, for example, discussion in Ref. [53]. The constraints derived from this experiment are shown by the black dotted line labeled Tullney et al. [45] in Fig. 8.6.

A different experiment was performed with a dual-species comagnetometer employing ^{129}Xe and ^{131}Xe and using Rb as an optical magnetometer for readout of the nuclear spin precession [46]. The Rb atoms are optically spin polarized and, through spin-exchange collisions, polarize the Xe nuclei parallel to a dc magnetic field. The Rb atoms serve as a magnetometer that detects the Xe precession since the transverse magnetic fields of the polarized Xe produce an oscillating transverse spin polarization of the Rb atoms. This is detected optically as a rotation of the polarization of a linearly polarized sense laser [46]. The setup is shown in Fig. 8.7 and limits on scalar-pseudoscalar interactions are derived in the distance range of

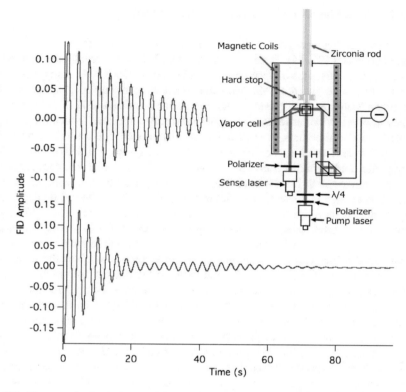

Fig. 8.7 The setup for the dual-species free-induction decay (FID) comagnetometer used to search for mm-scale monopole-dipole interactions. Sample FID data, where Xe spin precession was optically detected via laser light probing Rb spins co-located with the Xe spins, are shown (FID oscillations are frequency down-converted to $\approx 0.3\,\mathrm{Hz}$ from 45 Hz or 152 Hz). The unpolarized source is the movable zirconia rod, which can induce frequency shifts of the Xe spin precession if a monopole-dipole interaction of the form given by Eq. (8.21) exists. Figure from Ref. [46]

approximately 1 mm. NMR frequency shifts in polarized ^{129}Xe and ^{131}Xe could arise due to a monopole-dipole interaction when a zirconia rod was moved back and forth near the NMR cell. By comparing the simultaneous frequencies of the two Xe isotopes, magnetic field changes are distinguished from frequency shifts due to the monopole-dipole coupling using the same principle of comagnetometry discussed in the previous paragraphs. Using prior calculations of the neutron spin contribution to the nuclear angular momentum in ^{129}Xe and ^{131}Xe, a new upper bound on the product $g_s^N g_p^n$ for was obtained for ranges at the millimeter scale as shown by the red dashed line labeled Bulatowicz et al. [46] in Fig. 8.6.

8.4.1.2   Noble Gas: Alkali Comagnetometer Searches

Searches for new spin-dependent interactions have also been performed with K-^{3}He comagnetometers [37, 55]. Recent work was done with a moveable unpolarized Pb source mass at a distance of approximately 15 cm from the K-^{3}He comagnetometer in order to search for both electron- and nuclear-spin-coupled interactions that arise when there are both pseudoscalar and scalar couplings [37]. The experiment employed overlapping ensembles of spin-polarized K and ^{3}He, which are strongly coupled via Fermi-contact interactions during spin-exchange collisions when the resonant frequencies of K and ^{3}He are matched. This allows effective cancellation of magnetic fields and fast transient response [37]. In particular, the cancellation is for fields transverse to the external field due to the adiabatic following of the ^{3}He spins. (The system is insensitive to fields along the applied external field, because those do not create torques on the spins initially aligned along the leading field.)

Rubidium is used for spin-exchange optical pumping (SEOP), which polarizes the K by collisions, and Rb and K-^{3}He spin-exchange collisions polarize ^{3}He to approximately 2%. The comagnetometer signal is proportional to the difference of the anomalous magnetic-like field couplings to the nuclear spin in ^{3}He and the electron spin in K. The He-3 adiabatically follows the magnetic field and the leading field is tuned so that its effect on the K spins exactly balances that of the He-3 magnetization, making the comagnetometer insensitive to regular magnetic fields to first order. The rotation rate of the apparatus about the y axis represents an example of a non-magnetic coupling to spin that does not cancel in the comagnetometer. A challenge in this experiment was that the motion of the Pb source masses produced a subtle mechanical effect due to temperature changes correlated with the positions of the masses [37]. Constraints derived from this experiment are shown in both Figs. 8.3 and 8.6 by the dashed blue line labeled Lee et al. [37].

8.4.2   NMR-Based Spin-Dependent Searches

A sub-mm-range search was performed using room temperature polarized ^{3}He gas in a cell with a 250 μm thick window and unpolarized source mass [54]. The experimental diagram is shown in Fig. 8.8. The cylindrical ^{3}He cell is located in a uniform magnetic field. Correction coils compensate for residual leading field gradients. ^{3}He is polarized using spin-exchange optical pumping in the spherical pumping chamber, and polarized ^{3}He atoms diffuse into the lower 40-cm long cylindrical chamber, which has two hemispherical glass windows at both ends. Two pick-up coils are used: pick-up coil A is mounted below the window to measure the precession frequency shift of the polarized ^{3}He nuclei due to spin-dependent short-range interactions with the unpolarized mass. Pick-up coil B is positioned farther away to be insensitive to short-range interactions, and its signal is used to monitor the leading field drift and background fields. The frequencies measured in both coils are subtracted for each measurement. The ^{3}He cell position is adjusted

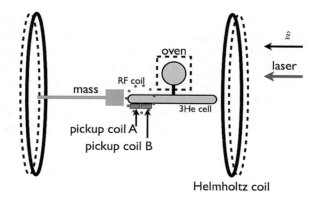

Fig. 8.8 Experimental setup for the NMR measurement with a ^{3}He sample cell, polarizing cell (spherical), polarizing laser, Helmholtz coils, and source mass. Precession of the polarized ^{3}He nuclei is measured by the induced EMF in the pick-up coils. Figure from Ref. [54]

to optimize the transverse spin relaxation time measured from coils A and B. The leading field is tuned to produce a ^{3}He Larmor frequency near 23.8 kHz and the authors apply a 24 kHz radiofrequency (RF) pulse to tip the spins by a small angle with negligible polarization loss. The precessing polarized ^{3}He nuclei induce an electromotive forces (EMF) in the pick-up coils which is recorded. Two source masses for the experiment are used having differing nucleon densities and low magnetic impurities: a ceramic mass block and a liquid mixture of 1.02% MnCl$_2$ in pure water. Constraints derived from this experiment are shown in Fig. 8.6 by the long-dashed red line labeled Chu et al. [54].

8.4.3   Resonant NMR-Based Spin-Dependent Interaction Search: ARIADNE

The Axion Resonant InterAction DetectioN Experiment (ARIADNE) aims to detect axion-mediated spin-dependent interactions between an unpolarized source mass and a spin-polarized ^{3}He low-temperature gas [56]. As previously noted in the discussion around Eq. (8.21), the axion can mediate an interaction between fermions (e.g., nucleons) with a potential given by

$$V_{sp}(r) = \frac{\hbar^2 g_s^N g_p^N}{8\pi m_f} \left(\frac{1}{r\lambda_a} + \frac{1}{r^2} \right) e^{-\frac{r}{\lambda_a}} (\boldsymbol{\sigma} \cdot \hat{\boldsymbol{r}}), \tag{8.77}$$

where m_f is their mass, $\boldsymbol{\sigma}$ is the Pauli spin matrix, $\boldsymbol{r}$ is the vector between them, and $\lambda_a = h/(m_a c)$ is the axion Compton wavelength. For the QCD axion the scalar and dipole coupling constants g_s^N and g_p^N are related to the axion mass. Since the axion couples to $\boldsymbol{\sigma}$, which is proportional to the nuclear magnetic moment, the axion coupling can be treated as an effective "magnetic field" $\boldsymbol{B}_{\text{eff}}$ (i.e., a pseudo-magnetic field). This effective field is used to resonantly drive spin precession in a laser-polarized cold ^{3}He gas. This is accomplished by spinning an unpolarized tungsten mass sprocket near the ^{3}He vessel. As the teeth of the sprocket pass by the sample at the nuclear Larmor precession frequency, the magnetization in the longitudinally polarized He gas begins to precess about the axis of an applied field. This precessing transverse magnetization is detected with a SQUID. The ^{3}He sample acts as an amplifier to transduce the small effective magnetic field into a larger real magnetic field detectable by the SQUID.

Integrating over the source mass, via Eq. (8.77), an axion with $\lambda_a < R$ will generate a potential a distance r from the surface of the source mass

$$V_a(r) \approx \frac{g_s^N g_p^N}{2m_N} \lambda_a^2 n_N e^{-\frac{r}{\lambda_a}} , \tag{8.78}$$

where m_N and n_N are the nucleon mass and density of the material, respectively. Here we assume the NMR sample thickness is of order λ_a and the source mass surface is effectively flat. A spin-polarized nucleus near this rotating sprocket will feel an effective magnetic field of approximately

$$\boldsymbol{B}_{\text{eff}} \approx \frac{1}{\hbar \gamma_N} \nabla V_a(r)(1 + \cos(n\omega_{\text{rot}}t)) , \tag{8.79}$$

where γ_N is the nuclear gyromagnetic ratio and n is the number of segments, for a sample thickness of order λ_a. $\boldsymbol{B}_{\text{eff}}$ is parallel to the radius of the sprocket.

The NMR sample with net polarization M_z parallel to the axis of the sprocket (and a Larmor frequency $2\boldsymbol{\mu}_N \cdot \boldsymbol{B}_{\text{ext}}/\hbar = \omega$ determined by an axial field $\boldsymbol{B}_{\text{ext}}$) will develop a time-varying perpendicular magnetization M_x in response to the resonant effective axion field B_{eff} given by

$$M_x(t) \approx \frac{1}{2} n_s p \mu_N \gamma_N B_{\text{eff}} T_2 \left(e^{-t/T_1} - e^{-t/T_2} \right) \cos(\omega t) , \tag{8.80}$$

where p is the polarization fraction, n_s is the spin density in the sample, and μ_N is the nuclear magnetic moment. $M_x(t)$ grows approximately linearly with time until $t \sim T_2$, the transverse relaxation time, and then decays at the longer longitudinal relaxation time T_1. $M_x(t)$ can be detected by a SQUID with its pick-up coil axis oriented radially. Note the SQUID detects the changing magnetization of the sample, not the axion field itself (which is not a "real" magnetic field and thus does not affect the SQUID reading directly).

Superconducting shielding is needed around the sample to screen it from ordinary magnetic field noise which would otherwise limit the sensitivity of the measurement. The ultimate limit is set by spin-projection noise (SPN) in the sample itself [56], given as

$$\sqrt{M_N^2} = \sqrt{\frac{\hbar \gamma n_s \mu_{\text{He}} T_2}{2V}} \tag{8.81}$$

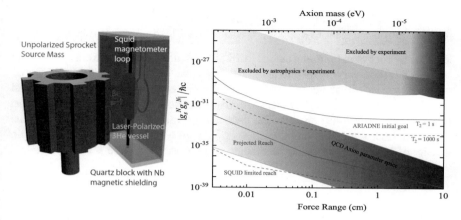

Fig. 8.9 (left) Setup: a sprocket-shaped source mass is rotated so its "teeth" pass near an NMR sample at its resonant frequency. (right) Projected reach for monopole-dipole axion-mediated interactions. The band bounded by the red (dark) solid line and dashed line denotes the limit set by transverse magnetization noise, depending on achieved T_2. Constraints and expectations for the QCD axion also are shown, adapted from Refs. [37, 56]

and the minimum transverse magnetic resonant field detectable with this setup is given by:

$$B_{\min} \approx p^{-1} \sqrt{\frac{2\hbar b}{n_s \mu_{He} \gamma V T_2}} = 3 \times 10^{-19} \text{ T} \qquad (8.82)$$

$$\times \left(\frac{1}{p}\right) \sqrt{\left(\frac{b}{1\,\text{Hz}}\right) \left(\frac{1\,\text{mm}^3}{V}\right) \left(\frac{10^{21}\,\text{cm}^{-3}}{n_s}\right) \left(\frac{1000\,\text{s}}{T_2}\right)}.$$

Here V is the sample volume, γ is the gyromagnetic ratio for ^{3}He $= (2\pi) \times 32.4$ MHz/T, b is the measurement bandwidth, and $\mu_{He} = -2.12 \times \mu_n$ is the ^{3}He nuclear moment, where μ_n is the nuclear Bohr magneton. The estimated SQUID magnetometer limited sensitivity is shown in Fig. 8.9.

The experiment sources the axion in the lab, and can explore all mass ranges in our sensitivity band simultaneously, unlike experiments which must scan over the allowed axion oscillation frequencies (masses) by tuning a cavity (e.g., as described in Chap. 4) or magnetic field (e.g., as described in Chap. 6). Distinct from other magnetometry experiments [37, 46, 54], the experiment uses a resonant enhancement technique. Assuming sources of systematic error and noise can be mitigated, the approach is expected to be spin-projection noise limited, and in principle allows several orders of magnitude improvement, yielding sufficient sensitivity to detect the QCD axion (Fig. 8.9). In principle, an experiment like ARIADNE could be adapted to use a polarized source mass, in order to search for anomalous dipole-dipole interactions [56]. Using a polarized source mass, however, increases the need for the screening of ordinary magnetic interactions.

? Problem 8.4 Magnetic Field "amplification factor" for a Magnetized NMR Sample Subject to an Effective Axion-Induced "magnetic field"

Within an order of magnitude, using Eq. (8.80), calculate the approximate amplitude of the (real) time-varying magnetic field B_{SQUID} that would be detected by a SQUID pick-up loop at a distance of 2 mm from the center of a 1-mm-radius spherical sample. Assume the induced transverse magnetization M_x is driven for a duration of T_2 by an axion with an effective field B_{eff}. Evaluate your expression for the dimensionless "amplification factor" ($B_{\mathrm{SQUID}}/B_{\mathrm{eff}}$) with $T_2 = 1000$ seconds, a spin density of 10^{21} spins per cubic centimeter, and unity polarization $p = 1$.

Solution on page 345.

8.5   Spectroscopic Measurements of Spin-Spin Coupled Interactions

Comparison of precision spectroscopy of atoms [57] and molecules [58] with theoretical expectations allows one to place stringent constraints on new exotic interactions at atomic scales. As an example, we consider here a recent experiment that has resulted in orders of magnitude improvement for constraints on the existence of anomalous dipole-dipole forces on angstrom length scales [58]. Constraints were obtained by comparison of NMR measurements and theoretical calculations of J-coupling in deuterated molecular hydrogen (HD). Such couplings have the form $J\boldsymbol{I} \cdot \boldsymbol{S}$ (here, $\boldsymbol{I}$ and $\boldsymbol{S}$ are nuclear spin operators) and arise due to a second-order hyperfine interaction. Exotic spin-spin interactions mediated by new bosons, described, for example, by Eqs. (8.28), (8.40), and (8.47), also contain terms proportional to $\boldsymbol{I} \cdot \boldsymbol{S}$ that can lead to a shift ΔJ of the J-coupling. Experimentally measured J-coupling is in good agreement with theoretical calculations [58], ruling out novel angstrom-range anomalous spin-dependent forces at a level several orders of magnitude better than prior constraints from molecular beam measurements [59] (Fig. 8.10).

8.6   Outlook

Future prospects for improvements in the search for novel spin-dependent interactions are promising with new cryogenic and quantum technologies. Cryogenic torsion balance technology could provide substantial gains beyond the thermal noise limit for spin-dependent torque experiments. Spin squeezing or coherent collective modes could offer prospects for improved sensitivity beyond the standard quantum limit of spin-projection noise in experiments such as ARIADNE, potentially

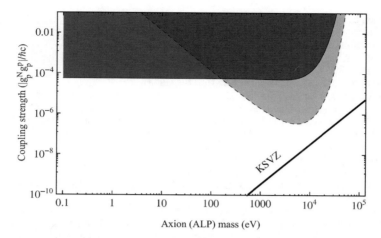

Fig. 8.10 Limits on dipole-dipole couplings between protons and nucleons $|g_p^N g_p^P|/(\hbar c)$, described by Eq. (8.28), derived from comparison of NMR measurements and theory for J-coupling in deuterated molecular hydrogen [58], shown by the light gray shaded region, along with limits from molecular beam experiments [59], shown by the dark gray shaded region. The prediction for dipole-dipole couplings mediated by the QCD axion in the Kim-Shifman-Vainshtein-Zakharov (KSVZ) model [18] is also shown for comparison. Figure adapted from Ref. [58]

allowing sensitivity all the way down to the SQUID-limited sensitivity (dashed-dotted line in Fig. 8.9). This would allow one to rule out the axion over a wide range of masses, and when combined with other promising techniques [60–62], and existing experiments [63, 64] already sensitive to QCD axions, could, in principle, allow the QCD axion to be searched for over its entire allowed mass range.

Acknowledgments AG is supported in part by NSF grants PHY-1806686 and PHY-1806671, the Heising-Simons Foundation, the John Templeton Foundation, the W. M. Keck foundation, and ONR Grant N00014-18-1-2370. Yun Chang Shin acknowledges support from the Institute for Basic Science under Grant No. IBS-R017-D1-2021-a00. We thank Derek F. Jackson Kimball, W. M. Snow, Younggeun Kim and DongOk Kim for discussions.

References

1. B.A. Dobrescu, I. Mocioiu, J. High Energy Phys. **11**, 5 (2006)
2. G. Bertone, D. Hooper, J. Silk, Phys. Rep. **405**, 279 (2005)
3. J.A. Frieman, M.S. Turner, D. Huterer, Annu. Rev. Astron. Astrophys. **46**, 385 (2008)
4. S. Perlmutter, et al., Astrophys. J. **517**, 565 (1999)
5. S. Weinberg, Rev. Mod. Phys. **61**, 1 (1989)
6. J.G. de Swart, G. Bertone, J. van Dongen, Nat. Astron. **1**, 0059 (2017)
7. S. Weinberg, Phys. Rev. Lett. **40**, 223 (1978)
8. A.A. Anselm, ZhETF Pisma Redaktsiiu **36**, 46 (1982)
9. A. Vilenkin, A.E. Everett, Phys. Rev. Lett. **48**, 1867 (1982)

10. J. Kim, Universe **3**, 68 (2017)
11. Y. Chikashige, Y. Fujii, K. Mima, Prog. Theo. Phys. **59**, 274 (1978)
12. P. Langacker, R.D. Peccei, T. Yanagida, Mod. Phys. Lett. A **01**, 541 (1986)
13. F. Wilczek, Phys. Rev. Lett. **49**, 1549 (1982)
14. R.D. Peccei, H.R. Quinn, Phys. Rev. Lett. **38**, 1440 (1977)
15. R.D. Peccei, H.R. Quinn, Phys. Rev. D **16**, 1791 (1977)
16. C. Haddock, J. Amadio, E. Anderson, L. Barrón-Palos, B. Crawford, C. Crawford, D. Esposito, W. Fox, I. Francis, J. Fry, H. Gardiner, A. Holley, K. Korsak, J. Lieffers, S. Magers, M. Maldonado-Velázquez, D. Mayorov, J.S. Nico, T. Okudaira, C. Paudel, S. Santra, M. Sarsour, H.M. Shimizu, W.M. Snow, A. Sprow, K. Steffen, H.E. Swanson, F. Tovesson, J. Vanderwerp, P.A. Yergeau, Phys. Lett. B **783**, 227 (2018)
17. S. Weinberg, Phys. Rev. Lett. **29**, 1698 (1972)
18. J.E. Moody, F. Wilczek, Phys. Rev. D **30**, 130 (1984)
19. D.J. Wineland, J.J. Bollinger, D.J. Heinzen, W.M. Itano, M.G. Raizen, Phys. Rev. Lett. **67**, 1735 (1991)
20. G. Vasilakis, J.M. Brown, T.W. Kornack, M.V. Romalis, Phys. Rev. Lett. **103**, 261801 (2009)
21. A.G. Glenday, C.E. Cramer, D.F. Phillips, R.L. Walsworth, Phys. Rev. Lett. **101**, 261801 (2008)
22. A.N. Youdin, J. Krause, D., K. Jagannathan, L.R. Hunter, S.K. Lamoreaux, Phys. Rev. Lett. **77**, 2170 (1996)
23. S. Aldaihan, D.E. Krause, J.C. Long, W.M. Snow, Phys. Rev. D **95**, 096005 (2017)
24. P. Fadeev, Y.V. Stadnik, F. Ficek, M.G. Kozlov, V.V. Flambaum, D. Budker, Phys. Rev. A **99**, 022113 (2019)
25. F. Ficek, D. Budker, Annalen der Physik **531**, 1800273 (2018)
26. T. Lancaster, S.J. Blundell, *Quantum Field Theory for the Gifted Amateur* (Oxford University Press, Oxford, 2014)
27. A. Zee, *Quantum Field Theory in a Nutshell*, vol. 7 (Princeton University Press, Princeton, 2010)
28. W.T. Ni, S.s. Pan, H.C. Yeh, L.S. Hou, J. Wan, Phys. Rev. Lett. **82**, 2439 (1999)
29. G.D. Hammond, C.C. Speake, C. Trenkel, A.P. Patón, Phys. Rev. Lett. **98**, 081101 (2007)
30. G. Hammond, A.P. Patón, C. Speake, C. Trenkel, G. Rochester, D. Shaul, T. Sumner, Phys. Rev. D **77**, 036005 (2008)
31. A.N. Youdin, D. Krause, Jr., K. Jagannathan, L.R. Hunter, S.K. Lamoreaux, Phys. Rev. Lett. **77**, 2170 (1996)
32. S.A. Hoedl, F. Fleischer, E.G. Adelberger, B.R. Heckel, Phys. Rev. Lett. **106**, 041801 (2011)
33. B.R. Heckel, E.G. Adelberger, C.E. Cramer, T.S. Cook, S. Schlamminger, U. Schmidt, Phys. Rev. D **78**, 092006 (2008)
34. D.J. Wineland, J.J. Bollinger, D.J. Heinzen, W.M. Itano, M.G. Raizen, Phys. Rev. Lett. **67**, 1735 (1991)
35. M.S. Safronova, D. Budker, D. DeMille, D.F. Jackson Kimball, A. Derevianko, C.W. Clark, Rev. Mod. Phys. **90**, 025008 (2018)
36. N. Crescini, C. Braggio, G. Carugno, P. Falferi, A. Ortolan, G. Ruoso, Phys. Lett. B **773**, 677 (2017)
37. J. Lee, A. Almasi, M. Romalis, Phys. Rev. Lett. **120**, 161801 (2018)
38. S. Baessler, V.V. Nesvizhevsky, K.V. Protasov, A.Y. Voronin, Phys. Rev. D **75**, 075006 (2007)
39. A.P. Serebrov, O.M. Zherebtsov, Astro. Lett. **37**, 181 (2011)
40. A.K. Petukhov, G. Pignol, D. Jullien, K.H. Andersen, Phys. Rev. Lett. **105**, 170401 (2010)
41. C.B. Fu, T.R. Gentile, W.M. Snow, *CPT and Lorentz Symmetry* (2011), pp. 244–248
42. C.B. Fu, T.R. Gentile, W.M. Snow, Phys. Rev. D **83**, 031504 (2011)
43. B.J. Venema, P.K. Majumder, S.K. Lamoreaux, B.R. Heckel, E.N. Fortson, Phys. Rev. Lett. **68**, 135 (1992)
44. D.F. Jackson Kimball, J. Dudley, Y. Li, D. Patel, J. Valdez, Phys. Rev. D **96**, 075004 (2017)
45. K. Tullney, F. Allmendinger, M. Burghoff, W. Heil, S. Karpuk, W. Kilian, S. Knappe-Grüneberg, W. Müller, U. Schmidt, A. Schnabel, F. Seifert, Y. Sobolev, L. Trahms, Phys. Rev. Lett. **111**, 100801 (2013)

46. M. Bulatowicz, R. Griffith, M. Larsen, J. Mirijanian, C.B. Fu, E. Smith, W.M. Snow, H. Yan, T.G. Walker, Phys. Rev. Lett. **111**, 102001 (2013)
47. T. Wu, J.W. Blanchard, D.F. Jackson Kimball, M. Jiang, D. Budker, Phys. Rev. Lett. **121**, 023202 (2018)
48. J.M. Brown, S.J. Smullin, T.W. Kornack, M.V. Romalis, Phys. Rev. Lett. **105**, 151604 (2010)
49. N.F. Ramsey, Phys. Rev. **100**, 1191 (1955)
50. I.I. Rabi, N.F. Ramsey, J. Schwinger, Rev. Mod. Phys. **26**, 167 (1954)
51. F. Bloch, A. Siegert, Phys. Rev. **57**, 522 (1940)
52. C. Gemmel, W. Heil, S. Karpuk, K. Lenz, C. Ludwig, Y. Sobolev, K. Tullney, M. Burghoff, W. Kilian, S. Knappe-Grüneberg, et al., European Phys. J. D **57**, 303 (2010)
53. D.F. Jackson Kimball, New J. Phys. **17**, 073008 (2015)
54. P.H. Chu, A. Dennis, C.B. Fu, H. Gao, R. Khatiwada, G. Laskaris, K. Li, E. Smith, W.M. Snow, H. Yan, W. Zheng, Phys. Rev. D **87**, 011105 (2013)
55. G. Vasilakis, J.M. Brown, T.W. Kornack, M.V. Romalis, Phys. Rev. Lett. **103**, 261801 (2009)
56. A. Arvanitaki, A.A. Geraci, Phys. Rev. Lett. **113**, 161801 (2014)
57. F. Ficek, D.F. Jackson Kimball, M.G. Kozlov, N. Leefer, S. Pustelny, D. Budker, Phys. Rev. A **95**, 032505 (2017)
58. M.P. Ledbetter, M.V. Romalis, D.F. Jackson Kimball, Phys. Rev. Lett. **110**, 040402 (2013)
59. N.F. Ramsey, Physica A **96**, 285 (1979)
60. D. Budker, P.W. Graham, M. Ledbetter, S. Rajendran, A.O. Sushkov, Phys. Rev. X **4**, 021030 (2014)
61. J.L. Ouellet, C.P. Salemi, J.W. Foster, R. Henning, Z. Bogorad, J.M. Conrad, J.A. Formaggio, Y. Kahn, J. Minervini, A. Radovinsky, N.L. Rodd, B.R. Safdi, J. Thaler, D. Winklehner, L. Winslow, Phys. Rev. Lett. **122**, 121802 (2019)
62. M. Silva-Feaver, S. Chaudhuri, H. Cho, C. Dawson, P. Graham, K. Irwin, S. Kuenstner, D. Li, J. Mardon, H. Moseley, R. Mule, A. Phipps, S. Rajendran, Z. Steffen, B. Young, IEEE Trans. Appl. Superconductivity **27**, 1 (2017)
63. N. Du, N. Force, R. Khatiwada, E. Lentz, R. Ottens, L.J. Rosenberg, G. Rybka, G. Carosi, N. Woollett, D. Bowring, A.S. Chou, A. Sonnenschein, W. Wester, C. Boutan, N.S. Oblath, R. Bradley, E.J. Daw, A.V. Dixit, J. Clarke, S.R. O'Kelley, N. Crisosto, J.R. Gleason, S. Jois, P. Sikivie, I. Stern, N.S. Sullivan, D.B. Tanner, G.C. Hilton, Phys. Rev. Lett. **120**, 151301 (2018)
64. L. Zhong, S. Al Kenany, K.M. Backes, B.M. Brubaker, S.B. Cahn, G. Carosi, Y.V. Gurevich, W.F. Kindel, S.K. Lamoreaux, K.W. Lehnert, S.M. Lewis, M. Malnou, R.H. Maruyama, D.A. Palken, N.M. Rapidis, J.R. Root, M. Simanovskaia, T.M. Shokair, D.H. Speller, I. Urdinaran, K.A. van Bibber, Phys. Rev. D **97**, 092001 (2018)

Chapter 9
Light-Shining-Through-Walls Experiments

Aaron D. Spector

Abstract The light-shining-through-walls (LSW) method of searching for ultra-light bosonic dark matter (UBDM) uses lasers and strong dipole magnets to probe the coupling between photons and UBDM in the presence of a magnetic field. Since these experiments take place entirely in the laboratory, they offer a unique opportunity to perform a model independent measurement of this interaction. This involves shining a high-power laser through a magnetic field toward a wall which blocks the light. The interaction between the laser and the magnetic field generates a beam of UBDM that passes through the wall. Beyond the wall is another region of strong magnetic field that reconverts the UBDM back to photons that can then be measured by a single photon detection system. The sensitivity of these kinds of experiments can be improved further by implementing optical cavities before and after the wall to amplify the power of the light propagating through the magnetic fields. This chapter gives an introduction to LSW experiments and discusses a number of interesting challenges associated with the technique.

9.1 Introduction

Light-shining-through-walls (LSW) experiments offer the unique ability to measure the coupling between photons and the UBDM field over a wide range of masses in a purely laboratory setting. As Fig. 9.1 shows, these experiments work by shining a high-power source of light through a static magnetic field toward an opaque wall. While the wall blocks the light, the interaction between the photons and the magnetic field will generate a UBDM field which travels through it. Past the wall is another region of static magnetic field where the UBDM field converts back to photons which can then be measured with a detector. One of the strengths of LSW experiments is that since they do not rely on model-dependent astrophysical

A. D. Spector (✉)
Deutsches Elektronen-Synchrotron (DESY), Hamburg, Germany
e-mail: aaron.spector@desy.de

© The Author(s) 2023
D. F. Jackson Kimball, K. van Bibber (eds.), *The Search for Ultralight Bosonic Dark Matter*, https://doi.org/10.1007/978-3-030-95852-7_9

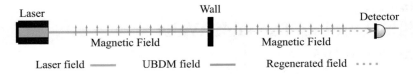

Fig. 9.1 A simple layout for an LSW experiment. The laser field is the red solid line, while the blue line is the UBDM field. A wall then blocks the light from the laser, while the UBDM field passes directly through it. The regenerated field is then shown as a red dotted line. The detector measures the regenerated field and does not interact with the UBDM field

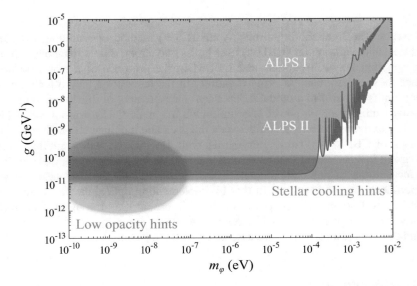

Fig. 9.2 Limits on g set from ALPS I [1] in orange and the projected sensitivity of ALPS II in blue. The hints from the transparency of the universe for TeV photons are shown in red [2], while the range of g that could cause stars to cool faster than their models predict is shown in pink [3]. ALPS II will be the first experiment to search for the UBDM coupling to photons over the mass range that could cause these phenomena in a purely laboratory setting

sources to produce UBDM fields, their systematic uncertainty is related only to the experimental apparatus itself.

In this chapter, we will refer to the region of magnetic field before the wall as the "production area" and the magnetic field after the wall as the "regeneration area." The electromagnetic field reconverted from the UBDM field in the regeneration area will be called the "regenerated field" or "regenerated photon signal."

We will also use the Any Light Particle Search II (ALPS II) [4, 5] as a reference point for the design of these experiments. From Fig. 9.2, we can see that ALPS II will be able to probe the coupling constant g between photons and the UBDM field down to $g < 2 \times 10^{-11}\,\mathrm{GeV}^{-1}$ for masses below 0.1 meV. This will allow ALPS II to explore a very important region of the parameter space where there are several hints of the existence of UBDM fields from astronomical observations mentioned earlier in Chap. 3. These include measurements of highly energetic photons from distant sources that indicate the universe is more transparent at these energies than predictions of the standard model would suggest [2]. This is shown in Fig. 9.2 as the circular region in the lower left corner. In addition to this, UBDM fields could also explain why stellar cooling rates uniformly exceed the expectations of their models [3]. This is shown as the band from $10^{-11} < g < 10^{-10}$. While other experiments may have investigated parts of these regions before, ALPS II will be the first to measure this range of g without relying on any astrophysical models of the production–regeneration process of the UBDM fields or the interstellar magnetic fields.

9.1.1 UBDM Interaction with Photons in a Magnetic Field

As we saw in Chap. 2, the term in the Lagrangian that defines the interaction between the photons and the UBDM field is

$$\mathcal{L}_{\text{UBDM}} = -\frac{1}{4} g \varphi F_{\mu\nu} \tilde{F}^{\mu\nu} . \tag{9.1}$$

In this equation, g is the coupling constant mentioned in the previous section. For LSW experiments, we can define the amplitude of the pseudoscalar UDBM fields φ_p generated in the production area as an integral of the dot product between an oscillating electric field E, supplied by the laser, and a static magnetic field B over an interaction length x,

$$\varphi_p(x, t) = e^{-i(\omega t - k_\varphi x)} \frac{ig}{2k_\varphi} \int dx' E(x') \cdot B(x') e^{-ik_\varphi x'} , \tag{9.2}$$

where ω is the angular frequency of the electric field, while k_φ is the wavenumber of the UBDM field described by the following equation:

$$k_\varphi = \sqrt{\omega^2 - m_\varphi^2} . \tag{9.3}$$

The maximum amplitude in Eq. (9.2) will occur when $E \parallel B$, while no field is produced if $E \perp B$. Likewise, opposite is true of scalar fields and the amplitude will be largest when $E \perp B$ and zero when $E \parallel B$. Therefore, LSW experiments can search for pseudoscalar fields by aligning the polarization of the laser to the magnetic field, while tuning the polarization of the laser orthogonal to the magnetic field when searching for scalar fields.

If we assume that B is static in time and uniform over a length L, the amplitude of the UBDM field can be simplified using plane wave approximations for E and B:

$$\varphi(x,t) = \frac{ig}{2k_\varphi} B E_0 e^{-i(\omega t - k_\varphi x)} \int dx' e^{iqx'} . \tag{9.4}$$

In this equation, q is a parameter that helps quantify the phase matching between the UBDM field generated at different points along the static magnetic field and is described by

$$q = n\omega - \sqrt{\omega^2 - m_\varphi^2} \approx \omega(n-1) + \frac{m_\varphi^2}{2\omega} . \tag{9.5}$$

From Eq. (9.4), it is apparent that when the mass is large enough or the interaction length is long enough that $qL > 1$, the experiment will lose some sensitivity as the UBDM field generated in the production area does not sum coherently. In the case where qL is some integer multiple of 2π greater than zero, the destructive interference prevents any field from being generated at all.

By evaluating the integral in Eq. (9.4), we can find the probability $\mathcal{P}_{\gamma \to \varphi}$ that a photon in the production area will convert to a UBDM field:

$$\mathcal{P}_{\gamma \to \varphi} = \frac{1}{4} \frac{\omega}{k_\varphi} (gBL)^2 |F(qL)|^2 . \tag{9.6}$$

In the equation above, $F(qL)$ represents the form factor for the magnetic field and can be simplified to

$$\left| F_{\text{single}}(qL) \right| = \left| \frac{2}{qL} \sin\left(\frac{qL}{2}\right) \right| . \tag{9.7}$$

Here, we can again see the effect of destructive interference in the generation of the UBDM field as the form factor goes to zero when $qL = 2\pi N$ for positive integer values of N. This effect, along the $1/qL$ factor outside the sine function, limits the sensitivity of LSW experiments at higher masses. This is apparent in the sensitivity curves for ALPS I and ALPS II shown in Fig. 9.2. There we can also see how a more detailed model of the magnetic field, which considers the gaps between the magnets, can produce patterns in the sensitivity at higher masses [6].

As it happens, the probability $\mathcal{P}_{\varphi \to \gamma}$ of the reverse process occurring and the UBDM field reconverting back to a photon in the regeneration area is the same as $\mathcal{P}_{\gamma \to \varphi}$. Therefore, a simple LSW experiment with the same magnetic field before and after the wall using a laser that travels only a single pass through the generation area will produce the following number of photons N_γ, in the regenerated field over a measurement time τ, for masses in which $qL \ll 1$:

$$N_\gamma = \frac{1}{16} (gBL)^4 \tau P_i . \tag{9.8}$$

We should make note of the fact here that these regenerated photons will have an identical energy to those that were used to generate the UBDM field. The magnetic field and length are obviously critical to the sensitivity of the experiment as the regenerated power is proportional to $(BL)^4$. The input power P_i, shown in units of photons per second, matters as well, but in this case the number of regenerated photons "only" scales linearly with it.

Plugging in the ALPS II parameters of 560 T·m of magnetic field length and an input power of 50 W gives an interesting result. For couplings down to $g < 2 \times 10^{-11}\,\mathrm{GeV}^{-1}$, this would only produce 1 regenerated photon over the course of 700 000 years. This is no mistake, remember that the N_γ above is only the number of regenerated photons for our simple example of an LSW experiment. This helps illustrate the importance of the additional techniques that LSW experiments like ALPS II can use to boost the power of the regenerated signal. These systems and how they impact the sensitivity are discussed in the later sections.

? Problem 9.1 Measuring the Mass of the UBDM Field

Suppose we build a simple LSW experiment with using a laser with an angular frequency of ω and a uniform magnetic field of length L for the production area and regeneration area. What is the lowest mass that the experiment is *insensitive* to? If we inject an inert gas into both the production and regeneration areas to give the optical path an index of refraction, at what value of n will the experiment achieve maximum sensitivity to that mass? How could this be used to find the mass of the UBDM field?

Solution on page 346.

9.1.2 Magnets

As we just discussed, LSW experiments rely on strong magnetic fields to promote an interaction between photons and the UBDM field. The magnets used for LSW experiments can be evaluated based on three critical parameters: (1) the strength and orientation of their static magnetic dipole field, (2) their length, and (3) the size of the bore. While it should be obvious why the length and magnetic field strength are important, a sufficient bore diameter is also crucial for LSW experiments so that light is not lost due to clipping as the lasers propagate through the beam tube. We will discuss later how this is essential when cavities are used to amplify the power of the input laser field and the reconverted field.

Fortunately, superconducting dipole magnets that were originally developed for particle accelerators are well suited for LSW experiments as they can produce strong dipole fields over very long distance with bore diameters sufficient to accommodate the accelerator beams. These magnets are constructed by coiling a superconducting thread to form many layers of wire, increasing the total current to induce very strong magnetic fields. They can also be connected in strings to produce magnetic fields with km lengths.

Reaching and maintaining superconductivity of course requires a cryogenic system that constantly supplies liquid He to cool the thread. Nevertheless, since they are a core element in modern accelerators, there are facilities all around the world that possess the cryogenic infrastructure necessary for operating them. Furthermore, after many years of development, the technology is very mature and magnets that can produce static dipole fields as high as 9 T with a uniform polarization over long distances and sufficient free apertures for LSW experiments are even currently in use at the LHC [7].

9.1.3   Light-Tightness

For LSW experiments to reach their optimal sensitivity, background signals must be suppressed below the sensitivity of the detection system. You may remember from earlier that the regenerated photons will have the same energy of those used to generate the UBDM field. Therefore, one source of background that all LSW experiments must cope with is light leaking through the wall from the production area to the regeneration area, as this would create a signal at the detector that is indistinguishable from one induced by an interaction with the UBDM field.

While sufficiently suppressing this light may seem like a trivial task, it is complicated by two points. First, current detection systems are capable of sensitivities on the order of a single photon per week. Second, as we will see later, the optical systems for LSW experiments are very sophisticated and need to transfer light between the production and regeneration areas of the experiment while a measurement is taking place. This interface is a particularly vulnerable point in terms of light-tightness. On top of that, there also needs to be systems that can verify the sensitivity of the experiment by checking parameters such as the alignment of the laser that generates the UBDM field. This involves having a shutter in the wall that can be opened to allow light to propagate directly from the production area to the regeneration area. This is another vulnerable point as stray light can find some scattering path through the shutter. The chance of a laser field actually transmitting through the bulk material of the shutter, on the other hand, is not a major concern. Even a few μm of material is enough to prevent any light from reaching the regeneration area, and in reality, the shutter will be substantially thicker than this.

9.2   Boosting Sensitivity with a Production Cavity

One way to increase the power in the reconverted signal is to amplify the power of the circulating light in the production area. Optical cavities or resonators can be very useful in this regard as state-of-the-art mirror coatings will allow cavities on the order of 100 m to amplify their circulating power by four orders of magnitude or more. As we saw in the previous section, the power in the regenerated photon signal scales linearly with the power circulating through the static magnetic field in the production area. Therefore, installing a production cavity (PC) there will amplify the power of the regenerated photon signal by the power build-up factor of the cavity β_p, if full power build-up can be achieved. Because of this caveat, in practice, it is more precise to quantify this in terms of circulating power in the PC P_c. With this, we can calculate the number of photons in the regenerated signal over a measurement time of τ from the following:

$$N_\gamma = \frac{1}{16}\left(g_{a\gamma\gamma}BL\right)^4 P_c \tau \ . \tag{9.9}$$

Therefore, a PC with power build-up of 10,000 will boost the sensitivity of the experiment with respect to g by a factor of ten.

9.2.1 Linear Cavity

While a variety of resonator designs exist, two mirror linear optical cavities are the most relevant for LSW experiments. As Fig. 9.3 shows, these types of cavities use two partially transmissive mirrors separated by a distance L and aligned with their surfaces normal to each other. The laser field enters the cavity through the input mirror M_1 and exits through the output mirror M_2. Let us suppose that these mirrors have a reflectivity of R_1 and R_2, transmissivity of T_1 and T_2. For lossless mirrors, these quantities are related by the following expression:

$$1 = R + T. \tag{9.10}$$

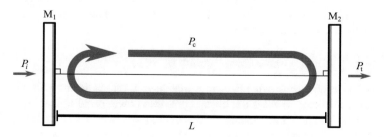

Fig. 9.3 Diagram of a two-mirror linear optical cavity with an input mirror M_1 and an output mirror M_2. The input power is given by P_i, while the circulating power is P_c, the transmitted power is P_t, and the length of the cavity is L

While these can also be expressed as field coefficients, in this chapter we will work in the convention that these are in terms of power.

In the 1D example, a laser field that is incident on the input mirror will be resonant if the length of the cavity is some integer multiple of the wavelength of the laser. The frequency spacing between the resonances of the cavity is known as its free spectral range (FSR) and can be found from $f_{\text{FSR}} = c/2L$. If the resonance condition is satisfied, the ratio of circulating power to input power can then be approximated by the following equation when $T_1, T_2, \rho \ll 1$:

$$\beta_{\text{P}} \equiv \left(\frac{P_{\text{c}}}{P_i} \right)_{\text{max}} \approx \frac{4T_1}{(T_1 + T_2 + \rho)^2} \, . \tag{9.11}$$

This is what we referred to earlier as the power build-up factor of the cavity. In this equation, ρ is the power losses that the circulating field accrues after each round trip. To reemphasize a point we made on the previous page, if the combined mirror transmissivities and losses are on the order of 100 ppm, the PC can amplify the power converted to the UBDM field by more than four orders of magnitude.

The dependence of circulating power on the laser frequency can be found from the cavity Lorentzian:

$$P_{\text{c}} = \frac{\beta P_i}{1 + \left(\frac{2\mathcal{F}}{\pi} \sin \left(\pi \frac{\Delta f}{f_{\text{FSR}}} \right) \right)^2} \, . \tag{9.12}$$

In this equation, the *finesse* of the cavity, $\mathcal{F}$, is defined as the ratio between the linewidth (full width half maximum, FWHM) of the cavity resonance f_c and the FSR:

$$\mathcal{F} \equiv \frac{f_{\text{FSR}}}{f_c} \approx \frac{2\pi}{(T_1 + T_2 + \rho)} \qquad (\text{when} \quad T, \rho \ll 1) \, . \tag{9.13}$$

Therefore, in order to achieve the maximum circulating power in the cavity, the input laser must be controlled such that the difference between its frequency and the cavity resonance is much less than f_c or $f_{\text{FSR}}/\mathcal{F}$.

? Problem 9.2 Maximum Power Build-Up

Let us suppose we are provided mirrors that have total scattering and absorption losses of ρ per mirror. What is the highest possible power build-up achievable for a two-mirror cavity and what transmissivities of the mirrors need to be used? Suppose we need 1% of the input power in transmission of the cavity. What is the highest power build-up that we can achieve and what mirror transmissivities are necessary under these conditions?

Solution on page 347.

9.2.2 Cavity Spatial Modes

In the previous section, we only discussed the longitudinal mode of the cavity, and however it is important to also consider the spatial profile of the cavity eigenmodes. This is especially true for longer baseline LSW experiments since the modes are constrained by the magnet bore and the power build-up factor can be limited by its diameter. In addition to this, in the next section, we will discuss dual cavity LSW experiments that also use a resonator in the regeneration area. When this technique is used, it is also important to ensure that the cavities share nearly the same transversal mode.

The transversal modes of the cavity resonances are commonly expressed in a basis set of Hermite–Gauss or Laguerre–Gauss modes, and the higher order modes can therefore be described as the product of the fundamental mode and Hermite or Laguerre polynomials. For LSW experiments, it is advantageous to operate with the fundamental mode since it will provide the smallest beam sizes over the longest baseline, thus reducing the clipping losses on the magnet bore. Therefore, while there are cases where the Hermite–Gauss and Laguerre–Gauss description can be very useful, we will not explore it further.

The field, when in the fundamental mode, follows a Gaussian distribution that can be described by the following equation when using the paraxial approximation:

$$E(r, z) = E_0 \frac{w_0}{w(z)} \exp\left(-\frac{r^2}{w(z)^2}\right) \exp\left(i\left[kz - \psi(z) + \frac{kr^2}{2R(z)}\right]\right). \qquad (9.14)$$

The intensity distribution for a beam with a power of P is then given by

$$I(r, z) = \frac{2P}{\pi w(z)^2} \exp\left(-2\frac{r^2}{w(z)^2}\right). \qquad (9.15)$$

In these equations, $w(z)$ represents the $1/e^2$ radius of the intensity distribution as the following function of z:

$$w(z) = w_0 \sqrt{1 + \left(\frac{z}{z_r}\right)^2}. \qquad (9.16)$$

Figure 9.4 shows a visual representation of the spatial mode of a Gaussian beam as it propagates through its waist position. As we can see, the radius of the distribution, shown as the thick black line, has the minimum waist w_0 at the position $z = 0$. In the near field ($z \ll z_r$), the beam is collimated and its size remains relatively constant, while in the far field ($z \gg z_r$) the waist expands linearly with z. The parameter z_r is known as the Rayleigh length and is defined as the distance from the waist position at which $w(z_r) = \sqrt{2}w_0$. It depends only on the minimum waist size and the laser wavelength:

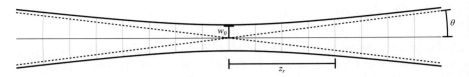

Fig. 9.4 Profile of a Gaussian beam with a minimum waist size of w_0, a Rayleigh length of z_r, and a divergence half-angle of θ. The waist size is shown as the thick black line, while wavefronts at different positions are shown as gray lines. As the beam propagates further into the far field, it will asymptotically approach the dotted lines that illustrate the divergence angle

$$z_r = \frac{\pi w_0^2}{\lambda}. \tag{9.17}$$

From this equation, we can see that the Rayleigh length is proportional to the area of the beam at the minimum waist position and inversely proportional to the wavelength. Therefore, producing a beam that is well collimated over long distances requires using a larger beam as smaller beams will only remain collimated for shorter distances. Furthermore, shorter wavelength lasers can produce the same Rayleigh length as longer wavelength lasers using smaller beam sizes. This is a critical point for LSW experiments as the diameter of the magnet bore will typically limit the length of the experiment, and once the waist size approaches some fraction of the bore diameter, clipping losses will limit the possible power build up factor of the cavities.

This can also be seen by looking at the divergence half angle θ, of the beam in the far field. In this regime, the dependence of the waist size on z can be approximated by the linear relation $w(z) \approx w_0 z / z_r$, and by plugging in for z_r we can define θ as

$$\theta = \frac{\lambda}{\pi w_0}. \tag{9.18}$$

There are also several terms in the complex exponential of Eq. (9.14). The first term, kz, is just the product of the wavenumber and the longitudinal position. The second term, $\psi(z)$, is known as the Gouy phase and represents the natural phase shift that the field will experience passing through the waist. The fields will also have spherical wavefronts, shown as light gray lines in Fig. 9.4, and the final term in the exponential, $kr^2/2R(z)$, introduces the curvature $R(z)$ of the wavefronts. The dependence of the wavefront curvature on the longitudinal position is given by

$$R(z) = z \left[1 + \left(\frac{z_r}{z} \right)^2 \right]. \tag{9.19}$$

Here, it is apparent that at the minimum waist position the wavefronts are flat ($R(0) = \infty$), while at the Rayleigh length $R(z_r) = 2z_r$ and in the far field $R(z) \approx z$. This is important as the radius of curvature of the mirrors sets the wavefront curvature at their position, and this will determine the shape of the transversal mode throughout the cavity.

In order for the cavity to achieve full power build-up, the input beam must be in the same spatial mode as the eigenmode of the cavity. The coupling efficiency between the laser and the cavity can be found by calculating the spatial overlap η, between the input laser field, shown in this equation as E, and the cavity eigenmode, expressed here as E':

$$\eta = \frac{\left|\int E^* E' dA\right|^2}{\int |E|^2 dA \int |E'|^2 dA} . \tag{9.20}$$

In this equation, we evaluate the overlap integral between the two normalized fields over an area A and then take its absolute square. Any spatial dependence in the differential phase between the wavefronts of the two fields or mismatch in the spatial distribution of their amplitudes will lead to a loss in the coupling of the field to the cavity. We should note the fact that the spatial overlap is independent of the longitudinal position of the plane it is evaluated over.

? Problem 9.3 Eigenmode Waist Size Versus Length

Derive the relationship between the minimum waist size w_0 of the cavity eigenmode and its length L for a two-mirror cavity in which both mirrors have a radius of the curvature equal to L.

Solution on page 347.

9.2.3   Stabilization of Optical Cavities

So far we have only considered static cavities using an input laser with a fixed frequency. In reality, though, both the laser frequency and cavity length will have some noise and a control system is needed to maintain the frequency of the laser with respect to the length of the cavity or vice versa. Much of the pioneering work in this field was done by the gravitational wave community as these types of control systems are critical to sensing the tiny phase fluctuations that gravitational waves introduce into detectors such as Advanced LIGO [8] and LSW experiments benefit considerably from this.

One of the most well-known and widely used techniques is the Pound–Drever–Hall (PDH) laser frequency stabilization [9, 10]. PDH takes advantage of the fact that close to resonance, the phase of the field reflected by the cavity will be linearly proportional to the frequency difference between the input laser and the cavity. Using phase modulation sidebands, the reflected phase can be measured to generate an electronic signal that can then be fed back to the laser frequency or the cavity length to maintain the resonance condition.

A similar technique known as differential wavefront sensing is capable of sensing the alignment of the laser with respect to the spatial eigenmode [11–13]. Here, a quadrant photodetector (QPD) measures the four quadrants of reflected power distribution. Again with phase modulation sidebands, misalignments between the wavefronts of the incident laser field and the circulating field that is leaking out of the input mirror of the cavity can be measured. By feeding this signal back to alignment actuators, the spatial overlap between the laser and the cavity eigenmode can be maintained.

9.2.4 Achieving High Finesse

Of course, achieving a high finesse or power build-up comes with its own set of challenges. As Problem 9.2 illustrates, the highest possible power build can be achieved when the losses (ρ) and the output mirror transmissivity (T_2) are as low as possible, while the input mirror transmissivity (T_1) obeys the condition $T_1 = T_2 + \rho$.

While we have some control on the transmissivities of the mirrors, there is a limit to how much we can suppress the losses. Three of the most common causes of the intracavity losses that LSW experiments must consider are scattering and absorption from the cavity mirrors and clipping on the free aperture of the magnet bore.

The scattering losses for mirrors used in high-finesse long-baseline cavities are typically driven by the surface roughness of the substrates themselves. In this case, the total integrated scatter (TIS) at normal incidence for smooth surfaces can be approximated by

$$\text{TIS} = \left(\frac{4\pi\sigma}{\lambda}\right)^2 . \tag{9.21}$$

In this equation, σ is the integrated RMS deviation from a perfect spheroid of the reflective surface of the mirror evaluated over the area of the beam. Therefore, larger beams will be exposed to features at lower spatial frequencies than smaller beams. Since the amplitude of these features tends to increase as the spatial frequency decreases, larger beams will typically experience higher scattering losses than smaller beams. For longer baseline cavities, this can limit in the maximum possible power build-up factor.

For LSW experiments, the clipping losses from the magnet aperture must also be considered. For this purpose, the magnet strings can be thought of as series of connected pipes. If we form a cavity by placing mirrors at the ends of the string, the clipping losses will be the percentage of light lost from scattering off of the walls of the pipes. When straight magnets are used, this can be approximated by integrating the power of the Gaussian beam at the position of its widest radius inside the magnets over the free aperture of the string. For long-baseline high-finesse cavities, this can create a problem as longer cavities will have larger eigenmodes. Unless magnets with extremely large bore diameters are used, clipping losses on the magnet bore will actually limit the maximum length of the string. To help compensate for this, the production cavity can be designed such that the Rayleigh length of the eigenmode is roughly equal to half the length of magnet string that contains it. If a cavity is also used in the regeneration area, then we must consider the clipping losses there as well. In this case, the Rayleigh length of both cavities should be half the combined length of the two strings. This will allow for the smallest possible beams at the exits of the string.

? Problem 9.4 Clipping Losses and Cavity Length

Suppose you have a site and supply of magnets where you are free to make an LSW experiment as long as you want using magnets with a bore diameter of 50 mm and a laser operating at a wavelength of 1064 nm. You would like to use a production cavity with a power build-up of 10,000 using a flat mirror and a concave mirrors with a radius of curvature that you may choose. How long can you make the magnet string in the production area before you can no longer reach a power build-up of 10,000? (Hint: use the relationship you derived in Problem 9.2 on the maximum power build-up factor.)

Solution on page 348.

9.2.5 High-Power Operation

As we have already discussed, the higher the power the PC can support, the more sensitive the experiment will be. This can be complicated, though, by a number of effects that degrade the performance of the cavity as higher powers are used due to absorption in the optical coatings.

One of the issues with absorption in the optical coatings is that areas with a higher incident intensity will reach a higher steady-state temperature than the areas of the mirror with a lower incident intensity [14]. This will cause larger thermal expansion in the central region of the mirror creating a change in its effective radius of curvature that makes it more convex. With this, the Rayleigh length of the cavity eigenmode will increase resulting in a larger beam size of the circulating field. If the beam size increases enough, this can lead to power losses due clipping on the free aperture of the magnets or a reduction in the spatial mode matching.

High-power operation can also lead to an increase in the cavity losses if there are point absorbers on the surface of the optical coating or embedded in it [15]. These points will also reach higher steady state-temperatures than the rest of the mirror and can introduce low-frequency spatial features through thermal expansion. If the spatial features are smaller than the beam size, this can lead to additional scattering losses.

9.3 Dual Cavity LSW Experiments

The sensitivity of LSW experiments can be increased further by using a regeneration cavity (RC) after the wall to amplify the regenerated field by a resonant enhancement factor. While this factor is nearly identical to the expression we used earlier for the power build-up, the concepts are somewhat distinct as the regenerated signal is injected without its power being attenuated by one of the mirrors. There it is amplified and then attenuated only as it leaves the cavity. In this way, the resonant enhancement factor is an expression of the amplification, in power, of the regenerated field that is actually incident on the detector. The power build-up in the PC, on the other hand, is the amplification of the power of the input laser while that field is still circulating in the PC.

With this, the resonant enhancement factor, β_R, can be expressed as the following approximation:

$$\beta_R \approx \frac{4T_{\text{out}}}{(T_1 + T_2 + \rho)^2}. \tag{9.22}$$

In this equation, T_{out} refers to the transmissivity of the mirror at the detector port of the RC and could be either T_1 or T_2. As we discussed in the previous section, if we say mirror "1" is the mirror at the detector, the highest possible power build-up will be achieved when $T_1 = T_2 + \rho$ and T_2 is chosen to be as low as possible.

Figure 9.5 shows a simplified optical setup where the flat mirrors of the PC and RC are coupled to a central optical bench (COB) in the middle of the experiment. In this setup, the radius of curvature of the curved mirrors at the end stations can be chosen to be roughly the length of the entire such that the Rayleigh length of the cavity eigenmodes is half the length of the entire magnet string, to minimize clipping losses on the magnet bore. It is important that the Rayleigh length is not the exact length of the cavities as this could lead to higher order mode degeneracies that will interfere with their performance. The wall is then located in between the

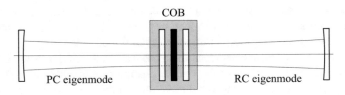

Fig. 9.5 Standard layout for a dual cavity LSW experiment with a COB at the center that houses the flat mirrors and wall, with curved mirrors located at the end stations

mirrors on the COB. We should note that the distance between these mirrors is much smaller than the length of the cavities.

When the dual cavity configuration is used, the number of regenerated photons at the detector will be

$$N_\gamma = \frac{1}{16}\left(g_{a\gamma\gamma}BL\right)^4 \eta\beta_R P_c \tau, \qquad (9.23)$$

where η is the spatial overlap between the two cavity eigenmodes and P_c is the total circulating power in the PC. With a spatial overlap on the order of one, 150 kW circulating in the PC, an RC resonant enhancement factor of 20,000, and BL of 560 T·m, a two-week measurement will produce roughly 50 photons at the detector for a $g_{a\gamma\gamma}$ of $2 \times 10^{-11} \text{GeV}^{-1}$. From this, we see that the product of the PC power build-up and RC resonant enhancement factor can help LSW experiments gain more than 8 orders of magnitude in the signal strength in the regenerated field. This can increase their sensitivity in terms of $g_{a\gamma\gamma}$ by a factor of 100.

The regenerated field can be treated like a weak input field and thus will need to be resonant with the length of the RC and in its spatial eigenmode. Since the PC transmitted field should be an accurate representation of the regenerated field, it can be used to verify the resonance condition and spatial overlap.

9.3.1 Dual Resonance

Remember that the regenerated field will be in the same spatial mode and have the same frequency as the field circulating in the RC. Therefore, for the regenerated field to be resonant with the RC, the field circulating in the PC must also be resonant with the RC. For this to occur, the frequency of the PC circulating field f_{PC} must meet the condition

$$f_{PC} = N\frac{c}{2L_{RC}}, \qquad (9.24)$$

where the right side of the equation gives the corresponding resonance of the RC. Here, N is some whole integer number and $c/2L_{RC}$ is the FSR of the RC. Any static offset from the resonance condition will lead to a loss in the resonant enhancement factor that follows the cavity Lorentzian expressed earlier in Eq. (9.12). As the input laser to the PC is frequency stabilized to its length, this tuning can actually be done by adjusting the length of either of the cavities.

Once the cavities are set to the correct length, the resonance condition must then be maintained in the presence of environmental noise. To do this, the frequency changes of the PC circulating field must somehow track the length changes of the RC or vice versa. This requires a sensing system capable of comparing these two parameters. This can be done by stabilizing the frequency of a reference laser (RL) to the length of the RC using PDH and interfering it with the light transmitted from the PC. With this system, a direct measurement can be made of the frequency difference between the PC circulating field and the RC resonance. This is important as this information can then be fed back to stabilize the length of one of the cavities with respect to the other.

This transfer of the frequency information between the RC resonance and the PC circulating field must be done while still preventing the light circulating in the PC from entering the RC. This would create background signals that are indistinguishable from the regenerated signal. Therefore, it is clear that we cannot use the same frequency for RL as the light circulating in the PC. The limits on the available frequency that we can choose for RL are actually dependent on the energy resolution of our detection method. This system must be able to tell the difference between the light we are using to sense the length of the RC and the actual regenerated signal we are trying to measure.

As we will see in the next section, the COB is one of the critical design features of dual cavity LSW experiments and its passive stability can be used to maintain the alignment between the flat mirrors. Therefore, it makes sense to actuate on the length of the cavity via one of the curved mirrors at the end stations instead.

Stabilizing the length of one of the cavities requires an actuator capable of moving the mirror fast enough and with enough dynamic range to overcome the differential length noise between the cavities over the course of a measurement. One way to do this is to mount the mirror to a piezo-electric actuator which can expand or contract based on an input voltage. The information from the measurement of the frequency difference between the PC field and the RC can be used to stabilize the length of one of the cavities by feeding back to the piezo actuator.

Without additional seismic isolation, these systems will require control bandwidths in the kHz range. This can be difficult as the mass of the mirror, internal resonances of the piezo, and the rigidity of the mount can limit the speed with which actuation is possible. As we saw from the previous section, the longer the length of the cavities is, the larger their mirrors must be to avoid clipping losses. As the mirror gets larger, it quickly becomes difficult to actuate with the necessary speed as their mass typically increases nonlinearly with the active area. Therefore, if the cavities in future LSW experiments are much longer than 100 m, they may also require more sophisticated systems that use passive isolation to suppress the seismic noise in addition to actively controlling the lengths of the cavities.

9.3.2 Spatial Overlap

Just as it is important to maintain the resonance condition of the regenerated field with respect to the RC length, this field must also be spatially coherent with the eigenmode of the RC. Any lateral displacement or angular misalignment between the modes will lead to a reduction in the coupling efficiency of the regenerated field to the RC. Just as we discussed in the previous section, the spatial mode of the regenerated field will be a replica of the PC circulating field. Therefore. the spatial overlap η [Eq. (9.20)] between the PC and RC eigenmodes can be used to estimate the coupling of the regenerated field to the RC.

For small mode-matching errors, η can be approximated by Eq. (9.25), where Δx is a transversal offset in the minimum waist position between the input field and the cavity eigenmode, $\Delta\theta$ is an angular offset between their optical axes, Δw is a difference in the minimum waist size, and Δz is a difference in the position of the waist along the optical axis:

$$\eta = 1 - \left(\frac{\Delta x}{w_0}\right)^2 - \left(\frac{\Delta\theta}{\theta}\right)^2 - \left(\frac{\Delta w_0}{w_0}\right)^2 - \left(\frac{\Delta z}{2z_r}\right)^2. \tag{9.25}$$

Due to the length of the cavities being much larger than the separation between their waist positions, $\Delta z/2z_r$ is much less than one and should not cause a significant reduction in coupling efficiency. Likewise, $\Delta w_0/w_0$ is also insignificant as the waist size in cavities with this geometry is determined only by their length and the radius of curvature of the end mirrors, and these values should be nearly identical for the PC and RC.

The angular misalignment and transversal displacement of the eigenmodes, on the other hand, can cause a significant loss in the spatial overlap and Fig. 9.6 shows examples of how each of these effects can occur in dual cavity setups. The angular misalignment of the cavity eigenmodes will be determined by the alignment error between the flat cavity mirrors since the optical axes of the cavities must be perpendicular to them. To ensure that there is less than 1% loss in the spatial coupling from this effect, their alignment must be within one-tenth of the cavity divergence half-angle. To put this in context, for ALPS II, this is 57 μrad and the requirement on the misalignment between the central mirrors is < 5 μrad. In ALPS II, this is achieved by rigidly mounting the mirrors to a COB. The COB is effectively just a large metallic plate which has demonstrated the necessary alignment stability in tests of prototypes.

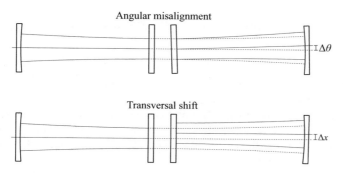

Fig. 9.6 Examples of an angular misalignment (top) and lateral shift (bottom) between the cavity spatial eigenmodes. The angle of the optical axes is determined by the alignment of flat mirrors, while their transversal positions are controlled by the alignment and position of the curved mirrors relative to the flat mirrors

The transversal waist position of each of the cavity eigenmodes will be determined by the position and alignment of the curved mirror. Since the optical axis will be normal to both mirrors, a transversal displacement of the curved cavity mirror will lead to an equal shift in the position of its optical axis. Changes in the alignment of the curved mirrors will produce a shift in the optical axis equal to the product of the angular displacement of the curved mirror and the radius of curvature of the mirror:

$$\Delta x = R \Delta \theta_{\text{curved}} . \tag{9.26}$$

The lower diagram in Fig. 9.6 shows an example of this effect. To put some numbers on this, in ALPS II, $w_0 = 6 \, \text{mm}$ and the requirements on the transversal shift between the cavities eigenmodes are $< 1 \, \text{mm}$ or a $< 3\%$ power loss. With a radius of curvature of $214 \, \text{m}$ for the curved mirrors, this means that their alignment must be controlled with better than $5 \, \mu\text{rad}$ precision.

The relative transversal shift between the cavity eigenmodes can be sensed by measuring cavity fields transmitted from flat mirrors with QPDs on the COB. The changes in the transversal position of the cavity eigenmodes relative to the COB will then lead to changes in the differential power level measured between the quadrants. By feeding back to alignment actuators on the curved cavity end mirrors, a control loop can be used to stabilize the positions of the eigenmodes.

Like cavity length stabilization, alignment stabilization can also be tricky. A difference here is that the requirements on the alignment stability of the cavities are usually much more forgiving relative to the environmental noise than the length stability requirements. Because of this, the cavity alignment control can be much slower (control bandwidths on the order of $1 \, \text{Hz}$) than the length control (control bandwidths on the order of $1 \, \text{kHz}$) while still sufficiently suppressing the noise. If the alignment noise is low enough, it can even be the case that no active alignment stabilization is necessary for the cavities.

9.3.3 Verification of the Resonance Condition and Spatial Overlap

When using a dual cavity setup to amplify the regenerated photon signal, it becomes all the more important to verify that the optical system is aligned and properly tuned. In particular, the resonance condition and spatial overlap must be checked. As we mentioned earlier, one of the design features of LSW experiments is a shutter in the wall that will allow light to freely propagate from the production area to the regeneration area when it is opened.

In dual cavity LSW experiments, the shutter must be located in between the flat cavity mirrors on the COB. When the control systems are sufficiently suppressing the environmental noise and the shutter is open, the PC transmitted field should couple directly to the RC. With prior knowledge of the reflectivities of the RC mirrors, the total coupling efficiency of the regenerated field to the RC can be estimated by measuring the ratio of the PC power incident on the RC to the power that transmits through it.

9.4 Detection Techniques

As we mentioned in the introduction, LSW experiments require detection systems capable of measuring single photons over time scales of weeks in order to reach their target sensitivity. This section will discuss two different detection schemes which are capable of this, both of which will be implemented in ALPS II. These are heterodyne interferometry and transition edge sensors. We should emphasize that each of these systems places distinct constraints on the optical setup, and therefore they cannot be operated in parallel.

9.4.1 Heterodyne Interferometry

Heterodyne interferometry works by optically mixing a laser, which we will refer to as the local oscillator (LO), with the regenerated field to create an interference beat note in the power which we can then measure. By using the coherence between the two fields, we can distinguish between this low power signal and noise. Figure 9.7 shows how a detection system using heterodyne interferometry could be implemented in a dual cavity LSW experiment. In this diagram, the high-power laser (red) is coupled to the PC and has an angular frequency of ω_s, while the local oscillator (blue), with an angular frequency of ω_{LO}, is injected to the RC through a Faraday isolator (FI). As the regenerated field, shown as the dotted red line, circulates in the RC, it naturally mixes with the LO field. This produces an interference beat note at the difference frequency $\Delta\omega$, between the two fields. This beat note on the LO power can then be measured at the science photodetector PD_S.

As we discussed in the previous section, for optimal resonant enhancement of the regenerated signal, the cavity lengths must be tuned such that the frequency difference between the two fields is held at some integer number of FSRs of the RC. Furthermore, since heterodyne detection systems rely on the absolute phase coherence of the local oscillator to the regenerated signal, any drift in the relative phase between these signals will lead to a reduction in sensitivity. We should note that this goes beyond even the requirements of dual resonance.

This necessitates some additional system that can sense the phase relationship between the local oscillator and the PC transmitted field. As we can see in Fig. 9.7, the simplest way to do this would be to interfere the fields transmitted by the cavities at a beam splitter on the COB. The phase of the interference beat note can then be monitored by a photodetector PD_M. This system must also be capable

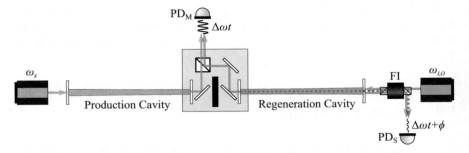

Fig. 9.7 Simplified design of a heterodyne detection system for a dual cavity LSW experiment

of sensing the optical path-length changes between the flat cavity mirrors on the COB [16], although the components that perform this function are not shown in the figure. The technical challenges become even more significant when considering that all of this must be accomplished without compromising the light-tightness of the experiment.

We can see how difficult it is to measure the power of the regenerated field by looking at the expression for the expected power at PD$_S$:

$$P(t) = P_{LO} + P_S + 2\sqrt{P_{LO} P_S} \cos(\Delta\omega t - \phi) + \chi_{SN} \ (\Delta\omega = \omega_{LO} - \omega_s). \tag{9.27}$$

The static terms P_{LO} and P_s represent the DC power of the local oscillator laser and the weak signal field. In LSW experiments, these powers differ by over 20 orders of magnitude effectively making a measurement of the P_S term impossible. The third term shows the interference beat note between the local oscillator and the regenerated field. The beat note has an amplitude of $2\sqrt{P_{LO} P_S}$, which corresponds to sub-pW amplitudes for a regenerated signal of one photon per day when a 10 mW local oscillator is used. This means we need to measure an oscillation in the power with an amplitude that is over 10 orders of magnitude lower than its mean value.

Furthermore, the beat note will also be embedded within what is known as shot noise due to photon counting statistics. This will make it impossible to identify the interference beat note simply from a time series of the power. Instead, we can calculate the power spectral density (PSD) to find the signal. The PSD is a measure of the density of power in each of the frequency components that make up the signal. As we will see, heterodyne interferometry takes advantage of the fact that the PSD of coherent signals will increase with the measurement time, while the PSD of incoherent signals will remain the same over time.

The PSD of shot noise measured by the photodetector will be equal to

$$PSD_{SN} = P_{LO} h\nu, \tag{9.28}$$

where $h\nu$ is the energy per photon of the laser. The single-sided PSD of the interference beat note at the difference frequency in the absence of noise is given by the following equation for a measurement time τ:

$$\text{PSD}_{\text{IB}}(\Delta\omega) = P_\text{S} P_{\text{LO}} \tau . \tag{9.29}$$

The signal-to-noise ratio (SNR) can then be found by calculating the ratio of these PSDs [17],

$$\text{SNR} = \frac{S_{\text{IB}}(\Delta\omega)}{S_{\text{SN}}(\Delta\omega)} = \frac{P_\text{S}}{h\nu}\tau = \langle N_\text{S}\rangle. \tag{9.30}$$

As expected, the signal-to-noise ratio is proportional to the measurement time, but what is also apparent is that the signal-to-noise ratio is equal to the expected number of regenerated photons. This is, of course, contingent on several factors such as the shot noise being at a level well above the technical noise of the photodetector. Also, the beat note between the regenerated signal and LO remaining coherent with the oscillator used to perform the PSD. Additionally, for this condition to be valid stray light and other sources of background signals must be sufficiently suppressed.

The power in the regenerated field can then be calculated by dividing the PSD at $\Delta\omega$ by the measurement time and LO power:

$$P_\text{S} = \frac{\text{PSD}_{\text{IB}}(\Delta\omega)}{P_{\text{LO}}\tau} . \tag{9.31}$$

In principle, heterodyne interferometry should not be limited by any fundamental backgrounds. Nevertheless, the system must be well designed such that the various electronic signals used to maintain the coherence of the fields only experience a limited coupling to the detection electronics. Otherwise, this will create background signals that cannot be distinguished from the regenerated field.

9.4.2   Transition Edge Sensors

An entirely different technology that can also be used to measure the regenerated field is transition edge sensors (TES) [18]. These devices are capable of measuring the heat induced by the incidence of single photons on an absorptive chip. They are well equipped to face the challenges posed by LSW experiments due to their low noise and high efficiency.

The diagram to the left of Fig. 9.8 shows a simplified version of the TES electrical circuit. TESs work by holding a small chip, typically made of tungsten, at the temperature threshold to superconductivity. When a photon is absorbed by the chip, it will cause a sudden spike in its temperature that provokes a change in its resistance and thus the current passing through the sensor. An inductive coil (L) in series allows the pulse in the current to be measured using a SQUID.

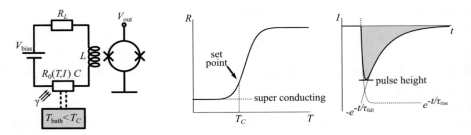

Fig. 9.8 On the left is a simplified diagram of the TES circuit. Here, the chip has a resistance R_0 when held at a temperature T with a current I passing through it. The current passing through the inductive coil L is measured with a SQUID. The center plot shows an R versus T curve of the chip at the superconducting transition, with the set point T_c. The plot on the right shows an ideal pulse and how the rise time τ_{rise} and fall time τ_{fall} effect the pulse shape

A bias current can be introduced which, when traveling over a shunt resistor R_L, can be treated as a constant voltage source (V_{bias}) . This configuration is critical to the stability of the system as when $R_L \ll R_0$ the electrothermal feedback is negative and the sensor operates at a steady state between heat introduced by the flow of current through it and the heat dissipated by a thermal link to a cold bath held at a lower temperature T_{bath}.

Since the bias current puts an additional heat load on the chip, it can also be used to tune and maintain the working point of the system. The R versus T curve in the middle of Fig. 9.8 shows how the chip transitions from a normal state to the superconducting state as its temperature drops. The set point is chosen at some temperature T_C along this curve, below the point where the derivative $\partial R / \partial T$ is at a maximum, to optimize the dynamic range of the system.

As the right-hand plot in Fig. 9.8 shows, pulses in current will have several defining features that help identify whether or not they were indeed the result of an incident photon, and if so how much energy was transferred to the chip. One of these is the rise time τ_{rise}, a measure of the time constant of the initial leveling off of the change in current after the photon is absorbed. The rise time is dependent on the inductance of the coil and the total dynamic resistance of the circuit. Then, there is the fall time τ_{fall} or the time constant of the decay of the current back to its steady-state value. The fall time will also be determined by the same parameters which set the rise time along with several others. These additional parameters are the derivative of the resistance with respect to the bias current $\partial R / \partial I$ and temperature of the chip $\partial R / \partial T$, along with the temperature of the chip itself T_C, all at the working point, as well as the thermal conductivity of the link to the cold bath. Finally, there is the height of the pulse, which will depend on the energy introduced by the incident photon.

The following expression can be used as a simple model for a pulse with A as a scaling constant:

$$\delta I(t) = A\left(e^{-t/\tau_{\text{rise}}} - e^{-t/\tau_{\text{fall}}}\right). \tag{9.32}$$

All incident photons will produce pulses with the same rise time and fall time, with the energy of the photon determining the pulse height. Simply integrating the pulse will give energy induced by the photon. However, for robustness a pulse fitting algorithm is typically applied to the measured data, which not only provides data on the photon energy, but can also help distinguish whether or not the source of the pulse was actually an incident photon, rather than the intrinsic noise of the system.

The energy resolution of the TES can be determined by measuring many photons from a single frequency source and constructing a histogram of energies with the template fitting routine. A perfect energy resolution would result in the same measured energy for all of the incident photons. The noise of the system will, however, lead to spreading of the histogram with the energy resolution of the TES being the width of the distribution. This is a critical parameter for LSW experiments as the better the energy resolution is, the better the TES can distinguish background events from signal photons. Energy resolutions down to 5% have been demonstrated [19].

One of the limiting sources of background events when using TESs for LSW experiments is black-body radiation. The primary concern is not actually events at the signal energy, since the lasers typically operate at energies outside the black-body spectrum at room temperature. Instead, the main issues arise from events called "pile-ups", in which two pulses occur so close close together in time that is is impossible to distinguish them from a single event. If the energies of the two black-body photons sum to an energy close to that of the regenerated field, they can be mistaken as a signal.

One way to mitigate this problem is to filter out the black-body photons before they are incident on the chip. This is complicated by the fact that the filter must be operated in a cryogenic environment, and any optics after the filter that couple the light to the TES must also be cold. Otherwise, the filter itself along with the warm optics would generate their own black-body spectrum creating a background. This is further complicated by the fact that the regenerated field is normally coupled to the TES via an optical fiber.

We should note here that the black-body spectrum does not need to be completely eliminated, only reduced to the point where the background rate no longer effects the sensitivity of the experiment. If the black-body pile-up can be sufficiently suppressed, it is possible for TESs to achieve background rates below to $1 \times 10^{-5}\,\text{s}^{-1}$ before other backgrounds, such as the radioactivity of the materials in the vicinity of the chip, become limiting.

? Problem 9.5 Black-Body Pile-Ups

Let us assume we are performing a 10^6 s measurement using a TES with a rise time of 0.1 μs, and we have a background rate due to black-body radiation for photons at energies from $0.45\,\text{eV} < h\nu < 0.55\,\text{eV}$ of 100 photons per second. Assuming that the photons obey Poissonian statistics, what is the expectation value for the number of unresolvable pile-ups? For simplicity, assume that "unresolvable" means that the photons arrive at the chip within one rise time of each other.

Solution on page 348.

9.5   Conclusion

As we have discussed, LSW experiments are capable of measuring the coupling between a UBDM field and electromagnetic fields without relying on model-dependent astrophysical sources. Instead, the UBDM fields are generated in the laboratory with a laser and a string of magnets. Using such a well understood mechanism of production for the UBDM field is a major advantage of LSW experiments over other types of searches.

In order to increase further their sensitivity, these experiments can use optical cavities on both sides of the wall to increase the power of the regenerated field at the detector. This, however, requires a sophisticated optical system to stabilize the length and alignment of the cavities. Furthermore, the system must have the ability to verify that it is properly tuned, all while suppressing background signals below the sensitivity of the detectors. With detection systems capable of sensitivities on the order of one photon per day, modern LSW experiments such as ALPS II will be able to probe the electromagnetic–UBDM field interaction down to couplings $\sim 2 \times 10^{-11}\,\text{GeV}^{-1}$.

Acknowledgments The author would like to thank Axel Lindner and Jan H. Põld for the illuminating conversations.

References

1. K. Ehret, M. Frede, S. Ghazaryan, M. Hildebrandt, E.A. Knabbe, D. Kracht, A. Lindner, J. List, T. Meier, N. Meyer, D. Notz, J. Redondo, A. Ringwald, G. Wiedemann, B. Willke, Phys. Lett. B **689**, 149 (2010)
2. M. Meyer, D. Horns, M. Raue, Phys. Rev. D **87**, 035027 (2013)
3. M. Giannotti, I. Irastorza, J. Redondo, A. Ringwald, J. Cosmol. Astropart. P. **2016**, 057 (2016)
4. R. Bähre, B. Döbrich, J. Dreyling-Eschweiler, S. Ghazaryan, R. Hodajerdi, D. Horns, F. Januschek, E.A. Knabbe, A. Lindner, D. Notz, et al., J. Instrum. **8**, T09001 (2013)

5. M.D. Ortiz, J. Gleason, H. Grote, A. Hallal, M.T. Hartman, H. Hollis, K.S. Isleif, A. James, K. Karan, T. Kozlowski, A. Lindner, G. Messineo, G. Mueller, J.H. Poeld, R.C.G. Smith, A.D. Spector, D.B. Tanner, L.W. Wei, B. Willke, Phys. Dark Universe **35**, 100968 (2022)
6. P. Arias, J. Jaeckel, J. Redondo, A. Ringwald, Phys. Rev. D **82**, 115018 (2010)
7. L. Bottura, G. De Rijk, L. Rossi, E. Todesco, IEEE Trans. Appl. Supercond. **22**, 4002008 (2012)
8. J. Aasi, B. Abbott, R. Abbott, T. Abbott, M. Abernathy, K. Ackley, C. Adams, T. Adams, P. Addesso, R. Adhikari, et al., Classical Quantum Gravity **32**, 074001 (2015)
9. R.W.P. Drever, J.L. Hall, F.V. Kowalski, J. Hough, G.M. Ford, A.J. Munley, H. Ward, Appl. Phys. B **31**, 97 (1983)
10. E.D. Black, Am. J. Phys. **69**, 79 (2001)
11. E. Morrison, B.J. Meers, D.I. Robertson, H. Ward, Appl. Opt. **33**, 5041 (1994)
12. G. Heinzel, A. Rüdiger, R. Schilling, K. Strain, W. Winkler, J. Mizuno, K. Danzmann, Optics Communications **160**, 321 (1999)
13. H. Grote, G. Heinzel, A. Freise, S. Goßler, B. Willke, H. Lück, H. Ward, M.M. Casey, K.A. Strain, D.I. Robertson, J. Hough, K. Danzmannx, Classical Quantum Gravity **21**, S441 (2004)
14. W. Winkler, K. Danzmann, A. Rüdiger, R. Schilling, Phys. Rev. A **44**, 7022 (1991)
15. L. Glover, M. Goff, J. Patel, I. Pinto, M. Principe, T. Sadecki, R. Savage, E. Villarama, E. Arriaga, E. Barragan, et al., Phys. Lett. **382**, 2259 (2018)
16. A. Hallal, G. Messineo, M.D. Ortiz, J. Gleason, H. Hollis, D.B. Tanner, G. Mueller, A.D. Spector, Phys. Dark Universe **35**, 100914 (2022)
17. Z.R. Bush, S. Barke, H. Hollis, A.D. Spector, A. Hallal, G. Messineo, D. Tanner, G. Mueller, Phys. Rev. D **99**, 022001 (2019)
18. K. Irwin, G. Hilton, *Transition-Edge Sensors* (Springer, Berlin, Heidelberg, 2005), pp. 63–150
19. J. Dreyling-Eschweiler, N. Bastidon, B. Döbrich, D. Horns, F. Januschek, A. Lindner, J. Mod. Optic. **62**, 1132 (2015)

Chapter 10
Global Quantum Sensor Networks as Probes of the Dark Sector

Andrei Derevianko and Szymon Pustelny

Abstract Most dark matter searches to date employ a single sensor for detection. In this chapter, we explore the power of distributed networks in dark matter searches. Compared to a single sensor, networks offer several advantages, such as the ability to probe spatiotemporal signatures of the putative signal and, as a result, an improved rejection of false positives, better sensitivity, and improved confidence in the dark matter origin of the sought-after signal. We illustrate our general discussion with two examples: (1) the Global Network of Optical Magnetometers for Exotic physics searches (GNOME) and (2) the constellation of atomic clocks on board satellites of the Global Positioning System (GPS).

10.1 Introduction

The goal of this chapter is to give an introduction to direct searches for ultralight dark matter (UBDM) using *networks* of precision sensors. This chapter reviews a meta-technique as it combines individual direct searches described in preceding chapters. A single apparatus couples to a dark matter (DM) field at its specific location, while a geographically distributed network can probe DM constituents at multiple locations. Thus a network approach enables testing additional signatures based on spatiotemporal correlation properties of putative DM signals. This leads to both an enhanced sensitivity and to a greater confidence in the DM origin of the sought-after signal (Fig. 10.1).

To reiterate the preceding chapters, our galaxy, the Milky Way, is embedded in a DM halo and rotates through the halo. The Sun moves through the DM halo towards

A. Derevianko (✉)
Department of Physics, University of Nevada, Reno, NV, USA
e-mail: andrei@unr.edu

S. Pustelny
Institute of Physics, Jagiellonian University, Kraków, Poland
e-mail: pustelny@uj.edu.pl

Fig. 10.1 A network of atomic clocks in the "sea" of wavy dark matter. The confidence level in the dark matter origin of the sought signal can be improved because of the specific spatiotemporal correlation properties of virialized DM fields

the Cygnus constellation at galactic velocities $v_g \approx 230$ km/s. Further, in the DM halo reference frame, the velocity distribution of DM objects is nearly Maxwellian with the dispersion of $v_{vir} \sim 270$ km/s (virial velocity) and a cut-off at the galactic escape velocity $v_{esc} \approx 650$ km/s. The DM energy density ρ_{dm} in the vicinity of the Solar system is estimated at the level of 0.3 GeV/cm^3, corresponding to about one hydrogen atom per three cubic cm.

All the evidence for dark matter (galactic rotation curves, gravitational lensing, peaks in the cosmic microwave background spectra, etc.—see discussion in Chap. 1) comes from galactic scale observations. The challenge in planning a laboratory experiment lies in extrapolating down from the 10-kpc characteristic galactic length scales to laboratory scales. These are truly vast extrapolation scales and a large number of theoretical models can fit the observations (as discussed in Chap. 3). For the goals of this chapter, we broadly classify DM candidates as either being "wavy" or "clumpy." As with the ocean, one may distinguish between either a relatively calm surface with characteristic ripples or solitary perturbations such as tsunami that preserve their shape while traveling across many miles. The former is an example of the wavy DM (nearly uniform field composed of many interfering waves) and the latter of the clumpy DM candidates.

The "wavy" DM is typically composed of non-self-interacting ultralight DM candidates. Due to the large mode occupation numbers (see Chaps. 1–3), such fields behave as classical entities coherent on a scale of individual detectors. At a single node, these fields would drive a signal oscillating at the DM field Compton frequency. An important point is that such candidates are *waves*, and while they do induce an oscillating-in-time signal at a given spatial location, DM signals at different locations have a fixed phase relation, i.e., the signals at distinct nodes are correlated. Thereby, a discovery reach can be improved by sampling the DM wave at multiple locations. In the wavy DM models, the DM field is composed of numerous waves traveling at different velocities and in different directions. Interference of DM waves results in a stochastic field, characterized by the coherence length and coherence time (see Chaps. 1–3). Namely, the coherence properties of the DM field determine space-time correlations of the DM signal measured at different nodes. We will discuss the relevant correlation properties of wavy DM fields and network performance in Sect. 10.4.2.

"Clumpy" DM is another distinct theoretical possibility. Here, DM is not distributed uniformly but rather occurs in the form of clumps: massive, large-scale, composite DM objects. Formation of clumps generically requires some form of interaction (self-interaction) between the elementary DM constituents, but even the ever-present gravitational interaction leads to instabilities and clumping (see discussion in Chap. 3). Examples of "clumpy" objects include "dark stars" [1], Q-balls [2, 3], solitons, and clumps formed by dissipative interactions in the DM sector. Alternatively, a significant fraction of the DM mass-energy could be stored in "topological defects" manifesting as monopoles, strings, or domain walls [4]. Self-interacting fields can include bosonic and fermionic DM candidates. The characteristic spatial extent of topological defects is determined by the Compton wavelength of the underlying DM field. For an Earth-sized object, this translates into a characteristic mass of DM field quanta of $\sim 10^{-14}$ eV, which places such DM fields in the category of ultralight candidates.

If DM takes such a "clumpy" form, sensors would not register a continuous oscillating signal associated with the "wavy" DM but rather would observe transient events associated with a DM clump sweeping through the detector [5–7]. Network-based searches seek patterns of synchronous propagation of DM-induced perturbations ("glitches") in sensor data streams; the perturbation is expected to sweep through the network at galactic velocities. The value of the network in searches for DM clumps lies in a much suppressed rate of false positives, as inevitable intrinsic noise (especially flicker noise) of a single-node sensor can mimic an encounter with a DM clump. Moreover, even if the DM-induced glitches are large, an unsuspecting experimentalist is likely to discard the event and attribute it to something perhaps unexplained but mundane (see blog post [8]). An appearance of the same glitch at all the nodes substantially raises confidence level in the detection of the sought-after signal. This strategy is analogous to that of gravitational wave observatories [9], where the same waveform is registered by multiple geographically separated detectors with the prescribed time delays. We will discuss network-based searches for clumpy DM in Sect. 10.4.3.

There are several networks of precision quantum sensors in existence. The authors are involved in the DM searches with atomic clocks and atomic magnetometers and, for concreteness, we focus on networks comprised of these two sensor types. We illustrate our general discussion with two examples: (1) the Global Network of Optical Magnetometers for Exotic physics searches (GNOME) and (2) the constellation of atomic clocks onboard satellites of the Global Positioning System (GPS). Section 10.2 introduces couplings (portals) of ultralight DM fields to the clocks and magnetometers. Essentially, we are interested in interactions that either vary fundamental constants or lead to fictitious magnetic fields coupled to atomic or nuclear spins. Section 10.3 introduces basics of atomic clocks and magnetometers. Section 10.4.1 reviews existing networks of quantum sensors. Network detection of wavy dark matter is discussed in Sect. 10.4.2 and of clumpy dark matter in Sect. 10.4.3. Some of the recent results are presented in Sect. 10.5 and conclusions are drawn in Sect. 10.6. Since the intended audience includes broader physics community, we restore $\hbar$ and c in the formulae in favor of using natural or atomic units.

10.2 Portals Into Dark Sector

Quantitative studies of interaction between the DM and Standard Model (SM) particles/fields require specification of how the two sectors interact. We follow a phenomenological approach of the so-called portals, when the gauge invariant operators of the SM fields are coupled to the operators that contain fields from the dark sector (see, e.g., Ref. [10] and Sect. 2.4.1 of Chap. 2). While a large number of Lorentz-invariant portals can be constructed, here we focus on those that can affect atomic clocks and magnetometers. In this section, we spell out these portals and discuss existing, DM-model independent, constraints on the portals. Proposals for direct DM searches should be more sensitive to new interactions than these established constraints.

In the following, we focus on either scalar or pseudoscalar DM fields ϕ. We consider interaction Lagrangians that are linear, $\mathcal{L}^{(1)}$, and quadratic, $\mathcal{L}^{(2)}$, in ϕ. While linear interactions invariably arise in perturbative treatments, quadratic interactions naturally appear for scalars possessing either $\mathbb{Z}_2$[1] or $\mathbb{U}(1)$ intrinsic symmetries.

For atomic clocks,

$$\mathcal{L}_{\text{clk}}^{(1)} = \left(-\sum_f \Gamma_f^{(1)} m_{f,0} c^2 \bar{\psi}_f \psi_f + \frac{\Gamma_\alpha^{(1)}}{4} F_{\mu\nu} F^{\mu\nu} \right) \sqrt{\hbar c}\, \phi \,, \qquad (10.1)$$

$$\mathcal{L}_{\text{clk}}^{(2)} = \left(-\sum_f \Gamma_f^{(2)} m_{f,0} c^2 \bar{\psi}_f \psi_f + \frac{\Gamma_\alpha^{(2)}}{4} F_{\mu\nu} F^{\mu\nu} \right) \hbar c\, \phi^2 \,. \qquad (10.2)$$

The structure of these portals is such that various parts of the SM Lagrangian are multiplied by DM fields, with Γ's being the associated coupling constants (to be determined or constrained). In the above interactions, f runs over all the SM fermions (fields ψ_f and masses m_f), and $F_{\mu\nu}$ is the electromagnetic Faraday tensor. Here we used the Lorentz-Heaviside system of electromagnetic units that is common

---

[1] Qualitatively, $\mathbb{Z}_2$ symmetry means that for a real-valued field ϕ, the Lagrangian remains invariant under sign swap operation, $\phi \to -\phi$. Thus, ϕ and $-\phi$ obey the same equation of motion.

in particle physics. In these expressions, the combination $\sqrt{\hbar c}\,\phi$ is measured in units of energy, $[E]$, i.e., $\Gamma_X^{(1)}$ are measured in $[E]^{-1}$ and $\Gamma_X^{(2)}$ in $[E]^{-2}$. The $\mathcal{L}_{\text{clk}}$ portals effectively alter fundamental constants [7], such as the electron mass m_e and the fine-structure constant $\alpha = q^2/\hbar c$.

? Problem 10.1 Dark matter-induced variation of fundamental constants

Show that the portals (10.1) and (10.2) lead to the effective redefinition of fermion masses m_f and the fine-structure constant α:

$$m_f(\mathbf{r}, t) = m_{f,0} \times \left[1 + \Gamma_f^{(n)}\left(\sqrt{\hbar c}\,\phi(\mathbf{r}, t)\right)^n\right], \qquad (10.3)$$

$$\alpha(\mathbf{r}, t) \approx \alpha_0 \times \left[1 + \Gamma_\alpha^{(n)}\left(\sqrt{\hbar c}\,\phi(\mathbf{r}, t)\right)^n\right], \qquad (10.4)$$

for the linear ($n = 1$) and quadratic ($n = 2$) portals, where $m_{f,0}$ and α_0 are the nominal (unperturbed) values, i.e., demonstrate that fundamental constants become both space and time dependent.

Solution on page 349.

It is conventional to recast the linear coupling strengths $\Gamma_X^{(1)}$ in terms of dimensionless "moduli" [11] (see Sect. 2.5.3)

$$d_X \equiv \left(\frac{E_{\text{Pl}}}{\sqrt{4\pi}}\right)\Gamma_X^{(1)}, \qquad (10.5)$$

with $E_{\text{Pl}} = \sqrt{\hbar c^5/G_N}$ being the Planck energy and G_N being the Newtonian constant of gravitation. We focus on the electron mass modulus d_{m_e} and the electromagnetic gauge modulus d_e, where $X = \alpha$ in this case. The most stringent limits on these moduli come from the tests of Einstein's equivalence principle violation (see Fig. 1 of Ref. [11]). For the parameter space relevant to atomic clocks, the excluded regions are $d_e \gtrsim 10^{-3}$ and $d_{m_e} \gtrsim 10^{-2}$.

For quadratic couplings, for consistency with prior literature, we work with energy scales

$$\Lambda_X \equiv \frac{1}{\sqrt{|\Gamma_X^{(2)}|}}. \qquad (10.6)$$

The most stringent (DM-model independent) constraints on the energy scales, $\Lambda_{m_e,\alpha} \gtrsim 3\,\text{TeV}$ and $\Lambda_{m_p} \gtrsim 10\,\text{TeV}$, come from the bounds on the thermal emission rate from the cores of supernovae [12]. The authors of Ref. [12] estimated emissivity of ϕ quanta due to the pair annihilation of photons and other processes such as the bremsstrahlung-like emission. They also considered tests of the gravitational force which resulted in similar constraints; compared to the linear Lagrangians $\mathcal{L}_{\text{clk}}^{(1)}$ these are milder, because the quadratic Lagrangians lead to the interaction potentials that scale as an inverse *cube* of the distance between the test bodies as only the exchange of pairs of ϕ's are allowed (for linear Lagrangians, the ϕ-mediated interaction potentials scale as the inverse distance). There are additional limits on quadratic couplings arising from Big Bang nucleosynthesis, black-hole superradiance and other mechanisms which are beyond the scope of this chapter (see Ref. [13] for details).

For magnetometers, we consider the following interaction Lagrangians [5]

$$\mathcal{L}_{\text{mag}}^{(1)} = \frac{1}{f_l} J^{\mu} \partial_{\mu} \phi \,, \tag{10.7}$$

$$\mathcal{L}_{\text{mag}}^{(2)} = \frac{1}{f_q^2} J^{\mu} \partial_{\mu} \phi^2 \,. \tag{10.8}$$

In these expressions, $J^{\mu} = \bar{\psi} \gamma^{\mu} \gamma_5 \psi$ is the axial-vector current for SM fermions and f_l and f_q are the characteristic energy scales (decay constants) associated with the linear and quadratic spin portals. These Lagrangians give rise to the effective spin-dependent interactions

$$\mathcal{H}_{\text{mag}}^{(1)} \approx -\frac{2(\hbar c)^{3/2}}{f_l} \mathbf{S} \cdot \nabla \phi \,, \tag{10.9}$$

$$\mathcal{H}_{\text{mag}}^{(2)} \approx -\frac{2(\hbar c)^2}{f_q^2} \mathbf{S} \cdot \nabla \phi^2 \,, \tag{10.10}$$

where $\mathbf{S}$ is the atomic or nuclear spin.

> **? Problem 10.2 Dark matter-induced pseudo-magnetic field**
>
> Starting from the portal (10.7), derive the spin-dependent interaction Hamiltonian (10.9).
>
> *Solution on page 350.*

Similar to the clocks, the most stringent limits on axion spin couplings f_l and f_q come from astrophysical observations, in particular, supernova 1987A (see discussion in Chap. 3). The basic framework for setting up the constraints comes from analysis of axion production through $N + N \rightarrow N + N + a$, where N is the nucleon and a is the axion. If such a reaction occurs, it would lead to core emission of axions and increased supernova cooling rate. In turn, this would result in shortening of neutrino pulses from the supernova explosion, which was not observed with detectors such as Kamiokande, IMB, and Baksan [14]. However, the axion production would occur under conditions difficult to fully describe, thus a rather conservative limit on the decay constant f_l at 2×10^8 GeV is derived from the observations [15]. Alternatively, the constraint can be formulated based on the kaon decay $K \rightarrow \pi a$, which gives comparable value of 10^8 GeV [16]. For the quadratic coupling f_q the limit is much weaker yielding 10^4 GeV [5, 12]. It is important to note, however, that there do exist theoretical scenarios where these astrophysical bounds can be circumvented [17], and therefore laboratory-based detection experiments as described here play a crucial role.

10.3   How Do Atomic Clocks and Magnetometers Work?

Although atomic clocks and atomic magnetometers measure different physical quantities, at the most fundamental level, both devices effectively measure the energy/frequency splitting between atomic states. For clocks, the atomic levels are chosen in such a way that the transition frequency, ideally, remains independent of external fields and environmental dynamics. Thereby, measurements of the transition frequency provide a reference that can be used for telling time. In contrast, the measured energy-level splitting in atomic magnetometers depends on the spin state and hence the applied magnetic field. In such a way, measurement of the splitting provides information about the strength (and often direction) of the magnetic field. Additionally, a common feature of atomic clocks and magnetometers is that they both employ photons for preparation and monitoring, and sometimes also manipulation, of the atoms used for the measurement.

10.3.1   Atomic Clocks

Measuring time requires observation of a stable periodic process. The elapsed time is simply a product of the number of counted periods and the fixed duration of each period. A grandfather clock is a mechanical realization of this formula: each swing of the pendulum is counted by the escapement mechanism, which advances the clock's hands. In atomic clocks, an atomic transition serves as a frequency reference for an external source of electromagnetic radiation, referred to as the local oscillator (LO). The frequency source is tunable and once its frequency is in resonance with the atomic transition, the period of oscillation is fixed and one counts the number of oscillations at the source. The simple formula "time = number of oscillations × known oscillation period" applies once again.

   One may generally distinguish between two types of atomic clocks: *microwave* and *optical* clocks. This dichotomy is based on the frequency band of the reference atomic transition. In the microwave clocks, two hyperfine levels, associated with a state of a given electronic angular momentum, are used. In the case of alkali-metal atoms, which are often used in atomic-physics experiments, the splitting of two ground-state hyperfine levels ranges between hundreds of MHz (in lithium) to nearly $10\,\text{GHz}$ (in cesium). In fact, the ground-state hyperfine level splitting in ^{133}Cs defines the SI unit of time, the second. Alternatively, in optical clocks, it is the energy splitting between two different electronic states that serves as the frequency reference. As these are typically separated by hundreds of THz, compared to GHz in microwave clocks, optical clocks have better fractional frequency accuracies than their microwave counterparts.

We focus on microwave clocks, as these are used in GPS. The atoms (quantum oscillators) are interrogated with light and microwave radiation. The microwave field is driven from the LO referenced to a microwave cavity. The cavity frequency is tunable and a feedback (servo) loop drives the LO frequency to be in resonance with the reference atomic transition. Technically, atomic clocks measure the quantum phase Φ of an atomic oscillator with respect to that of the LO. The accumulated phase and thereby the quantum probability of a resonant transition is determined by a time integral of the difference in frequencies between the clock atom and the LO. Both the atomic oscillator and the LO can be affected by the DM fields. The DM-induced accumulated-phase difference over measurement time $t_0 = t_j - t_{j-1}$ is

$$\Phi_j^{\text{DM}} = 2\pi \int_{t_{j-1}}^{t_j} \left[\nu_{\text{atom}}^{\text{DM}}(t') - \nu_{\text{LO}}^{\text{DM}}(t') \right] dt' . \qquad (10.11)$$

This DM-induced phase is interpreted by the servo-loop logic as if the cavity frequency ν_{LO} has drifted away from its nominal value ν_{clock}. Technically, the servo-loop would introduce a correction $\Phi_j^{\text{DM}}/(2\pi t_0)$ to the LO frequency ν_{LO}. In other words, DM affects the time as measured by the clocks.

One can simplify the DM-induced phase in two practically relevant cases. If we assume that the characteristic duration τ of DM field action on the clock is much longer than t_0 (slow regime), $\Phi_j^{\text{DM}} \approx 2\pi [\nu_{\text{atom}}^{\text{DM}}(t_j) - \nu_{\text{LO}}^{\text{DM}}(t_j)]t_0$. In the opposite limit of a short transient of duration $\tau \ll t_0$ occurring at time $t' \in (t_{j-1}, t_j)$, $\Phi_j^{\text{DM}} \approx 2\pi [\nu_{\text{atom}}^{\text{DM}}(t_j) - \nu_{\text{LO}}^{\text{DM}}(t_j)]\tau$. Then DM leads to fractional frequency excursions

$$s^{\text{DM}} \equiv \frac{\nu_{\text{atom}}^{\text{DM}} - \nu_{\text{LO}}^{\text{DM}}}{\nu_{\text{clock}}} \times \frac{\min(\tau, t_0)}{t_0} , \qquad (10.12)$$

where the second factor accounts for both the slow and fast regimes.

As discussed in Sect. 10.2, we are interested in portals that lead to variation of fundamental constants, such as the fine-structure constant α or electron mass m_e. Atomic frequencies are primarily affected by the induced variation of the Rydberg constant, $R_\infty = m_e c^2 \alpha^2$. Optical clocks can exhibit additional α dependence due to relativistic effects for atomic electrons. Microwave clocks operate on hyperfine transitions and are additionally affected by the variation in the quark masses, m_q and the strong coupling constant. The reference cavity is also a subject to the DM influence. For example, the variation in the Bohr radius $a_0 = \alpha^{-1}\hbar/(m_e c)$ affects cavity length $L \propto a_0$ and thus the cavity resonance frequencies. Conventionally, one introduces coefficients $\kappa_X = \partial \ln \nu / \partial \ln X$ quantifying sensitivity of a resonance frequency ν to the variation in the fundamental constant X. Then

$$\kappa_{m_e}^{\text{atom}} \approx 1 \, , \kappa_\alpha^{\text{atom}} \approx 2 \, , \kappa_{m_e}^{\text{cavity}} \approx -1 \, , \kappa_\alpha^{\text{cavity}} \approx -1 \, .$$

It is worth noting that there are exceptional cases of enhanced sensitivity to variation of fundamental constants, for example, in the actively pursued ^{229}Th nuclear clock ($\kappa_\alpha \approx 10^4$) [18], and clocks based on highly charged ions ($\kappa_\alpha \lesssim 10^2$) [19].

Since DM portals pull on the fundamental constants [Eqs. (10.3) and (10.4)] and thus on the LO and atomic frequencies, the putative DM signal [Eq. (10.12)] can then be expressed as

$$s(t)^{\text{DM}} = \Gamma_{\text{eff}}^{(n)} \left(\sqrt{\hbar c} \, \phi(t) \right)^n \times \frac{\min(\tau, t_0)}{t_0} \, , \tag{10.13}$$

where $n = 1$ or 2 for the linear and quadratic portals, respectively. Here we also introduced the effective coupling constants

$$\Gamma_{\text{eff}}^{(n)} \equiv \sum_X K_X \Gamma_X^{(n)} \, , \tag{10.14}$$

where $K_X = \kappa_X^{\text{atom}} - \kappa_X^{\text{LO}}$ is the differential sensitivity coefficient and the summation runs over all relevant fundamental constants.

As with any device, there are two issues that must be addressed in experiments with atomic clocks: systematic errors (accuracy) and statistical uncertainties (stability). Systematic errors quantify how well the quantum oscillator is protected from external perturbations. Although the community of physicists working on atomic clock development devotes significant efforts to characterizing clock accuracies, these are not relevant to the goal of detecting DM signals (unless conventional physics perturbations mimic the sought-after DM signatures). Sensitivity to DM portals is determined by the clock stability which quantifies statistical uncertainties. As with most statistical errors, these are reduced by increasing measurement time τ_{meas}. The clock stabilities are characterized using the Allan variance $\sigma_y(\tau_{\text{meas}})$, quantifying the statistical error in fractional clock frequency as a function of the measurement time. Typically, the Allan variance scales as $1/\sqrt{\tau_{\text{meas}}}$. One can interpret $\sigma_y(\tau_{\text{meas}})$ as the error in the determination of the mean fractional clock frequency. In other words, namely $\sigma_y(\tau_{\text{meas}})$ determines the non-DM noise component of the fractional clock excursions (first factor) entering the DM signal (Eq. (10.12)). At $\tau_{\text{meas}} = 1$ s, modern atomic clocks have Allan deviations of $\sim 10^{-12}$ for GPS clocks and 10^{-16} for optical clocks.

10.3.2 Atomic Magnetometers

Magnetic field measurement requires monitoring of a physical quantity that depends on the magnetic field. In the case of (optical) atomic magnetometers, this quantity may be the intensity or polarization of light propagating through a medium subjected to an external magnetic field. If the medium is spin polarized, e.g., by interaction with polarized light (optical pumping), the magnetic field changes the initial spin polarization, which affects the characteristics of the transmitted light.

At the microscopic level, interaction of the field with atomic magnetic moments leads to the precession of the moments at the Larmor frequency $\nu_L = \gamma B/(2\pi)$, where γ is the gyromagnetic ratio for the atom. Periodic evolution of the system (i.e., precession of spins around the magnetic field direction) enables synchronous pumping of the atoms (e.g., by modulating light frequency at the Larmor frequency or its harmonic), leading to a resonant response of the atoms and hence stronger optical signals. Tracking the position of the resonance, by modifying the frequency of the LO driving the modulation, enables accurate magnetic field measurements. Alternatively, atomic magnetometers may be subjected to continuous perturbation, e.g., continuous-wave (CW) light, when competition between such processes as optical pumping, Larmor precession, and spin-polarization relaxation results in appearance of quasi-static optical signals. While this scheme allows measurements of relative field changes (unless the signal is calibrated, in which case absolute measurements can be made) and typically leads to smaller dynamic range (up to about 100 nT), the magnetometers typically have better sensitivities (1 fT/$\sqrt{\text{Hz}}$ or below), than their dynamically driven counterparts.

As shown in Eqs. (10.9) and (10.10) the axion-field gradient acts as a pseudo-magnetic field. Generally, this pseudo-magnetic field differs for electrons and nucleons. By rewriting the Hamiltonians using the total angular momentum $\boldsymbol{F}$ of an atom ($\boldsymbol{F}$ is a sum of electronic spin, electronic angular momentum, and nuclear spin), one obtains

$$\mathcal{H}^{(1)} \approx -\frac{(\hbar c)^{3/2}}{f_l^{\text{eff}}}\frac{\boldsymbol{F}\cdot\nabla\phi}{F}, \tag{10.15}$$

$$\mathcal{H}^{(2)} \approx -\frac{(\hbar c)^2}{\left(f_q^{\text{eff}}\right)^2}\frac{\boldsymbol{F}\cdot\nabla\phi^2}{F}, \tag{10.16}$$

where f_l^{eff} and f_q^{eff} are linear and quadratic effective decay constants. The relation of the effective decay constants $f_{l,q}^{\text{eff}}$ to the electron f_e, proton f_p, and neutron f_n decay constants can be calculated as described, for example, in Ref. [20]. In the case of ^{3}He and ^{39}K, two atoms often used in atomic magnetometers, the linear coupling constants take the form

$$^3\text{He:} \qquad \frac{1}{f_l^{\text{eff}}} = \frac{1}{f_n}, \tag{10.17}$$

$$^{39}\text{K}: \quad \frac{1}{f_l^{\text{eff}}} = \frac{1}{4 f_e} - \frac{3}{20 f_p}, \tag{10.18}$$

where we assumed that the angular momentum F is mostly due to an unpaired neutron in ^{3}He and the $d_{3/2}$ valence proton in ^{39}K.

Since the exotic spin couplings are orders of magnitude weaker than the conventional Zeeman interaction, suppression of Zeeman coupling to any uncontrollable (e.g., environmental) magnetic fields becomes of prime importance. Therefore, most experiments searching for exotic spin couplings are housed inside magnetic shields. The shields, commonly made of high permeability material (e.g., mu-metal or ferrites), passively attenuate stray magnetic fields by a factor on the order of 10^6. As shown in Ref. [21], for many experimental geometries and conditions, magnetic shields do not substantially reduce the sensitivity to exotic spin couplings. To further reduce the sensitivity to the magnetic fields, the magnetometers can be operated in the so-called comagnetometer arrangement, where two distinct atoms or nuclei are used for field sensing (see Chap. 8 for more information). Often, a noble gas and alkali atoms are used, as they sense the field through the coupling to the nuclear spin in the first case and predominantly through the coupling to the electronic spin in the second case. Comparison of the responses of the species to the magnetic field removes sensitivity to the field, leaving system sensitive to other, particularly exotic, couplings. Moreover, due to principally different coupling of exotic physics to different atomic species, comagnetometry also allows disentangling individual couplings to electrons, protons, and neutrons.

At the most fundamental level, the sensitivity of atomic magnetometers is determined by the quantum nature of atoms and light, i.e., by the spin-projection and photon shot noise. In an optimized system, the magnetometric sensitivity is determined as described in, e.g., Auzinsh et al. [22], Ledbetter et al. [23]

$$\delta B_{\text{opt}} = \frac{\hbar}{g \mu_B} \sqrt{\frac{1}{N_{\text{at}} T_2 \tau_{\text{meas}}}}, \tag{10.19}$$

where g is the Landé factor, N_{at} is the number of atoms involved in field sensing, T_2 is the transverse spin relaxation time, and τ_{meas} is the duration of the measurement. This gives the fundamental sensitivity limit between 0.01 and 1 fT/$\sqrt{\text{Hz}}$ for a typical magnetometer.

? Problem 10.3 Atomic-projection noise limit on magnetometric sensitivity

Derive the atomic-projection limit on the sensitivity of an atomic magnetometer.

Solution on page 350.

10.4   DM Searches with Network of Sensors

10.4.1   Overview of Existing Networks

Networks are ubiquitous in our life, with one of the most well-known examples being the internet. In telecommunication settings, the utility of the network is proportional to the square of the number of nodes (Metcalfe's law). This scaling law reflects the total number of unique connections $N_s(N_s - 1)/2$ for the number of nodes N_s. If one considers that the price of the network increases as N_s, there is a certain critical number of nodes above which the network becomes economically viable. Similar considerations (with significant caveats) apply to networks of quantum sensors. One can argue that the sensitivity of a classical network to the exotic physics should improve generically as $\sqrt{N_s}$ since the same putative signal is measured independently by N_s sensors. However, such an argument neglects the vetoing power of the interconnected network that results in a reduced rate of false positives. In addition, the cost of deploying N_s sensors in research environment is vastly different from commercial settings. The reason is that the cost of developing a single table-top sensor in a university lab vastly exceeds the cost of the hardware. Thus the cost of the second identical sensor is mostly the cost of the hardware (economy of scale). Another possibility is an integration of already developed sensors, then the additional cost is the cost of synchronizing data acquisition or links.

Perhaps the most widely celebrated network in the physics community is LIGO (the Laser Interferometer Gravitational Wave Observatory)—a gravitational wave observatory initially consisted of two sites in the US. While this network is adding more locations, the original black-hole merger gravitational wave detection from 2015 used only two spatially separated interferometers in the waveform template matching [9]. The appearance of the same waveform in both interferometers, with the proper time delay between the two, greatly supported the credibility of the discovery claim.

There are several criteria [25] for a network to detect the signal pattern due to a macroscopic DM object sweeping through a network of N_s sensors:

(i) The network should be sufficiently dense so that the string and monopole-type DM objects can overlap with at least several geographically distinct nodes.

(ii) The network volume should be sufficiently large in order to increase the rate of encounters with string and monopole-type DM objects.

(iii) Per the standard halo model the DM objects sweep through the network at galactic velocities ($v_g \sim 300\,\text{km/s}$). Thus the sampling rate should be sufficiently high to enable tracking the propagation of the DM object through the network (see Fig. 10.2). The tracking enables reconstruction of the geometry and dynamics of the encounter.

Fig. 10.2 Simulated
response of an Earth-scale
network of atomic clocks to a
spherically symmetric
Gaussian-profiled dark matter
clump (a monopole or a
Q-ball). The traces in the
bottom panel show
time-evolution of the phase of
quantum oscillators for three
distinct locations. From
Ref. [7]

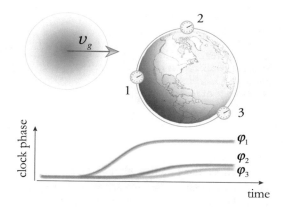

(iv) Although not necessary, it is desirable that the encounters of DM objects with
     the network are sufficiently rare, so that only a single DM object interacts with
     the network at any given time.

A particular example of a global network fulfilling these criteria is the GPS. The
network is nominally comprised of $N_s = 32$ satellites in a medium-Earth orbit
(altitude $\sim 20,000$ km). Microwave Rb or Cs atomic clocks onboard the satellites
drive microwave signals, which are broadcast to Earth. A network of specialized
Earth-based GPS receivers measures the carrier phase of these microwave signals,
which is then used to deduce the satellite clock data. The network can be extended
to incorporate clocks from high-quality Earth-based receiver stations and other
navigation systems, such as the European Galileo, Russian GLONASS, and Chinese
BeiDou, and networks of laboratory clocks [26]. An additional and important
advantage of the GPS network is the public availability of nearly two decades of
archival data enabling relatively inexpensive data mining. Such searches for dark
matter-induced transient variation of fundamental constants are the focus of the
GPS.DM collaboration [7, 27].

GNOME is a network of shielded optical atomic magnetometers specifically
targeting transient events associated with exotic physics [28]. To the best of
our knowledge, this is the first network ever constructed specifically to search
for physics beyond the SM. Presently, GNOME consists of $N_s = 12$ atomic
magnetometers located at stations throughout the world (six sensors in North
America, five in Europe, three in Asia, and one in Australia), with a number of new
stations under construction in Israel, India, and Germany [29]. Each magnetometer
is located within a multilayer magnetic shield to reduce the influence of magnetic
noise. The overall network sensitivity is close to 100 fT depending on the number of
stations active [30]. It is noteworthy that besides the traditional analysis technique
that takes advantage of the spatiotemporal pattern of the network signal to veto false
positives, GNOME offers further ability to limit the rate of false positives. Due to the
pseudoscalar character of the coupling in atomic magnetometers (Sect. 10.3.2) and
different directions of spin polarization in specific GNOME stations, the amplitude

Fig. 10.3 A map of existing low-energy precision measurement laboratories (red dots) around the globe. Such a network can serve as a global dark matter observatory. Adopted from Ref. [24]

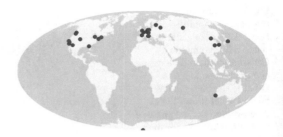

and sign of the putative signals also carry information about the coupling. This information can be used to further improve rejection of false positives. Additionally, several stations are implementing sensors employing a dense polarized noble gas in a comagnetometer configuration [31]. This arrangement has a reduced sensitivity to magnetic couplings and hence enhanced sensitivity to exotic spin couplings. This new network of noble-gas-based comagnetometers will form an Advanced GNOME with an anticipated sensitivity to spin couplings a hundred times better than the existing GNOME. As of the summer of 2020, there is about a year of GNOME data collected which can be analyzed to search for exotic physics signals, and results of the first search for dark matter using the GNOME data set has recently been completed as described in Ref. [32].

Let us finally reiterate a vision for a global dark matter observatory [24] (see Fig. 10.3), that is a natural extension of the GNOME architecture. Up to date, individual direct DM searches employ a broad range of sensors: atomic clocks, magnetometers, accelerometers, interferometers, cavities, resonators, permanent electric-dipole and parity-violation measurements, and extend to gravitational wave detectors (see Ref. [33] for a review). These distinct tools are typically located at geographically separated laboratories across several continents (see Fig. 10.3) or in space. These tools already form nodes of the network and only the links are missing. In the most basic version, even the physical links are not necessary as the synchronization can be implemented with a GPS-assisted time-stamping of data acquisition [6, 34]. Some of the enumerated instruments are sensitive to the same portals (e.g., atomic clocks, cavities, atom interferometers, and gravimeters are all sensitive to the DM-induced variation of fundamental constants), which would lead to an important complementarity of the searches at individual nodes.

? Problem 10.4 Noise suppression of false positive events with a sensor network

Determine the suppression of the false positive event rate by introduction of an additional sensor into a network searching for "clumpy" dark matter.

Solution on page 352.

10.4.2   Network-Based Searches for "Wavy" Dark Matter

In the wavy models, DM is composed of ultralight spin-0 bosonic fields, oscillating at their Compton frequency $\omega_\phi = m_\phi c^2/\hbar$, where m_ϕ is the boson mass, see Chaps. 1 and 2. Multiple proposals covered in this book focus on searching for an oscillating signal at the Compton frequency. Unfortunately, in a laboratory environment, an observation of an oscillating signal could be ascribed to some mundane ambient noise and it is desirable to establish additional DM signatures. Due to the DM virial velocity distribution, these DM fields are stochastic in nature (again, see Chap. 1) and we refer to them as virialized ultralight fields (VULFs). Their coherence times and coherence lengths are related to DM properties. An additional signature [24] relies on a VULF spatiotemporal correlations that can be probed with a network. Formally, the two-point field correlation function is defined as

$$g\left(\Delta t, \Delta \mathbf{r}\right) = \langle \phi\left(t' = t + \Delta t, \mathbf{r}' = \mathbf{r} + \Delta \mathbf{r}\right) \phi\left(t, \mathbf{r}\right) \rangle,$$

where averaging is over stochastic realizations of the DM field.

DM field correlations imprint correlations on the putative DM signal. Indeed, in the assumption of the linear portals, see Sect. 10.2, the measured quantity has a DM-induced admixture $s_X(t, \mathbf{r})$ that is proportional to the field value $\phi(t, \mathbf{r})$ at the device location. Then the correlation between DM signals at the two locations is related through the DM field correlation function

$$\langle s_X\left(t', \mathbf{r}'\right) s_X\left(t, \mathbf{r}\right) \rangle \propto \langle \phi\left(t', \mathbf{r}'\right) \phi\left(t, \mathbf{r}\right) \rangle.$$

The correlation function for spatiotemporal variations of fundamental constants is also expressed in terms of DM field correlation function, e.g.,

$$\frac{\langle \alpha\left(t', \mathbf{r}'\right) \alpha\left(t, \mathbf{r}\right) \rangle}{(\alpha_0)^2} = 1 + \hbar c \left(\Gamma_\alpha^{(1)}\right)^2 g\left(\Delta t, \Delta \mathbf{r}\right),$$

where we used the DM-induced variation (10.4) of the fine-structure constant.

The correlation function derived in Ref. [24] reads

$$g\left(\Delta t, \Delta \mathbf{r}\right) \approx \frac{1}{2} \Phi_0^2 \, \mathcal{A}\left(\Delta t, \Delta \mathbf{r}\right) \cos\left[\omega_\phi' \Delta t - \mathbf{k}_g \cdot \Delta \mathbf{r} + \Psi\left(\Delta t, \Delta \mathbf{r}\right)\right]. \quad (10.20)$$

Here ω_ϕ' is the Doppler-shifted value of the Compton frequency $\omega_\phi' = \omega_\phi + m_\phi v_g^2/(2\hbar)$ and $\mathbf{k}_g = m_\phi \mathbf{v}_g/\hbar$ is the "galactic" wave vector associated with the apparatus motion through the DM halo (towards the Cygnus constellation). The effective field amplitude Φ_0 is related to the DM energy density as $\Phi_0 = \frac{\hbar}{m_\phi c}\sqrt{2\rho_{\mathrm{dm}}}$, which comes from directly evaluating the time-like (00) component of the stress-energy tensor for the bosonic field. Further, the amplitude and phase are defined as

$$\mathcal{A}(\Delta t, \Delta \mathbf{r}) = \frac{\exp\left(-\frac{|\Delta \mathbf{r} - \mathbf{v}_g \Delta t|^2}{2\lambda_c^2} \frac{1}{1 + (\Delta t/\tau_c)^2}\right)}{\left[1 + (\Delta t/\tau_c)^2\right]^{3/4}}, \tag{10.21}$$

$$\Psi(\Delta t, \Delta \mathbf{r}) = -\frac{|\Delta \mathbf{r} - \mathbf{v}_g \Delta t|^2}{2\lambda_c^2} \frac{\Delta t/\tau_c}{1 + (\Delta t/\tau_c)^2} + \frac{3}{2} \tan^{-1}(\Delta t/\tau_c),$$

where the coherence time $\tau_c \equiv (\xi^2 \omega_\phi)^{-1} \approx 10^6/\omega_\phi$ and coherence length $\lambda_c \equiv \hbar/(m_\phi \xi c)$ are expressed in terms of the virial velocity $\xi c \approx 10^{-3} c$. The correlation function encodes the priors on VULFs and the DM halo, such as the DM energy density in the vicinity of the Solar system, motion through the DM halo at $\mathbf{v}_g$ and the virial velocity ξc. Thereby, the correlation function provides an improved statistical confidence in the event of an observation of a DM signal.

The *N-point* correlation function required for the multi-node network can be fully expressed in terms of the derived *two-point* correlation function since the DM field is Gaussian in nature. The statistical significance of the correlation function for a network was explored in Ref. [24]. If all N_s nodes are separated by distances larger than the coherence length λ_c, compared to a single apparatus, the network sensitivity improves as $N_s^{1/4}$. In the opposite limit of the node separations being much smaller than λ_c (fully coherent network), the statistical sensitivity is improved by the factor $\sqrt{N_s}$. Network searches for wavy DM are in their infancy, but are expected to gain in significance once the oscillating DM signal is discovered. The network will be necessary to confirm the DM origin of the signal.

10.4.3 Network-Based Searches for "Clumpy" Dark Matter

In the "clumpy" dark matter models, DM is postulated to be composed of macroscopic objects, such as topological defects (TDs). Monopoles (0D), strings (1D), and domain walls (2D) are all examples of TDs of various dimensionalities. Other examples of macroscopic DM candidates include "dark stars" [1], Q-balls [2, 3], solitons, and clumps formed due to dissipative interactions in the DM sector. A special case of clumpy DM are DM "blobs" [35], particle-like DM objects sourcing long-range Yukawa-type interactions with the SM sector.

As an illustration, we focus on topological defects. Inside the defect, the amplitude of the DM field $\mathcal{A}$ and the energy density of the defect is related by $\rho_{\text{inside}} = \mathcal{A}^2/(\hbar c\, d^2)$, where d is the width of the defect (we use the convention where the field has units of energy). The DM object width d is treated as a free observational parameter. For topological defects, this width may be linked to the mass of the DM field particles m_ϕ through the healing length which is on the order of the Compton wavelength, $d \sim \hbar/(m_\phi c)$. Further, the local DM energy density ρ_{dm} may be expressed in terms of d and $\mathcal{A}$ by assuming that these objects saturate the local DM energy density,

Fig. 10.4 While crossing
through the domain wall, wall
timings and amplitudes of the
signals recorded in different
GNOME nodes (red dots)
form a spatiotemporal pattern
that enables determination of
the properties of dark matter
and reduce a false positive
rate. Courtesy of Arne
Wickenbrock

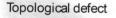

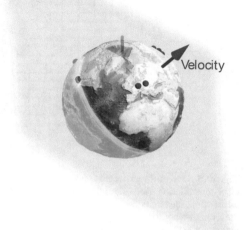

$$\mathcal{A}^2 = \hbar c \, \rho_{\mathrm{dm}} d^2 \frac{\mathcal{T}_e}{\tau}, \qquad\qquad (10.22)$$

where $\tau \sim d/v_g$ is the characteristic duration of crossing through a point-like
instrument and $\mathcal{T}_e$ is the average time between encounters of the sensor with DM
clumps. These relations hold for all types of defects.

So far both the GPS.DM and GNOME searches focused on domain walls
(Fig. 10.4). Their signature is especially simple as the wall would cross all the
sensors with the same amplitude of the DM signal. Domain wall-like signatures
can appear naturally in the context of bubbles, i.e., domain walls closed on
themselves [27]. Locally, one can neglect the bubble curvature as long as the bubble
radius is much larger than the spatial extent of the sensor network. Since bubbles are
spherically symmetric, gravitationally interacting ensembles of these DM objects
are a subject to the equation of state for pressureless cosmological fluid as required
by the ΛCDM paradigm, see Chap. 3.

An example of a DM signature for a "thin" domain wall ($d/v_g < t_0$) sweeping
through the GPS constellation is shown in Fig. 10.5. GPS clock data are reported
with respect to some other fixed (reference) clock. Thus the signal pattern would
involve DM-induced perturbations to both satellite clocks and the reference clock.
For identical types of reference and satellite clocks, the domain wall creates a
perturbation of the reference clock that leads to a "timing glitch" of equal magnitude

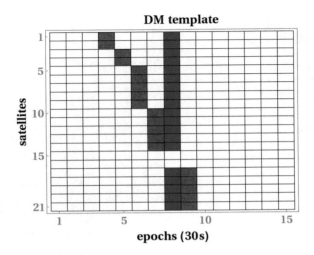

Fig. 10.5 One of the expected frequency signatures for a thin domain wall sweeping through the GPS constellation. Red (blue) tiles indicate positive (negative) DM-induced frequency excursions, while white tiles mark the absence of the signal. In this example, the satellites are listed in the order they were swept (though in general the order depends on the incident direction of the DM object and is not known a priori). The slope of the red line encodes the incident velocity of the wall. The reference clock was swept within the 30 s leading to epoch 8. Satellites 15 and 16 do not record any frequency excursions, since they are spatially close to (degenerate with) the reference clock and are swept within the same sampling period. Adopted from Ref. [27]

but opposite sign as compared to those appearing when the domain wall passes through satellite clocks.[2] Consider the pattern of "glitches" in GPS clocks shown in Fig. 10.5. The domain wall first sweeps through satellites 1 and 2 in epoch 4, causing a temporary positive frequency excursion, then encounters clocks 3 and 4 in the next epoch, and so on. When the domain wall passes through the reference clock in epoch 8, most of the clocks show a temporary negative frequency excursion, since the reference clock itself experiences a positive frequency excursion (the exceptions being satellites 15 and 16 which are spatially close to the reference clock). In the case of GNOME, however, there is no reference magnetometer, i.e., all the magnetometers are independent. Thus for identical magnetometers the sought-after pattern would involve only the "diagonal" (red tiles) in Fig. 10.5.

---

[2] This is simply a consequence of the fact that the timing glitches are determined by comparison between the reference and satellite clocks. If, for example, the glitch causes a clock to temporarily run fast, when the domain wall passes through the reference clock, satellite clocks appear to run slow with respect to the reference clock, whereas when the domain wall passes through a satellite clock, the satellite clock appears to run fast compared to the reference clock.

10.5   Putting It All Together

In this section, we illustrate the implementation of the described ideas with a search for clumpy DM [27] using archival GPS data. The archival GPS data is publicly available and the dataset includes atomic clock data and satellite positions sampled every $t_0 = 30$ s.

Returning to the discussion of Sect. 10.3.1, we focus on "thin" domain walls and quadratic couplings. Then, with Eq. (10.22) for the DM field amplitude, the DM signal (10.13) becomes

$$s_0^{DM} = \Gamma_{eff}^{(2)} (\hbar c)^2 \, \rho_{dm} d^2 \frac{\mathcal{T}_e}{\tau} . \tag{10.23}$$

This is the amplitude of the signal during the time interval when the wall overlaps with a sensor, otherwise there is no signal, $s_0^{DM} = 0$.

The key qualifier for Eq. (10.23) is that one must be able to distinguish between the clock noise and DM-induced frequency excursions. Discriminating between the two sources relies on measuring time delays between DM events at network nodes, see Fig. 10.5. The velocity of the sweep is encoded in the time delay between two DM-induced frequency excursion and it must lie within the boundaries predicted by the standard halo model (the distributed response of the network encodes the spatial structure and kinematics of the DM object and its coupling to the sensors).

To search for domain wall signals, the GPS.DM collaboration analyzed the GPS data streams in two stages [27]. At the first stage, they scanned all the data from October 2016 to May 2000 searching for the most general patterns associated with a domain wall crossing, without taking into account the order in which the satellites were swept. They required at least 60% of the clocks to experience a frequency excursion at the same epoch, which would correspond to when the wall crossed the reference clock (vertical blue line in Fig. 10.5). This 60% requirement is a conservative choice based on the GPS constellation geometry and ensures sensitivity to walls with relative speeds of up to 700 km/s. Then, the GPS.DM collaboration checked if these clocks also exhibit a frequency excursion of similar magnitude (accounting for clock noise) and opposite sign anywhere else within a given time window (red tiles in Fig. 10.5). Any epoch for which these criteria were met was counted as a "potential event." Above a certain threshold for the DM signal (10.23), no potential events were seen.

The second stage of the search involved analyzing the potential events in more detail, so that their status could be elevated to "candidate events" if warranted by the evidence. GPS.DM examined a few hundred potential events that had frequency excursions magnitudes just below the threshold values, by matching the data streams against the expected patterns, where the velocity vector and wall orientation were treated as free parameters. At this second stage, GPS.DM accounted for the ordering and time at which each satellite clock was affected. This was done by matching data against a bank of signal templates that was itself generated using DM halo

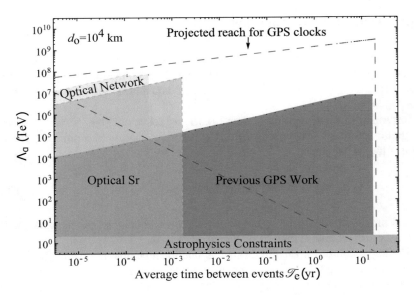

Fig. 10.6 Projected discovery reach for thin wall dark matter objects along with existing constraints. The red dashed lines represent the least stringent and most stringent discovery reaches for the 2010 GPS atomic clock network. The shaded cyan region are the constraints coming from astrophysics [12], while the salmon shaded regions are the constraints placed by the GPS.DM collaboration [27]. The green shaded region contains the constraints placed by optical clock experiments [36], while the yellow region—by a global terrestrial network of laboratory clocks [37]. Adopted from Ref. [25]

properties. As a result of this pattern matching, none of these events was consistent with domain wall signals.

Since GPS.DM did not find evidence for encounters with domain walls, there are two possibilities: either DM of this nature does not exist, or the DM signals are below the sensitivity. The derived limits on quadratic coupling energy scale Λ_α are presented in Fig. 10.6 (salmon shaded region). Here we assumed for simplicity that the DM-induced variation in the fine-structure constant α dominates. These limits represent a significant improvement over the astrophysical bounds, discussed in Sect. 10.2.

The ultimate discovery reach of the clock network is given by [25]

$$|\Gamma_X^{(2)}| \leq \frac{C}{\sqrt{N_E N_s}} \frac{\sigma_y(t_0)}{\hbar c \rho_{dm} \mathcal{T}_e d^2 K_X} . \qquad (10.24)$$

Again we assumed that a specific coupling constant Γ_X dominates so that $\Gamma_{eff} \rightarrow K_X \Gamma_X$. Here $C \sim O(1)$ is a constant that depends on the confidence level and details of the network implementation, $N_E = \mathcal{T}/\mathcal{T}_e$ is the total number of expected encounters with DM objects during total observation time $\mathcal{T}$ (as of 2020 $\sim$ 20 years for GPS archival data). This constraint translates into projected exclusion limits on the effective energy scale $\Lambda_X = 1/\sqrt{|\Gamma_X|}$, shown in Fig. 10.6. Notice that the derived constraints from the first GPS.DM search [27] are several orders of magnitude weaker than the projected limits. Reaching the full discovery potential requires implementing more advanced statistical approaches, such as Bayesian statistics [38] or matched filter techniques [25]; these are a subject of current efforts by the GPS.DM collaboration.

10.6   Summary

Distributed sensor networks, either at a local or a global scale, are one of the powerful strategies used in scientific research. Extension of these ideas to networks of precision quantum sensors opens novel opportunities, in particular searches for a variety of dark matter candidates. While these developments are in their early stages, we believe that additional spatiotemporal signatures offered by the networks are key to improving confidence in the DM origin of a putative signal (if discovered). Networks also enable a powerful rejection of false positives.

Searching for dark matter using sensor networks is a rapidly developing research area: soon after the initial proposals [5, 7] based on atomic magnetometers and atomic clocks, several new schemes involving terrestrial networks of optical clocks, atom interferometers, and optical cavities [37, 39] have been proposed. We also mention recently proposed space missions [40, 41] that involve rudimentary networks of atomic clocks and interferometers; while these missions focus on detection of gravitational waves, DM searches are one of their secondary goals.

Finally, once a network is built, it may find other applications. For example, quantum sensor networks can be used in searches for exotic low-mass fields (ELFs) emitted in cataclysmic astrophysical events such as black-hole or neutron-star mergers [42]. This idea opens up an intriguing "exotic physics" modality in multi-messenger astronomy, where the measured exotic physics signals are correlated with gravitational wave triggers and the progenitor position in the sky.

Acknowledgments We would like to thank members of GPS.DM and GNOME collaborations for numerous discussions. This work of A.D. was supported in part by the U.S. National Science Foundation under Grant Nos. PHY-1806672 and PHY-1912465 and the work of S.P. was supported in part by the National Science Centre, Poland.

References

1. E.W. Kolb, I.I. Tkachev, Phys. Rev. Lett. **71**, 3051 (1993)
2. S. Coleman, Nucl. Phys. B **262**, 263 (1985)
3. A. Kusenko, P.J. Steinhardt, Phys. Rev. Lett. **87**, 141301 (2001)
4. A. Vilenkin, Phys. Rep. **121**, 263 (1985)

5. M. Pospelov, S. Pustelny, M.P. Ledbetter, D.F. Jackson Kimball, W. Gawlik, D. Budker, Phys. Rev. Lett. **110**, 021803 (2013)
6. D. Budker, A. Derevianko, Phys. Today **68** (2015)
7. A. Derevianko, M. Pospelov, Nat. Phys. **10**, 933 (2014)
8. A. Derevianko, When would an unanticipated "new physics" event be apparent to an unsuspecting experimentalist? http://wp.me/p2Z9xm-8Z
9. B.P. Abbott, et al., Phys. Rev. Lett. **116**, 061102 (2016)
10. R. Essig, J. Jaros, W. Wester, arXiv:1311.0029 (2013)
11. A. Arvanitaki, S. Dimopoulos, K. Van Tilburg, Phys. Rev. Lett. **116**, 031102 (2016)
12. K.A. Olive, M. Pospelov, Phys. Rev. D **77**, 043524 (2008)
13. S. Sibiryakov, P. Sørensen, T.T. Yu, J. High Energ. Phys. **2020**, 75 (2020)
14. G.G. Raffelt, Annu. Rev. Nucl. Part. Sci. **49**, 163 (1999)
15. J.H. Chang, R. Essig, S.D. McDermott, J. High Energ. Phys. **2018**, 51 (2018)
16. G. Marques-Tavares, M. Teo, J. High Energy Phys. **2018**, 180 (2018)
17. W. DeRocco, P.W. Graham, S. Rajendran, Phys. Rev. D **102**, 075015 (2020)
18. V.V. Flambaum, Phys. Rev. Lett. **97**, 092502 (2006)
19. M.G. Kozlov, M.S. Safronova, J.R.C. López-Urrutia, P.O. Schmidt, Rev. Mod. Phys. **90**, 045005 (2018)
20. D.F. Jackson Kimball, New J. Phys. **17**, 073008 (2015)
21. D.F. Jackson Kimball, J. Dudley, Y. Li, S. Thulasi, S. Pustelny, D. Budker, M. Zolotorev, Phys. Rev. D **94**, 082005 (2016)
22. M. Auzinsh, D. Budker, D.F. Kimball, S.M. Rochester, J.E. Stalnaker, A.O. Sushkov, V.V. Yashchuk, Phys. Rev. Lett. **93**, 173002 (2004)
23. M.P. Ledbetter, I.M. Savukov, V.M. Acosta, D. Budker, M.V. Romalis, Phys. Rev. A **77**, 033408 (2008)
24. A. Derevianko, Phys. Rev. A **97**, 042506 (2018)
25. G. Panelli, B.M. Roberts, A. Derevianko, EPJ Quantum Technol. **7**, 5 (2020)
26. F. Riehle, Nat. Photon. **11**, 25 (2017)
27. B.M. Roberts, G. Blewitt, C. Dailey, M. Murphy, M. Pospelov, A. Rollings, J. Sherman, W. Williams, A. Derevianko, Nat. Comm. **8**, 1195 (2017)
28. S. Afach, D. Budker, G. DeCamp, V. Dumont, Z.D. Grujić, H. Guo, T.W. Jackson Kimball, D. F. Kornack, V. Lebedev, W. Li, H. Masia-Roig, S. Nix, M. Padniuk, C.A. Palm, C. Pankow, C. Penaflor, X. Peng, S. Pustelny, T. Scholtes, J.A. Smiga, J.E. Stalnaker, A. Weis, A. Wickenbrock, D. Wurm, Phys. Dark Univ. **22**, 162 (2018)
29. GNOME website. https://budker.uni-mainz.de/gnome/
30. H. Masia-Roig, J.A. Smiga, D. Budker, V. Dumont, Z. Grujic, D. Kim, D.F. Jackson Kimball, V. Lebedev, M. Monroy, S. Pustelny, T. Scholtes, P.C. Segura, Y.K. Semertzidis, Y.C. Shin, J.E. Stalnaker, I. Sulai, A. Weis, A. Wickenbrock, Phys. Dark Univ. **28**, 100494 (2020)
31. T.W. Kornack, M.V. Romalis, Phys. Rev. Lett. **89**, 253002 (2002)
32. S. Afach, B.C. Buchler, D. Budker, C. Dailey, A. Derevianko, V. Dumont, N.L. Figueroa, I. Gerhardt, Z.D. Grujić, H. Guo, et al. Nat. Phys. **17**, 1396 (2021)
33. M.S. Safronova, D. Budker, D. DeMille, D.F. Jackson Kimball, A. Derevianko, C.W. Clark, Rev. Mod. Phys. **90**, 025008 (2018)
34. P. Wlodarczyk, S. Pustelny, D. Budker, M. Lipinski, Nucl. Instrum. Methods **150**, 763 (2014)
35. D.M. Grabowska, T. Melia, S. Rajendran, Phys. Rev. D **98**, 115020 (2018)
36. P. Wcisło, P. Morzyński, M. Bober, A. Cygan, D. Lisak, R. Ciuryło, M. Zawada, Nat. Astron. **1**, 0009 (2016)
37. P. Wcislo, P. Ablewski, K. Beloy, S. Bilicki, M. Bober, R. Brown, R. Fasano, R. Ciurylo, H. Hachisu, T. Ido, J. Lodewyck, A. Ludlow, W. McGrew, P. Morzyński, D. Nicolodi, M. Schioppo, M. Sekido, R. Le Targat, P. Wolf, X. Zhang, B. Zjawin, M. Zawada, Sci. Adv. **4**, eaau486 (2018)

38. B.M. Roberts, G. Blewitt, C. Dailey, A. Derevianko, Phys. Rev. D **97**, 083009 (2018)
39. L. Badurina, et al., J Cosmol. Astropart. P. **2020**, 011 (2020)
40. G.M. Tino, et al., Eur. Phys. J. D **73**, 228 (2019)
41. Y.A. El-Neaj, et al., EPJ Quantum Technol. **7**, 6 (2020)
42. C. Dailey, C. Bradley, D.F. Jackson Kimball, I.A. Sulai, S. Pustelny, A. Wickenbrock, A. Derevianko, Nat. Astron. **5**, 150 (2021)

Correction to: The Search for Ultralight Bosonic Dark Matter

Derek F. Jackson Kimball and Karl van Bibber

Correction to:
D. F. Jackson Kimball, K. van Bibber (eds.), *The Search for Ultralight Bosonic Dark Matter*,
https://doi.org/10.1007/978-3-030-95852-7

This book was inadvertently published without updating the following corrections:

Chapter 2
The original version of this chapter was published with incorrect text on page 48 last paragraph 8th and 9th lines. Now, the changes have been updated in the chapter.

Solutions to Chapter Problems
The original version of "Solutions to Chapter Problems" was published with incorrect text on page 307 last paragraph 3rd line. Now, the changes have been updated in the chapter.

The updated original version for this book can be found at
https://doi.org/10.1007/978-3-030-95852-7_2
https://doi.org/10.1007/978-3-030-95852-7

Solutions to Chapter Problems

Since most of the galaxy's mass M is within the radius R of the star's circular orbit, the star's centripetal acceleration (v^2/R, where v is the star's rotational velocity) is equal to the gravitational pull of the galaxy divided by the star's mass:

$$\frac{v^2}{R} \approx \frac{G_N M}{R^2} \,, \tag{A.1}$$

and so

$$v \approx \sqrt{\frac{G_N M}{R}} \,. \tag{A.2}$$

Thus, $v \propto 1/\sqrt{R}$. If we assume that, in fact, the galaxy's mass is dominated by a spherical distribution of dark matter so that

$$M(R) \approx \int_0^R 4\pi \rho_{\rm dm}(r) r^2 dr \,, \tag{A.3}$$

where $\rho_{\rm dm}(r)$ is the dark matter density, to obtain the observed flat rotation curve, we demand that $\rho_{\rm dm}(r) \propto 1/r^2$. This yields $M(R) \propto R$, and thus based on Eq. (A.2), v is independent of R.

The original version of the chapter has been revised. A correction to this chapter can be found at https://doi.org/10.1007/978-3-030-95852-7_11

Problem 1.2: Minimum Mass of Fermionic Dark Matter

The spin–statistics theorem demands that only a single fermion can occupy a given quantum state, and so there is an upper bound on the possible fermion density in the dark matter halo. There is also an upper bound on the speed of the dark matter particles: to remain trapped within the gravitational potential of the galaxy, they cannot exceed the escape velocity v_{esc}. These two bounds conspire to set a lower limit m^* on the fermionic dark matter mass m.

The existence of a lower bound on m can be understood qualitatively in the following way. The maximum number density of fermions is capped at $\sim 1/\lambda_{dB}^3$, where λ_{dB} is the de Broglie wavelength of the fermions; this is the case when there is about one fermion per quantum state. This caps the mass density at about $\rho_{max} \sim m/\lambda_{dB}^3$. Since $\lambda_{dB} \propto 1/m$, $\rho_{max} \propto m^4$. Thus, if m is too small, ρ_{max} is smaller than ρ_{dm}, and the fermions cannot obtain the observed density of dark matter.

To derive a numerical value for m^*, we begin by considering the number of quantum states dN in a differential volume of phase space, which is given by dividing the phase space volume by h^3:

$$dN = 2\frac{d^3\mathbf{r} d^3\mathbf{p}}{h^3} , \qquad (A.4)$$

where we have included an additional factor of 2 to account for the spin degree of freedom for the spin-$1/2$ fermions. To find the maximum possible density, we assume that every possible quantum state is occupied, starting from the state with the smallest possible momentum up to the Fermi momentum p_F (the case of a zero-temperature Fermi gas). The density of quantum states $n_Q = dN/dV$, where $dV = d^3\mathbf{r}$ is the volume element, and so the maximum number density $n_{max} = n_Q$ is found by integrating over the possible momenta in spherical coordinates

$$n_{max} = \frac{2}{h^3} \int_0^{p_F} 4\pi p^2 dp = \frac{8\pi p_F^3}{3h^3} . \qquad (A.5)$$

Requiring that $p_F \leq m v_{esc}$ and also $\rho_{max} = m n_{max} \geq \rho_{dm}$, we obtain the relation

$$\frac{8\pi m^4 v_{esc}^3}{3h^3} \geq \rho_{dm} \qquad (A.6)$$

and find that the minimum mass of a fermionic dark matter particle is

$$m^* = \sqrt[4]{\frac{3h^3 \rho_{dm}}{8\pi v_{esc}^3}} . \qquad (A.7)$$

Numerically, given that $\rho_{dm} \approx 0.4 \text{ GeV/cm}^3$ and $v_{esc} \approx 2 \times 10^{-3}c$, we find

$$m^* \approx 10 \text{ eV}. \qquad (A.8)$$

Problem 1.3: Ultralight Bosonic Dark Matter Waves

The Compton frequency is given by

$$\omega_c = \frac{mc^2}{\hbar}, \tag{A.9}$$

and the Compton wavelength is given by

$$\lambda_c = \frac{2\pi c}{\omega_c} = \frac{2\pi \hbar}{mc} . \tag{A.10}$$

A useful numerical quantity to recall for such "back-of-the-envelope" estimates is $\hbar c \approx 200 \, \text{eV} \cdot \text{nm}$. For $m_b c^2 = 10^{-6} \, \text{eV}$, we find

$$\omega_c \approx \frac{\left(10^{-6} \, \text{eV}\right) \times \left(3 \times 10^{10} \, \text{cm/s}\right)}{200 \times 10^{-7} \, \text{eV} \cdot \text{cm}} \approx 2\pi \times 240 \, \text{MHz} , \tag{A.11}$$

and

$$\lambda_c \approx \frac{2\pi \times 3 \times 10^{10} \, \text{cm/s}}{1.5 \times 10^9 \, \text{s}^{-1}} \approx 130 \, \text{cm}. \tag{A.12}$$

The de Broglie wavelength is given by

$$\lambda_{\text{dB}} = \frac{2\pi \hbar}{mv} = \frac{\lambda_c c}{v} , \tag{A.13}$$

and so for virialized dark matter with $v \approx 10^{-3} c$ and $m_b c^2 = 10^{-6} \, \text{eV}$, $\lambda_{\text{dB}} \approx 10^5 \, \text{cm} \approx 1 \, \text{km}$. Since the axions are nonrelativistic, their kinetic energy is small compared to their rest energy, and we can estimate that each axion carries about $m_a c^2$ of energy. Based on the dark matter density ρ_{dm}, the average number of bosons $\langle N_b \rangle$ occupying a "quantum volume" λ_{dB}^3 can be estimated to be

$$\langle N_b \rangle \approx \frac{\rho_{\text{dm}} \lambda_{\text{dB}}^3}{m_b c^2} . \tag{A.14}$$

For $m_b c^2 = 10^{-6} \, \text{eV}$, $\langle N_b \rangle \approx 10^{30}$. Clearly, the mode density for such UBDM is quite large.

To see how these estimates change for lighter bosons, we note how each quantity scales with m_b: $\omega_c \propto m_b$, λ_c and λ_{dB} are $\propto 1/m_b$, and $\langle N_b \rangle \propto 1/m_b^4$. Thus, for $m_b c^2 = 10^{-12} \, \text{eV}$, we find $\omega_c \approx 2\pi \times 240 \, \text{Hz}$, $\lambda_c \approx 10^8 \, \text{cm}$, $\lambda_{\text{dB}} \approx 10^{11} \, \text{cm}$, and $\langle N_b \rangle \approx 10^{54}$.

Problem 1.4: Coherence of Ultralight Bosonic Dark Matter Fields

Since UBDM is cold (i.e., nonrelativistic), the energy of a UBDM particle is the sum of its rest energy and its kinetic energy:

$$E \approx m_b c^2 + \frac{1}{2} m_b v^2 , \tag{A.15}$$

and the spread in observed energies due to the virialized velocity distribution is

$$\Delta E \approx \frac{1}{2} m_b \Delta v^2 . \tag{A.16}$$

Thus, there is a spread of observed frequencies $\Delta \omega$ for UBDM:

$$\frac{\Delta \omega}{\omega} = \frac{\Delta E}{E} \approx \frac{\Delta v^2}{2c^2} . \tag{A.17}$$

Since $\Delta v^2 \approx 10^{-6} c^2$, the UBDM waves dephase after $\approx 10^6$ oscillations: this is when the accumulated phase differences become ~ 1. The Q-factor for UBDM in the Milky Way is thus $Q \approx 10^6$. This gives a coherence time of

$$\tau_{\text{coh}} \approx 10^6 \frac{2\pi \hbar}{m_b c^2} . \tag{A.18}$$

The coherence length is given by the product of τ_{coh} and the average boson velocity, $v \approx 10^{-3} c$, which is the de Broglie wavelength:

$$L_{\text{coh}} = v \tau_{\text{coh}} = \lambda_{\text{dB}} \approx 10^3 \frac{2\pi \hbar}{m_b c} . \tag{A.19}$$

For $m_b c^2 = 10^{-6}$ eV, $\tau_{\text{coh}} \approx 4$ ms and $L_{\text{coh}} \approx 10^5$ cm ≈ 1 km. For $m_b c^2 = 10^{-12}$ eV, $\tau_{\text{coh}} \approx 4000$ s ≈ 1 hour and $L_{\text{coh}} \approx 10^{11}$ cm.

Problem 1.5: Axion Dark Matter Field Amplitude

Since $\mathcal{L}$ has units of energy density and both ∂_μ and $m_a c / \hbar$ have units of inverse length, from Eq. (1.20), it is then evident that the axion field φ must have units of (energy/length)$^{1/2}$. The energy density of the axion field based on the Klein–Gordon equation,

$$\frac{1}{c^2} \frac{\partial^2}{\partial t^2} \varphi - \nabla^2 \varphi + \frac{m_a^2 c^2}{\hbar^2} \varphi = 0 , \tag{A.20}$$

can be equated to the dark matter energy density:

$$\rho_{\rm dm} \approx \frac{1}{2}\left(\frac{m_a c}{\hbar}\right)^2 \langle \varphi^2 \rangle \,, \tag{A.21}$$

and thus

$$\langle \varphi^2 \rangle \approx \frac{2\rho_{\rm dm}\hbar^2}{m_a^2 c^2} \,. \tag{A.22}$$

Problem 1.6: Dark Electromagnetic Fields

From Eq. (1.21), we have

$$\mathcal{E}' \approx \sqrt{8\pi\rho_{\rm dm}} \approx 40 \text{ V/cm} \,. \tag{A.23}$$

From Eq. (1.22), we have

$$\mathcal{B}' \approx \frac{v}{c}\mathcal{E}' \approx 10^{-4} \text{ G} \,. \tag{A.24}$$

Problem 2.1: Vacuum Field and Boson Mass in Spontaneous Symmetry Breaking

By taking the derivative of the potential with respect to ϕ and setting to zero, we find

$$\frac{\partial V}{\partial \phi} = -\mu^2\phi + \frac{\lambda}{3!}\phi^3 = 0 \,, \tag{A.25}$$

which gives us the roots 0, $+\sqrt{6\mu^2/\lambda}$ and $-\sqrt{6\mu^2/\lambda}$. By inspection of the potential (the plot on the right of Fig. 2.2) or by taking the second derivative,

$$\frac{\partial^2 V}{\partial \phi^2} = -\mu^2 + \frac{\lambda}{2}\phi^2 \,, \tag{A.26}$$

we see that the roots $\pm\sqrt{6\mu^2/\lambda}$ correspond to minima and thus the vacua of the field ϕ.

To find the mass of the associated boson, we Taylor expand the potential around one of the minima. Let us choose $\phi_0 = \sqrt{6\mu^2/\lambda}$:

$$V(\phi) = V(\phi_0) + \left[\frac{\partial V}{\partial \phi}\right]_{\phi=\phi_0} (\phi - \phi_0) + \frac{1}{2!}\left[\frac{\partial^2 V}{\partial \phi^2}\right]_{\phi=\phi_0} (\phi - \phi_0)^2 + \cdots$$

$$(A.27)$$

$$= V(\phi_0) + \mu^2(\phi - \phi_0)^2 + \cdots .$$

$$(A.28)$$

The constant offset of the potential, $V(\phi_0)$, has no physical consequences and can be subtracted. Keeping only the lowest order terms in $\bar{\phi} \equiv \phi - \phi_0$ (since we assume small perturbations around the minimum) and noting that $\partial_\mu \phi = \partial_\mu \bar{\phi}$, we find for the Lagrangian

$$\mathcal{L} \approx \frac{1}{2}\left(\partial_\mu \bar{\phi}\right)^2 - \mu^2 \bar{\phi}^2 .$$

$$(A.29)$$

By analogy with Eq. (2.13), we find that the boson mass is given by the relation

$$\frac{1}{2}m^2 = \mu^2 ,$$

$$(A.30)$$

so the mass $m = \sqrt{2}\mu$.

Problem 2.2: Lagrangian for Two Scalar Fields

Starting from the potential from Eq. (2.30),

$$V(\bar{\alpha}, \bar{\beta}) = -\frac{\mu^2}{2}\left[(\bar{\alpha} + \alpha_0)^2 + \bar{\beta}^2\right]$$
$$+ \frac{\lambda}{4!}\left[(\bar{\alpha} + \alpha_0)^2 + \bar{\beta}^2\right]^2 ,$$

$$(A.31)$$

we can expand the terms and substitute $\alpha_0 = \mu\sqrt{6/\lambda}$ to obtain

$$V(\bar{\alpha}, \bar{\beta}) = -\frac{1}{2}\mu^2\bar{\alpha}^2 - \mu^3\sqrt{\frac{6}{\lambda}}\bar{\alpha} - 3\frac{\mu^4}{\lambda} - \frac{1}{2}\mu^2\bar{\beta}^2$$
$$+ \frac{\lambda}{4!}\bar{\alpha}^4 + \mu\sqrt{\frac{6}{\lambda}}\bar{\alpha}^3 + \frac{3}{2}\mu^2\bar{\alpha}^2 + \mu^3\sqrt{\frac{6}{\lambda}}\bar{\alpha} + \frac{3}{2}\frac{\mu^4}{\lambda}$$
$$+ \frac{\lambda}{12}\bar{\alpha}^2\bar{\beta}^2 + \mu\sqrt{\frac{6}{\lambda}}\bar{\alpha}\bar{\beta}^2 + \frac{1}{2}\mu^2\bar{\beta}^2 + \frac{\lambda}{4!}\bar{\beta}^4 .$$

$$(A.32)$$

Combining like terms in Eq. (A.32)—noting in particular that (1) the terms linear in $\bar{\alpha}$, with no $\bar{\beta}$ dependence, cancel and also (2) the terms quadratic in $\bar{\beta}$, with no $\bar{\alpha}$

dependence, cancel—we arrive at

$$V(\bar{\alpha}, \bar{\beta}) = \mu^2\bar{\alpha}^2 + \mu\sqrt{\frac{\lambda}{6}}\bar{\alpha}^3 + \frac{\lambda}{4!}\bar{\alpha}^4$$

$$+ \frac{\lambda}{4!}\bar{\beta}^4 + \mu\sqrt{\frac{\lambda}{6}}\bar{\alpha}\bar{\beta}^3 + \frac{\lambda}{12}\bar{\alpha}^2\bar{\beta}^2 \tag{A.33}$$

$$- \frac{3}{2}\frac{\mu^4}{\lambda},$$

which yields Eq. (2.31).

Problem 2.3: Explicit and Spontaneous Symmetry Breaking

To verify that the minimum of the potential $V(\alpha, \beta)$ from Eq. (2.35), namely,

$$V(\alpha, \beta) = -\frac{\mu^2}{2}\left(\alpha^2 + \beta^2\right) + \frac{\lambda}{4!}\left(\alpha^2 + \beta^2\right)^2 - \epsilon\lambda\alpha_0^3\alpha, \tag{A.34}$$

occurs at

$$\alpha = \alpha_0(1 + 3\epsilon), \tag{A.35}$$

$$\beta = 0, \tag{A.36}$$

let us begin by taking the derivative of $V(\alpha, \beta)$ with respect to α:

$$\frac{\partial V}{\partial \alpha} = -\mu^2\alpha + \frac{\lambda}{6}\left(\alpha^3 + \alpha\beta^2\right) - \epsilon\lambda\alpha_0^3. \tag{A.37}$$

A bit of algebra shows that

$$\frac{\partial V}{\partial \alpha} = \frac{\lambda}{6}\left[-\frac{6\mu^2}{\lambda}\alpha + \left(\alpha^3 + \alpha\beta^2\right) - 6\epsilon\alpha_0^3\right],$$

$$= \frac{\lambda}{6}\left[\alpha^3 + \left(\beta^2 - \alpha_0^2\right)\alpha - 6\epsilon\alpha_0^3\right], \tag{A.38}$$

where we made the substitution $\alpha_0 = \sqrt{6\mu^2/\lambda}$. Setting $\alpha = \alpha_0' = \alpha_0(1 + 3\epsilon)$ and $\beta = 0$ in Eq. (A.38), we find

$$\left[\frac{\partial V}{\partial \alpha}\right]_{\alpha=\alpha_0',\beta=0} = \frac{\lambda}{6}\left[\alpha_0^3(1+3\epsilon)^3 - \alpha_0^3(1+3\epsilon) - 6\epsilon\alpha_0^3\right],$$

$$\approx \frac{\lambda}{6}\left[\alpha_0^3(1+9\epsilon) - \alpha_0^3(1+3\epsilon) - 6\epsilon\alpha_0^3\right], \tag{A.39}$$

$$\approx 0,$$

where in the second line of Eq. (A.39) we have kept only terms to first order in ϵ. Similarly, with respect to β,

$$\frac{\partial V}{\partial \beta} = -\mu^2\beta + \frac{\lambda}{6}\beta\left(\alpha^2 + \beta^2\right), \tag{A.40}$$

and so

$$\left[\frac{\partial V}{\partial \beta}\right]_{\beta=0} = 0. \tag{A.41}$$

This shows that $\alpha = \alpha_0(1+3\epsilon)$, $\beta = 0$ is an extremal point of the potential to first order in ϵ. Either by examining the second derivatives or by inspection of the potential plotted in Fig. 2.5, it is evident that this is a minimum.

Next, we rewrite the potential in Eq. (A.34) in terms of the suggested variable,

$$\bar{a} = \alpha - \alpha_0(1+3\epsilon), \tag{A.42}$$

which gives us

$$V(\bar{a}, \beta) = -\frac{\mu^2}{2}\alpha_0^2\left(1 + 3\epsilon + \frac{\bar{a}}{\alpha_0}\right)^2 - \frac{\mu^2}{2}\beta^2$$

$$+ \frac{\lambda}{4!}\alpha_0^4\left[\left(1 + 3\epsilon + \frac{\bar{a}}{\alpha_0}\right)^4 + 2\frac{\beta^2}{\alpha_0^2}\left(1 + 3\epsilon + \frac{\bar{a}}{\alpha_0}\right)^2 + \frac{\beta^4}{\alpha_0^4}\right] \tag{A.43}$$

$$- \epsilon\lambda\alpha_0^4\left(1 + 3\epsilon + \frac{\bar{a}}{\alpha_0}\right).$$

In order to simplify the above expression, we make the approximations that terms higher than first order in ϵ can be neglected as well as any terms higher than second order in the field variables $\bar{a}$ and β, yielding

$$V(\bar{a}, \beta) \approx -\frac{\mu^2}{2}\alpha_0^2 \left[1 + 6\epsilon + 2\frac{\bar{a}}{\alpha_0}(1 + 3\epsilon) + \left(\frac{\bar{a}}{\alpha_0}\right)^2 \right] - \frac{\mu^2}{2}\beta^2$$
$$+ \frac{\lambda}{4!}\alpha_0^4 \left[1 + 12\epsilon + 4\frac{\bar{a}}{\alpha_0}(1 + 9\epsilon) + 6\left(\frac{\bar{a}}{\alpha_0}\right)^2(1 + 6\epsilon) \right] \qquad \text{(A.44)}$$
$$+ \frac{\lambda}{12}\alpha_0^2\beta^2(1 + 6\epsilon)$$
$$- \epsilon\lambda\alpha_0^4 - \epsilon\lambda\alpha_0^3\bar{a} .$$

The above expression can be further simplified by eliminating constant terms in the potential (which do not affect the physics) and making the substitution $\lambda\alpha_0^2 = 6\mu^2$ to eliminate λ from the expression to help identify like terms that can be combined,

$$V(\bar{a}, \beta) \approx -\mu^2\alpha_0^2 \left[\frac{\bar{a}}{\alpha_0}(1 + 3\epsilon) + \frac{1}{2}\left(\frac{\bar{a}}{\alpha_0}\right)^2 \right] - \frac{\mu^2}{2}\beta^2$$
$$+ \mu^2\alpha_0^2 \left[\frac{\bar{a}}{\alpha_0}(1 + 9\epsilon) + \frac{3}{2}\left(\frac{\bar{a}}{\alpha_0}\right)^2(1 + 6\epsilon) \right] \qquad \text{(A.45)}$$
$$+ \frac{\mu^2}{2}\beta^2(1 + 6\epsilon) - 6\epsilon\mu^2\alpha_0\bar{a} ,$$

and canceling and combining like terms produces

$$V(\bar{a}, \beta) \approx \mu^2\bar{a}^2(1 + 9\epsilon) + 3\epsilon\mu^2\beta^2 . \qquad \text{(A.46)}$$

Since $\epsilon \ll 1$, the factor of $(1 + 9\epsilon)$ in the first term is ≈ 1, and so we obtain the sought-after Eq. (2.39), namely

$$V(\bar{a}, \beta) \approx \mu^2\bar{a}^2 + 3\epsilon\mu^2\beta^2 . \qquad \text{(A.47)}$$

Problem 2.4: Interactions Between Two Scalar Fields

The third-order terms of interest come from two terms in the second line of Eq. (A.43), namely

$$\frac{\lambda}{4!}\alpha_0^4 \left(1 + 3\epsilon + \frac{\bar{a}}{\alpha_0} \right)^4 \qquad \text{(A.48)}$$

and

$$\frac{\lambda}{4!}\alpha_0^4\left(\frac{2\beta^2}{\alpha_0^2}\right)\left(1 + 3\epsilon + \frac{\bar{a}}{\alpha_0}\right)^2 . \tag{A.49}$$

The expansion of the polynomial factor in (A.48), keeping only terms up to first order in ϵ and third order in the fields, gives

$$\left(1 + 3\epsilon + \frac{\bar{a}}{\alpha_0}\right)^4 \approx 1 + 4\left(\frac{\bar{a}}{\alpha_0}\right) + 6\left(\frac{\bar{a}}{\alpha_0}\right)^2 + 4\left(\frac{\bar{a}}{\alpha_0}\right)^3$$

$$12\epsilon + 36\epsilon\left(\frac{\bar{a}}{\alpha_0}\right) + 36\epsilon\left(\frac{\bar{a}}{\alpha_0}\right)^2 + 12\epsilon\left(\frac{\bar{a}}{\alpha_0}\right)^3 . \tag{A.50}$$

Since the terms lower than third order in the fields were already accounted for in Eq. (2.40), the new interaction term from (A.50) is

$$\frac{\lambda}{4!}\alpha_0^4\left[4\left(\frac{\bar{a}}{\alpha_0}\right)^3(1 + 3\epsilon)\right] \approx \frac{\lambda}{6}\alpha_0\bar{a}^3 . \tag{A.51}$$

Similarly, for the polynomial factor in (A.49),

$$\beta^2\left(1 + 3\epsilon + \frac{\bar{a}}{\alpha_0}\right)^2 \approx \beta^2\left[(1 + 6\epsilon) + 2\left(\frac{\bar{a}}{\alpha_0}\right)(1 + 3\epsilon)\right] \tag{A.52}$$

$$\approx 2\beta^2\left(\frac{\bar{a}}{\alpha_0}\right) , \tag{A.53}$$

giving a new interaction term

$$\frac{\lambda}{4!}\alpha_0^2\left[\frac{4\beta^2}{\alpha_0^2}\left(\frac{\bar{a}}{\alpha_0}\right)\right] \approx \frac{\lambda}{6}\alpha_0\beta^2\bar{a} . \tag{A.54}$$

Problem 2.5: Axion–Photon Interaction

The electromagnetic field tensor (Faraday tensor) $F^{\mu\nu}$ is given by [1]

$$F^{\mu\nu} = \partial^\mu A^\nu - \partial^\nu A^\mu \tag{A.55}$$

$$= \begin{pmatrix} 0 & -E_x & -E_y & -E_z \\ E_x & 0 & -B_z & B_y \\ E_y & B_z & 0 & -B_x \\ E_z & -B_y & B_x & 0 \end{pmatrix} , \tag{A.56}$$

where ∂^μ is the four-derivative, A^μ is the four-potential, and E_i and B_i are the electric and magnetic field components in the Cartesian basis. The dual field tensor is given by

$$\tilde{F}_{\alpha\beta} = \frac{1}{2}\varepsilon_{\alpha\beta\mu\nu}F^{\mu\nu} \tag{A.57}$$

$$= \begin{pmatrix} 0 & B_x & B_y & B_z \\ -B_x & 0 & E_z & -E_y \\ -B_y & -E_z & 0 & E_x \\ -B_z & E_y & -E_x & 0 \end{pmatrix}, \tag{A.58}$$

where $\varepsilon_{\alpha\beta\mu\nu}$ is the Levi-Civita totally antisymmetric tensor. Taking the trace of the product of the matrices (A.56) and (A.58) yields $F^{\mu\nu}\tilde{F}_{\mu\nu}$:

$$F^{\mu\nu}\tilde{F}_{\mu\nu} = 4\left(E_x B_x + E_y B_y + E_z B_z\right) = 4E \cdot B. \tag{A.59}$$

Substituting Eq. (A.59) into (2.63) yields Eq. (2.64).

Problem 3.1: Background Evolution of UBDM

The continuity equation, $\dot{\rho}/\rho = -3(w+1)\dot{a}/a$, can be integrated such that $\rho \propto a^{-3(w+1)}$. Note that this solution is also valid for the case $w = -1$, i.e., when ρ is constant. Substituting this into the Friedmann equation, we find

$$H^2 \equiv \left(\frac{\dot{a}}{a}\right)^2 = \frac{\rho}{3\,M_{pl}^2} \propto a^{-3(w+1)} \Rightarrow \dot{a}\,a^{\frac{3w+1}{2}} = c_1, \tag{A.60}$$

where c_1 is some constant. Integrating the equation above, we obtain

$$a = \begin{cases} c_2\,e^{c_1 t} & \text{if } w = -1 \\ (c_1 t + c_2)^{\frac{2}{3(w+1)}} & \text{else} \end{cases} \tag{A.61}$$

$$a^{\frac{3(w+1)}{2}} = c_1 t + c_2 \Rightarrow a \propto t^{\frac{2}{3(w+1)}}. \tag{A.62}$$

Consequently, we find via $H = \dot{a}/a$ that

$$H = \begin{cases} c_1 & \text{if } w = -1 \\ \frac{2c_1}{3(w+1)}\frac{1}{c_1 t + c_2} & \text{else.} \end{cases} \tag{A.63}$$

To change variables from conformal to physical time, note that

$$\bar{\phi}' \equiv \frac{\ddot{\bar\phi}}{d\tau} = \frac{dt}{d\tau}\frac{d\bar\phi}{dt} = a\,\dot{\bar\phi} \tag{A.64}$$

$$\bar{\phi}'' \equiv \frac{d}{d\tau}\bar{\phi}' = a'\dot{\bar\phi} + a^2\ddot{\bar\phi} = a\dot a\dot{\bar\phi} + a^2\ddot{\bar\phi} = a^2(H\dot{\bar\phi} + \ddot{\bar\phi}). \tag{A.65}$$

This yields the following field equation for $\bar\phi$:

$$0 = \ddot{\bar\phi} + 3H\dot{\bar\phi} + m^2\bar\phi = \ddot{\bar\phi} + \frac{2\,\dot{\bar\phi}}{3(w+1)\,t} + m^2\bar\phi. \tag{A.66}$$

The solutions of this equation can be expressed in terms of Bessel functions of first (J_ν) and second kind (Y_ν) with $\nu = (w-1)/2(w+1)$ and $\delta \equiv \sqrt{1-(2m/3H)^2}$

$$\bar\phi = \begin{cases} e^{-3H(1+\delta)t}\left(c_3 + c_4\,e^{3H\delta t}\right) & \text{if } w = -1 \\ t^\nu\left[c_3\,J_\nu(m\,t) + c_4 Y_\nu(m\,t)\right] & \text{if } w \neq -1, \end{cases} \tag{A.67}$$

where c_3 and c_4 are constants that depend on the initial conditions. While these may be used to derive expressions for the pressure and density, we note that the asymptotic behavior of Eq. (A.66) may be derived more generally. Note that, for early times, $t \to 0$,[1] we have $H \propto 1/t \to \infty$ and the corresponding term will dominate:

$$\ddot{\bar\phi} + 3H\dot{\bar\phi} \simeq 0 \Rightarrow \dot{\bar\phi} \propto a^{-3} \Rightarrow \bar\phi = c_5 + \begin{cases} c_6 e^{-3Ht} & \text{if } w = -1 \\ \ln(c_6\,t) & \text{if } w = 1 \\ c_6\,t^{\frac{w-1}{w+1}} & \text{else,} \end{cases} \tag{A.68}$$

where c_5 and c_6 are constants. In typical applications, e.g., a radiation-dominated universe with $w = 1/3$, we have $\bar\phi = c_5 + c_6\,t^{-1/2}$. The second term is divergent for $t \to 0$ and we also see that $\bar\phi \simeq c_5$ for $t \gg (c_6/c_5)^2$, which are two arguments usually used for ignoring the second term and saying that the background field is simply constant at early times, $\bar\phi = c_5$.

On the other hand, for late times $H(t) \ll m$, Eq. (A.66) is solved by a WKB-like solution,[2] and without loss of generality we take $\bar\phi \propto a^{-3/2}\cos(m\,t)$ and $\dot{\bar\phi} \propto a^{-3/2}[H\cos(m\,t) - m\sin(m\,t)]$. Note that $H \ll m$ also implies that the oscillation periods $T \sim 1/m \ll 1/H \sim t$ such that we can assume that H and a do not change much over each integration

---

[1] More precisely, this is the regime where $H(t) \gg m$.

[2] This can be checked by substituting this in Eq. (A.66) and noting that the non-vanishing terms are all small if $H \ll m$.

$$m^2 \dot{\bar{\phi}}^2 = P + \rho \Rightarrow m^2 \langle \dot{\bar{\phi}}^2 \rangle = \langle P \rangle + \langle \rho \rangle \equiv (w_{\text{eff}} + 1)\langle \rho \rangle \,. \tag{A.69}$$

Using $\langle \sin(m\,t) \cos(m\,t) \rangle = 0$ and $\langle \sin^2(m\,t) \rangle = \langle \cos^2(m\,t) \rangle = 1/2$, we find that

$$\langle \dot{\bar{\phi}}^2 \rangle \propto (H/m)^2/2 + 1/2 \,, \tag{A.70}$$

$$\langle \rho \rangle = m^2 \langle \dot{\bar{\phi}}^2 \rangle /2 + \langle V \rangle \propto H^2/4 + m^2/4 + m^2/4 \,, \tag{A.71}$$

$$\Rightarrow w_{\text{eff}} + 1 = 2 \frac{(H/m)^2 + 1}{(H/m)^2 + 2} \simeq 1 \Rightarrow w_{\text{eff}} \simeq 0 \,. \tag{A.72}$$

For the potential $V(\phi) = \lambda \phi^4$, Eq. (A.66) now becomes

$$0 = \ddot{\bar{\phi}} + \frac{2\dot{\bar{\phi}}}{3(w+1)\,t} + 4\lambda \bar{\phi}^3 \,. \tag{A.73}$$

For early times, there is no change to the previous argument since the potential is irrelevant.

For later times, the shape of Eq. (A.73) implies that we cannot rely on the WKB-like solutions anymore and instead follow the general approach presented in Refs [2, 3].[3] We repeat this derivation for a symmetric potential V, i.e., $V(-\bar{\phi}) = V(\bar{\phi})$, and $V(\bar{\phi}) = \lambda \bar{\phi}^n$ (n even). At the maximum $\hat{\bar{\phi}}$, the total energy density is given by the maximum potential value, $\rho(\hat{\bar{\phi}}) = V(\hat{\bar{\phi}}) \equiv \hat{V}$. Since the oscillations at late times are very rapid, the energy density of the field does not change much over one oscillation period T, and $\rho \approx \hat{V}$. The definition of ρ then implies that

$$\dot{\bar{\phi}}^2 = \frac{2}{m^2}(\rho - V) = \frac{2\hat{V}}{m^2}\left(1 - \frac{V}{\hat{V}}\right), \tag{A.74}$$

and we find for the oscillation period that

$$T = \int dt = \int \frac{dt}{d\bar{\phi}} d\bar{\phi} = \int_{-\hat{\bar{\phi}}}^{\hat{\bar{\phi}}} \frac{1}{|\dot{\bar{\phi}}|} d\bar{\phi} = \frac{2m}{\sqrt{2\hat{V}}} \int_0^{\hat{\bar{\phi}}} \frac{1}{\sqrt{1 - V/\hat{V}}} d\bar{\phi} \,, \tag{A.75}$$

which in turn implies that

$$w_{\text{eff}} + 1 = \frac{m^2 \langle \dot{\bar{\phi}}^2 \rangle}{\langle \rho \rangle} = \frac{m^2}{\hat{V}} \frac{1}{T} \int \dot{\bar{\phi}}^2 dt = \frac{m^2}{\hat{V}} \frac{1}{T} \int_{-\hat{\bar{\phi}}}^{\hat{\bar{\phi}}} \dot{\bar{\phi}}^2 / |\dot{\bar{\phi}}| d\bar{\phi} \tag{A.76}$$

---

[3] Note that, for the quartic potential $V(\bar{\phi}) = \lambda \bar{\phi}^4$, the solutions for $\bar{\phi}$ can be expressed in terms of so-called Jacobi elliptic functions times an oscillating function [3], which can be used to check the general solution presented here explicitly.

$$= \frac{2\sqrt{2}\,m}{\sqrt{\hat{V}}\,T} \int_0^{\hat{\phi}} \sqrt{1 - V/\hat{V}}\,\mathrm{d}\bar{\phi} \tag{A.77}$$

$$= 2\frac{\int_0^{\hat{\phi}} (1 - V/\hat{V})^{\frac{1}{2}}\,\mathrm{d}\bar{\phi}}{\int_0^{\hat{\phi}} (1 - V/\hat{V})^{-\frac{1}{2}}\,\mathrm{d}\bar{\phi}} = \frac{2n}{n+2} \Rightarrow w_{\mathrm{eff}} = \frac{n-2}{n+2}, \tag{A.78}$$

i.e., $w_{\mathrm{eff}} = 1/3$ for $n = 4$, which corresponds to the equation-of-state parameter of radiation.

We see that when Hubble friction dominates, i.e., typically at sufficiently early times, any scalar particles generically behave as dark energy with $w_{\mathrm{eff}} = -1$. Later on, however, (dominant terms of) the potential determine the behavior of the scalar field oscillations, which need not be that of dark matter (pressureless dust), and dark matter bounds hence do not apply.

Problem 3.2: Derivation of the Schrödinger–Poisson Equations for UBDM

First, let us write down the Klein–Gordon equation,

$$\Box\phi - \partial_\phi V = 0, \tag{A.79}$$

where the D'Alembertian is defined as

$$\Box = \frac{1}{\sqrt{-g}}\partial_\mu(\sqrt{-g}\,g^{\mu\nu})\partial_\nu, \tag{A.80}$$

and the potential is given by

$$V(\phi) = \frac{m^2}{2}\phi^2 + \frac{m^2}{2}\lambda\phi^4. \tag{A.81}$$

With $\Psi = \Phi$ and $a = 1$, the metric is given by

$$g = [g_{\mu\nu}] = \mathrm{diag}[-(1 + 2\Psi), 1 - 2\Psi, 1 - 2\Psi, 1 - 2\Psi]. \tag{A.82}$$

To first order, this gives

$$g^{-1} = [g^{\mu\nu}] = \mathrm{diag}[-(1 - 2\Psi), 1 + 2\Psi, 1 + 2\Psi, 1 + 2\Psi],$$
$$\sqrt{-g} = 1 - 2\Psi.$$

Thus, the D'Alembertian to first order is given by

$$\Box = 4\dot{\Psi}\partial_t - (1 - 2\Psi)\partial_t^2 + (1 + 2\Psi)\nabla^2, \tag{A.83}$$

and the Klein–Gordon equation (Eq. (A.79)) reads

$$- (1 - 2\Psi)\ddot{\phi} + 4\dot{\Psi}\dot{\phi} + (1 + 2\Psi)\nabla^2\phi - m^2\phi - 2m^2\lambda\phi^3 = 0. \tag{A.84}$$

We can rewrite the Klein–Gordon equation by multiplying with $-(1+2\Psi)$ (since this term goes to -1 in the nonrelativistic limit, the result remains unchanged save for an overall minus sign), which reduces the number of terms we need to consider later on:

$$\ddot{\phi} - 4\dot{\Psi}\dot{\phi} - (1 + 4\Psi)\nabla^2\phi + (1 + 2\Psi)m^2\phi + (1 + 2\Psi)2m^2\lambda\phi^3 = 0. \tag{A.85}$$

Let us take the ansatz for ϕ and write

$$\phi = \frac{1}{\sqrt{2m}}\left[\psi e^{imt} + \psi^* e^{imt}\right],$$

$$\dot{\phi} = \frac{1}{\sqrt{2m}}\left[e^{imt}\left(\dot{\psi} + im\psi\right) + e^{-imt}\left(\dot{\psi}^* - im\psi^*\right)\right],$$

$$\ddot{\phi} = \frac{1}{\sqrt{2m}}\left[e^{imt}\left(\ddot{\psi} + 2im\dot{\psi} - m^2\psi\right) + e^{-imt}\left(...\right)\right],$$

$$\nabla^2\phi = \frac{1}{\sqrt{2m}}\left[(\nabla^2\psi)e^{imt} + (\nabla^2\psi^*)e^{-imt}\right],$$

$$\phi^3 = \frac{1}{(\sqrt{2m})^3}\left[e^{imt}2|\psi|^2\psi + e^{-imt}2|\psi|^2\psi^* + \psi^3 e^{3imt} + \psi^{*3}e^{-3imt}\right].$$

Since terms for e^{-imt} are the complex conjugate, the terms in front of e^{+imt} need to vanish in order for the Klein–Gordon equation to be fulfilled. Thus, one needs to consider only the terms which go with an e^{+imt} oscillation (or e^{-imt}, respectively). This gives for Eq. (A.85)

$$\frac{1}{\sqrt{2m}}[-4\dot{\Psi}(\dot{\psi} + im\psi) + (\ddot{\psi} + 2im\dot{\psi} - m^2\psi) - (1 + 4\Psi)(\nabla^2\psi)$$

$$+ (1 + 2\Psi)m^2\psi + (1 + 2\Psi)2m^2\lambda\frac{2}{2m^2}|\psi|^2\psi] = 0$$

$$\Longleftrightarrow \quad \frac{1}{\sqrt{2m}}[-4\dot{\Psi}\dot{\psi} - 4\dot{\Psi}im\psi + \ddot{\psi} + 2im\dot{\psi} - \nabla^2\psi - 4\Psi\nabla^2\psi$$

$$+ 2\Psi m^2\psi + 2\lambda|\psi|^2\psi] + 4\Psi\lambda|\psi|^2\psi] = 0.$$

Now, let us take the nonrelativistic limit and consider either the limits given in the exercise or take $c \to \infty$. We calculated in natural units, where $c = 1$. Remember when reintroducing the c's that $\Psi \rightarrow \Psi/c^2$, that the time derivatives gain a factor of $1/c$ and that—for each m within the brackets—$m \rightarrow mc$. Hence, a number of terms vanish, and one is left with

$$i\dot{\psi} - \frac{1}{2m}\nabla^2\psi + m\Psi\psi + \frac{\lambda}{m}|\psi|^2\psi = 0. \tag{A.86}$$

Considering the complex conjugate, this equals Eq. (3.35) with $\lambda_{\text{GP}} = -\lambda$.

To calculate the energy density, ρ, to leading order, we recognize that $\rho = -T_0^0$. The stress energy tensor is given by

$$T_{\mu\nu} = \partial_\mu\phi\partial_\nu\phi - g_{\mu\nu}\left[\frac{1}{2}g^{\alpha\beta}\partial_\alpha\phi\partial_\beta\phi + V(\phi)\right], \tag{A.87}$$

and thus

$$T_0^0 = g^{00}T_{00}$$

$$= g^{00}\dot{\phi}^2 - g^{00}g_{00}\left(\frac{1}{2}g^{\alpha\beta}\partial_\alpha\phi\partial_\beta\phi + V(\phi)\right)$$

$$= \frac{1}{2}g^{00}\dot{\phi}^2 - \frac{1}{2}(\nabla\phi)^2 g^{ii} - V(\phi)$$

$$= -\frac{1}{2}\left[(1 - 2\Psi)\dot{\phi}^2 + (1 + 2\Psi)(\nabla\phi)^2 + 2V(\phi)\right].$$

The individual terms are given by

$$\phi^2 = \frac{1}{2m^2}\left[e^{2imt}\psi^2 + e^{-2imt}\psi^{*2} + 2|\psi|^2\right],$$

$$\dot{\phi}^2 = \frac{1}{2m^2}[e^{2imt}(\dot{\psi} + im\psi)^2 + e^{-2imt}(\dot{\psi}^* - im\psi^*)^2 +$$

$$2(|\dot{\psi}|^2 + \underbrace{im(\psi^*\dot{\psi} - \psi\dot{\psi}^*)}_{=2m\text{Im}(\dot{\psi}\psi^*)} + m^2|\psi|^2)],$$

$$(\nabla\phi)^2 = \frac{1}{2m^2}\left[(\nabla\psi)^2 e^{2imt} + (\nabla\psi^*)^2 e^{-2imt} + 2|\nabla\psi|^2\right].$$

Again, the leading order terms can be identified by either taking the limits given in the exercise or counting powers of c (remembering that the mass term in the potential has a factor of c^2). Neglecting the oscillatory terms, the remaining terms are $\dot{\phi}^2 \rightarrow |\psi|^2$ and $m^2\phi^2 \rightarrow |\psi|^2$. Thus,

$$\rho = \frac{1}{2}\left(|\psi|^2 + |\psi|^2\right) = |\psi|^2. \tag{A.88}$$

Problem 3.3: Relaxation of UBDM

Starting from

$$t_{\text{relax}} = 0.1\frac{R}{v}\frac{M}{m \log \Lambda}, \tag{A.89}$$

first one notes that the host of mass M with radius R and the quasi-particle of mass m follow from the same underlying density ρ. The host mass can be approximated by assuming it to be a sphere of constant density ρ, while the quasi-particle mass is effectively given by the size of the de Broglie wavelength:

$$M = \frac{4\pi}{3}\rho R^3 \tag{A.90}$$

$$m \sim \rho \left(\frac{\lambda_{\text{dB}}}{2}\right)^3. \tag{A.91}$$

The de Broglie wavelength is given by

$$\lambda_{\text{dB}} = \frac{h}{mv}, \tag{A.92}$$

where the velocity v approximately equals the quasi-particle velocity. The latter is justified since the dynamics of the quasi-particle are ultimately determined by the dynamics of the underlying axion particle and, thus, one can expect $v_{\text{qp}} \sim v_a$.

Plugging Eqs. (A.90), (A.91), and (A.92) into (A.89) yields

$$t_{\text{relax}} \sim 3.4(m/h)^3 v^2 R^4 \frac{1}{\log \Lambda}. \tag{A.93}$$

Plugging in the numbers, we obtain Eq. (3.47):

$$t_{\text{relax}} \sim \frac{10^{10}}{\log \Lambda}\left(\frac{m}{10^{-22}\,\text{eV}}\right)^3\left(\frac{v}{100\,\text{km/s}}\right)^2\left(\frac{R}{5\,\text{kpc}}\right)^4. \tag{A.94}$$

Problem 3.4: Estimating Superradiance Properties of UBDM

First, let us derive mass scale relevant for superradiance. In order to affect the action, the UBDM potential $m^2\phi^2/2$ should be comparable gravitational term induced by the Kerr BH, $M_{pl}^2 R/2$. Recalling that $8\pi G_N \equiv 1/M_{pl}^2$, the Ricci scalar in the Kerr geometry is given by

$$R^2 = 48 \, (G_N M)^2 \, \frac{\left(r^2 - a_J^2 \cos^2(\theta)\right)^2 \left[\left(r^2 + a_J^2 \cos^2(\theta)\right)^2 - 16 r^2 a_J^2 \cos^2(\theta)\right]}{\left(r^2 + a_J^2 \cos^2(\theta)\right)^6} .$$

$$(A.95)$$

Suppose, for simplicity, that we are at the equator ($\theta = \pi/2$) and that we are just inside the ergosphere at

$$r_{\text{ergo}} = G_N \left(M + \sqrt{M^2 - a_J^2 \cos^2(\theta)} \right) = 2 \, G_N M , \qquad (A.96)$$

which is also just the Schwarzschild radius of the black hole. We then find

$$R^2 \overset{\theta = \pi/2}{=} \frac{48 \, G_N^2 M^2}{r^6} \overset{r = r_{\text{ergo}}}{=} \frac{3}{4} \frac{1}{(G_N M)^4} . \qquad (A.97)$$

For typical field values $\phi \sim M_{pl}$ in the extreme environment surrounding the black hole, we find that

$$\frac{m^2\phi^2}{2} \sim \frac{m^2 M_{pl}^2}{2} \overset{!}{\sim} \frac{M_{pl}^2 R}{2} \sim \frac{M_{pl}^2}{2} \sqrt{\frac{3}{4} \frac{1}{(G_N M)^2}} \Rightarrow G_N M \sim 1 , \qquad (A.98)$$

where we ignored the $O(1)$ numerical factor.

Note that we might have used dimensional analysis (e.g., that $R \sim 1/R_S \sim 1/2GM$) instead of Eq. A.95 to arrive at a similar result.

To compute the Bosenova condition, the self-coupling of the UBDM particles, λ, needs to be large enough to play a role. We anticipate this to happen when the corresponding term in the action, $\lambda\phi^4/4!$, becomes of the same order as the potential term, $m^2\phi^2/2$. To facilitate this comparison, first note that the total energy density of the UBDM cloud can be equated to its total mass, M_{cloud}, divided by its volume, V_{cloud}, which are given by

$$M_{\text{cloud}} = Nm , \qquad (A.99)$$

$$V_{\text{cloud}} = \frac{4\pi}{3} R_{\text{cloud}}^3 \sim \frac{4\pi}{3} r_{\text{ergo}}^3 = \frac{32\pi}{3} (G_N M)^3 , \qquad (A.100)$$

where N is the number of UBDM particles. By equating $m^2\phi^2/2 = M_{\text{cloud}}/V_{\text{cloud}}$, we find that

$$\phi^2 = \frac{3}{16\pi} \frac{N}{(G_N M)^3 m} .$$
(A.101)

From $m^2\phi^2/2 \overset{!}{\sim} \lambda\phi^4/4!$ and using Eq. (A.101), we find further that

$$\phi^2 \sim \frac{12m^2}{\lambda} \Rightarrow N \sim 64\pi \frac{(G_N M m)^3}{\lambda} \sim \frac{64\pi}{\lambda} ,$$
(A.102)

where the last step made use of Eq. (A.98). To compare to Eq. (3.60), we use that $\lambda = m^2/f_a^2$ for an axion and eliminate m via the SR condition (A.98), such that $\lambda \sim 1/f_a^2(G_N M)^2$. Since $M_{pl} = 2.4 \times 10^{18}$ GeV $= 2.2 \times 10^{-39} M_\odot$, we arrive at

$$N \sim \frac{64\pi}{\lambda} \sim \frac{1}{\pi} \left(\frac{10\,M_\odot}{M_{pl}} \right)^2 \left(\frac{M}{10\,M_\odot} \right)^2 \left(\frac{f_a}{M_{pl}} \right)^2$$
(A.103)

$$\approx 6.6 \times 10^{78} \left(\frac{M}{10\,M_\odot} \right)^2 \left(\frac{f_a}{M_{pl}} \right)^2 ,$$
(A.104)

which is very close to Eq. (3.60) for the first energy level ($n = 1$).

Problem 3.5: Microlensing Constraints on UBDM

The Schrödinger–Poisson (SP) equation is given by

$$i\dot{\psi} + \frac{\nabla^2}{2m}\psi + \frac{\lambda_{\mathrm{GP}}}{m}|\psi|^2\psi = 0$$
(A.105)

and

$$\nabla^2\Phi = 4\pi G \left(|\psi|^2 - \int \mathrm{d}^3x\,|\psi|^2 \right) .$$
(A.106)

We can therefore write Eq. (A.105) using the given scaling relation $\{t, x, \psi, \Phi, \lambda_{\mathrm{GP}}\} \to \{\lambda^{-2}\hat{t}, \lambda^{-1}\hat{x}, \lambda^2\hat{\psi}, \lambda^2\hat{\Phi}, \lambda^{-2}\hat{\lambda}_{\mathrm{GP}}\}$.

For some constant and variable, c and q, respectively, we can write that $\frac{\partial}{\partial(cq)} = \frac{1}{c}\frac{\partial}{\partial(q)}$, and hence $\nabla \to \lambda\hat{\nabla}$. Therefore,

$$i\frac{\partial\lambda^2\hat{\psi}}{\partial(\lambda^{-2}\hat{t})} + \frac{\lambda^2\hat{\nabla}^2}{2m}\lambda^2\hat{\psi} + \frac{\lambda^{-2}\hat{\lambda}_{\mathrm{GP}}}{m}|\lambda^2\hat{\psi}|^2\lambda^2\hat{\psi} = 0,$$
(A.107)

$$\lambda^4 \left[i \frac{\partial \hat{\psi}}{\partial(\hat{t})} + \frac{\hat{\nabla}^2}{2m}\hat{\psi} + \frac{\hat{\lambda}_{GP}}{m}|\hat{\psi}|^2\hat{\psi} \right] = 0, \tag{A.108}$$

$$i \frac{\partial \hat{\psi}}{\partial(\hat{t})} + \frac{\hat{\nabla}^2}{2m}\hat{\psi} + \frac{\hat{\lambda}_{GP}}{m}|\hat{\psi}|^2\hat{\psi} = 0. \tag{A.109}$$

Therefore, Eq. (A.105) is invariant under the rescaling. This can similarly be shown for the second equation using the fact that for the solitons $|\psi|^2 \gg 1$.

For $\lambda = 1$, the soliton profile is given by

$$\frac{\rho_{sol}(r)}{m^2 M_{pl}^2} = \chi^2(mr) = \frac{1}{(1 + \alpha^2 m^2 r^2)^8}. \tag{A.110}$$

The mass of the soliton is then given by

$$M_{sol} = 4\pi \int_0^{r_s} \rho_{sol}(r)r^2 dr. \tag{A.111}$$

The rescaling relates to $\chi \rightarrow \lambda^2 \hat{\chi}$. Then, since $\rho_{sol}(r) = |\chi|^2$,

$$\begin{aligned} M_{sol} &= 4\pi \int_0^{r_c} |\chi|^2 r^2 dr \\ &= 4\pi \int_0^{r_c} \lambda^4 |\hat{\chi}|^2 \frac{1}{\lambda^3}\hat{r}^2 d\hat{r} \\ &= \lambda \hat{M}_{sol}. \end{aligned} \tag{A.112}$$

Similarly, $\rho_{sol}(r) = \lambda^4 \hat{\rho}_{sol}(\hat{r})$. We can therefore rescale the profile to

$$\frac{\lambda^4 \hat{\rho}_{sol}(r)}{m^2 M_{pl}^2} = \frac{1}{(1 + \alpha^2 \lambda^{-2}m^2\hat{r}^2)^8}. \tag{A.113}$$

We can define a scale radius $r_c = \frac{\lambda}{\alpha m}$ allowing us to write the density profile as

$$\rho_{sol}(r) = \frac{M_{pl}^2}{r_c^4 \alpha^4 m^2} \frac{1}{(1 + (\hat{r}/r_c)^2)^8}. \tag{A.114}$$

Integrating and making the change of variables, $u = \hat{r}/r_c$, we find that

$$\hat{M}_{sol} = \frac{4\pi M_{pl}^2}{r_c \alpha^4 m^2} \int_0^1 \frac{u^2}{(1 + u^2)^8} du. \tag{A.115}$$

The integral is now a constant (~ 0.246).[4] Fixing the units and dropping the hats, we then find

$$M_{\text{sol}} \sim 4 \times 10^8 \left(\frac{m}{10^{-22}\text{eV}}\right)^{-2} \left(\frac{r_c}{\text{kpc}}\right)^{-1} M_\odot.$$

(A.116)

The Einstein radius for such a lens is given by

$$R_E = 2 \times 10^{-7} \left(\frac{M_*}{M_\odot}\right)^{1/2} \text{kpc}.$$

(A.117)

Rearranging our mass–radius relation, we see

$$r_c \sim 4 \times 10^8 \left(\frac{m}{10^{-22}\text{eV}}\right)^{-2} \left(\frac{M_{\text{sol}}}{M_\odot}\right)^{-1} \text{kpc}.$$

(A.118)

For the object to lens like a point mass, we require that $r_c < R_E$, and therefore

$$4 \times 10^8 \left(\frac{m}{10^{-22}\text{eV}}\right)^{-2} \left(\frac{M_{\text{sol}}}{M_\odot}\right)^{-1} < 2 \times 10^{-7} \left(\frac{M_{\text{sol}}}{M_\odot}\right)^{1/2} \text{kpc},$$

(A.119)

which can be rearranged to

$$m < 5 \times 10^{-15} \left(\frac{M_{\text{sol}}}{M_\odot}\right)^{-3/4} \text{eV}.$$

(A.120)

This upper limit is maximized by being sensitive to masses as small as possible. Setting M_{sol} to the smallest mass detectable by HSC,

$$m < 1.4 \times 10^{-10}\text{eV}.$$

(A.121)

To calculate the range of the (T_{osc}, δ) parameter space, we neglect the activation function in Eq. (3.68) (since $S(x) \sim O(1)$). Requiring again that $R_{\text{MC}} < R_E$ and substituting our MC mass into the equation for the Einstein radius, we find the region range of (T_{osc}, δ) parameter space can be probed by microlensing to be

$$\frac{1}{\delta(1+\delta)^{1/3}} \left(\frac{T_{\text{osc}}}{2\text{GeV}}\right)^{1/2} \lesssim 7 \times 10^{-4}.$$

(A.122)

---

[4] This integral can be solved analytically. The result is insensitive to whether we define the mass to be $M(r < r_c)$ or $M(r < \infty)$.

Problem 4.1: Axion to Photon Production

An axion of mass equivalent to $3.3\,\mu eV = 5.29 \times 10^{-25}$ J corresponds to a photon with frequency $\nu = 5.29 \times 10^{-25}$ J$/h \approx 800$ MHz. To find the expected signal power, we can plug in the values listed in the problem into Eq. 4.1. We will use the standard assumption of $Q_a = 10^6$, which implies that the axion signal linewidth at 800 MHz is $\Delta\nu_a = \nu/Q_a = 800$ MHz$/10^6 = 800$ Hz. This is smaller than the cavity resonant linewidth given by $\Delta\nu_c = \nu_c/Q_L = (8 \times 10^8)/(6 \times 10^4) = 1.33 \times 10^4$ Hz.

$$P_{sig} = \left(\frac{(0.36)\alpha m_a}{\pi \cdot 0.006\,\text{GeV}^2}\right)^2 \left(\frac{\hbar^3 c^3 \rho_a}{(m_a)^2}\right)$$

$$\times \left(\frac{1}{\mu_0}(B_0)^2(\omega_c)(V)(C)(Q_0)\right) \tag{A.123}$$

$$\times \left(\frac{\beta}{(1+\beta)^2}\frac{1}{1 + ((2 \times \Delta\nu_a)/(\Delta\nu_c))^2}\right).$$

Canceling the factors of m_a^2 and plugging in the remaining terms including $\alpha \approx 1/137$, $\mu_0 = 4\pi \times 10^{-7}$ H m^{-1}, and $\hbar c = 1.97 \times 10^{-14}$ GeV cm,

$$P_{sig} \approx \left(\frac{0.36/137}{\pi \cdot 0.006\,\text{GeV}^2}\right)^2 \left((1.97 \times 10^{-14}\,\text{GeV cm})^3\,(0.45\,\text{GeV cm}^{-3})\right)$$

$$\times \left(\frac{1}{\mu_0}(7.6\,\text{T})^2 2\pi (8 \times 10^8\,\text{s}^{-1})(0.150\,\text{m}^3)(0.45)(180,000)\right)$$

$$\times \left(\frac{2}{9}\right) \times \left(\frac{1}{1 + \left(2 \times 800\,\text{Hz}/1.33 \times 10^4\,\text{Hz}\right)^2}\right). \tag{A.124}$$

One can see that the units (GeV, s, cm) in the top and bottom rows cancel out and that the middle row has units of W. Plugging these values into Eq. (4.1) yields $P_{sig} \approx 4.17 \times 10^{-23}$ W coming out of the cavity. For photons with 5.29×10^{-25} J of energy, this implies photon rates of 78 photons per second.

Problem 4.2: Cavity Resonance Frequencies

The frequency of the TM$_{010}$ mode of an empty cylindrical cavity of radius $r_{cavity} = 5.0$ cm is independent of the radius and is given by Eq. (4.8) to be

$$\nu TM_{010} = \frac{\omega TM_{010}}{2\pi} = 2.3\, \text{GHz}. \tag{A.125}$$

To find the number of TE modes between $1.3\, \text{GHz}$ and $3.3\, \text{GHz}$, we can use Eq. (4.9). The lowest frequency TE mode in a cavity of height $5.0\, \text{cm}$ is $\nu TE_{111} = 3.5\, \text{GHz}$, so there are zero TE modes within $1\, \text{GHz}$ of νTM_{010}. A cavity of height $10.0\, \text{cm}$ has the TE mode frequencies of $\nu TE_{111} = 2.3\, \text{GHz}$ and $\nu TE_{211} = 3.3\, \text{GHz}$, so there are two TE modes within $1\, \text{GHz}$ of νTM_{010}. Finally, a cavity of height $20.0\, \text{cm}$ has the TE mode frequencies of $\nu TE_{111} = 1.9\, \text{GHz}$, $\nu TE_{112} = 2.3\, \text{GHz}$, $\nu TE_{113} = 2.9\, \text{GHz}$, $\nu TE_{211} = 3.0\, \text{GHz}$, and $\nu TE_{212} = 3.3\, \text{GHz}$, so there are five TE modes within $1\, \text{GHz}$ of νTM_{010}.

Problem 4.3: Form Factor of an Annular Cavity

Since the external applied magnetic field is in the $\hat{z}$ direction, the dot product of the electric field of the modes of interest and the external applied magnetic field is

$$\mathbf{E} \cdot \mathbf{B_0} = E_z B_0. \tag{A.126}$$

The form factor of the TM_{0n0} mode in an annular cavity with the described dimensions is

$$C_{0n0} = \frac{\left(\int_V \sin k_0 \left(\rho - r_{\text{rod}} \right) B_0 \, dV \right)^2}{B_0^2 V \int_V \epsilon_r \sin^2 k_0 \left(\rho - r_{\text{rod}} \right) dV}. \tag{A.127}$$

Then, the volume integral can be separated into its cylindrical components and integrated in the $\hat{z}$ and $\hat{\phi}$ directions to give

$$C_{0n0} = \frac{2\pi h_{\text{cavity}} \left(\int_{r_{\text{rod}}}^{r_{\text{cavity}}} \rho \sin k_0 \left(\rho - r_{\text{rod}} \right) d\rho \right)^2}{V \epsilon_r \int_{r_{\text{rod}}}^{r_{\text{cavity}}} \rho \sin^2 k_0 \left(\rho - r_{\text{rod}} \right) d\rho}. \tag{A.128}$$

Integrating, we get $C_{010} = 0.81$ and $C_{030} = 0.09$. The results do not depend on the rod radius, cavity radius, or cavity height.

Problem 4.4: The Standard Quantum Limit

The noise temperature of the standard quantum limit is given by

$$T_{\text{SQL}} = \frac{h\nu}{k_B}, \tag{A.129}$$

where h is Plank's constant, ν is the frequency of interest, and k_B is the Boltzmann constant. Plugging in the frequencies of interest, $T_{SQL} \approx 34\,\text{mK}$ at 700 MHz, and $T_{SQL} \approx 290\,\text{mK}$ at 6 GHz.

Problem 5.1: Natural Lorentz–Heaviside Units

The speed of light c is given by

$$c = 299,792,458\,\frac{\text{m}}{\text{s}} = \{c\}[c], \tag{A.130}$$

where $c = 299,792,458$ is the numerical value of c and $[c]$ represents the unit, i.e., m/s. Equivalently, one obtains for the electric charge

$$e = 1.602176487 \times 10^{-19}\,\text{C} = \{e\}[e]. \tag{A.131}$$

Due to the fact that

$$1\frac{\text{GeV}}{c} = \frac{10^9\,\text{eV}}{c}$$

$$= \frac{10^9 \times \{e\}\,\text{J}}{\{c\}\frac{\text{m}}{\text{s}}}$$

$$= \frac{10^9 \times \{e\}\,\text{C}\,\frac{\text{J}}{\text{C}}}{\{c\}\frac{\text{m}}{\text{s}}}$$

$$= \frac{10^9 \times \{e\}\,\text{C}\,\frac{\text{J}}{\text{C}}}{\{c\}\frac{\text{m}}{\text{s}}}$$

$$= \frac{10^9}{\{c\}}e\,\frac{\text{J}\,\text{s}}{\text{C}\,\text{m}^2}\text{m}$$

$$= \frac{10^9}{\{c\}}e\,\text{T}\cdot\text{m}, \tag{A.132}$$

we find that

$$1\frac{\text{GeV}}{c} = 3.336\,e\,\text{T}\cdot\text{m}. \tag{A.133}$$

Using natural units, i.e., $c = 1$, Eq. (A.133) turns into

$$1\ \text{GeV} = 3.336\ e\ \text{T} \cdot \text{m}, \qquad (\text{A.134})$$

such that the relation between GeV and T·m depends on the definition of the electric units. In Gaussian units, the electric charge is chosen to be

$$e = \sqrt{\alpha} \approx 0.085, \qquad (\text{A.135})$$

while Heaviside units differ from the Gaussian units by a factor of $\sqrt{4\pi}$ and yield

$$e = \sqrt{4\pi\alpha} \approx 0.303. \qquad (\text{A.136})$$

In both systems, charge is dimensionless. Using Lorentz–Heaviside units in Eq. (A.134) yields

$$1\ \text{GeV} = 1.010\ \text{T} \cdot \text{m}. \qquad (\text{A.137})$$

Problem 5.2: Momentum Transfer

The axion energy is given by

$$E_a^2 = m_a^2 + p_a^2. \qquad (\text{A.138})$$

Thus, the momentum of the axion can be written as

$$p_a = E_a \sqrt{1 - \frac{m_a^2}{E_a^2}}, \qquad (\text{A.139})$$

where the development for $m_a \ll E_a$ yields

$$p_a \approx E_a - \frac{m_a^2}{2E_a}. \qquad (\text{A.140})$$

Analogously, one obtains

$$p_\gamma \approx E_\gamma - \frac{m_\gamma^2}{2E_\gamma}, \qquad (\text{A.141})$$

and with this, the momentum transfer q follows as

$$q = |p_a - p_\gamma| = \left| \frac{m_a^2}{2E_a} - \frac{m_\gamma^2}{2E_\gamma} \right|, \tag{A.142}$$

which is Eq. (5.22) if $E_a = E_\gamma$.

Problem 5.3: Effective Photon Mass in a Buffer Gas

The plasma frequency ω_p is given by

$$\omega_p^2 = 4\pi n_e \frac{e^2}{m} = 4\pi n_e r_0, \tag{A.143}$$

with n_e as the electron density and $r_0 = e^2/m$ the classical electron radius.

Using the fact that $n_e = N_e/V$, where N_e is the number of electrons in volume V, the plasma frequency can be expressed as

$$\omega_p^2 = 4\pi \frac{N_e}{V} r_0. \tag{A.144}$$

The effective photon mass follows then as

$$m_\gamma^2 = 4\pi \frac{N_e}{V} r_0. \tag{A.145}$$

In the case of helium, the number of electrons corresponds to twice the number of atoms N_a, i.e., $N_e = 2N_a$. And thus, applying the ideal gas law,

$$pV = nRT, \tag{A.146}$$

with pressure p, volume V, gas constant R, temperature T, and the amount of gas n given in mol, it follows that

$$\frac{N_e}{V} = \frac{2pN_A}{RT}. \tag{A.147}$$

In the above expression, we used the fact that $n = N_a/N_A$ and N_A is Avogadro's constant. Inserting Eq. (A.147) into Eq. (A.145) leads to

$$m_\gamma^2 = 8\pi \frac{r_0 N_A}{R} \frac{p}{T} = 5.130 \times 10^{11} \frac{p}{T} \frac{K}{\text{mbar m}^2} \tag{A.148}$$

or, using natural units ($\hbar c = 0.197$ GeV fm),

$$m_\gamma = \sqrt{0.020\frac{p/\text{mbar}}{T/\text{K}}}\,\text{eV},$$ (A.149)

which is the effective photon mass in helium.

Problem 5.4: Estimating the Focal Spot Size for Solar Axion Observations

To first order, the total angular spot size s_{total} is given by

$$
\begin{aligned}
s_{\text{total}} &= \sqrt{s_{\text{object}}^2 + s_{\text{optic}}^2} \\
&= \sqrt{(0.87\ \text{mrad})^2 + (0.58\ \text{mrad})^2} \\
&= 1.09\ \text{mrad}.
\end{aligned}
$$

The spatial diameter of the imaged focal spot can be calculated as focal length $f \times s_{\text{total}}$, and therefore the focal spot area a is

$$
\begin{aligned}
a &= \frac{\pi}{4}\,(s_{\text{total}} \times f)^2 \\
&= 0.23\ \text{cm}^2.
\end{aligned}
$$

Problem 5.5: Calculating an Exclusion Plot Using the Maximum Likelihood Method

The total expected number of counts μ_{ik} in the i-th energy bin E_i at the k-th pressure setting p_k can be expressed as

$$\mu_{ik} = b_{ik} + N_{ik},$$ (A.150)

where b_{ik} is the expected background in the i-th energy bin at density step k. It is assumed that this has been appropriately normalized. Due to low count numbers, Poissonian statistics have to be used and the maximum likelihood (ML) method can be applied (see, e.g., Chapter 40: Statistics in Ref. [4]). Generally, the standard likelihood function for Poissonian statistics is given by

$$\mathcal{L}_{\text{std}} = \prod_i e^{-\mu_{ik}}\frac{\mu_{ik}^{n_{ik}}}{n_{ik}!},$$ (A.151)

where n_{ik} is the number of counts measured during tracking in the i-th energy bin and the k-th pressure setting in the case of a CAST-like helioscope that we are considering here.

It is often convenient to work with a likelihood ratio of the kind:

$$\mathcal{L}_k = \frac{\mathcal{L}_{\text{std}}}{\mathcal{L}_{0k}} = \frac{\prod_i e^{-\mu_{ik}} \left(\mu_{ik}^{n_{ik}} / n_{ik}! \right)}{\prod_i e^{-n_{ik}} \left(n_{ik}^{n_{ik}} / n_{ik}! \right)}, \tag{A.152}$$

where $\mathcal{L}_0$ is merely the likelihood for which we have replaced μ_{ik} by the bin-by-bin model-independent ML estimator that is n_{ik}. Note that $\mathcal{L}_0$ does not depend on N_{ik} (thus neither on $g_{a\gamma\gamma}$), and therefore maximizing $\mathcal{L}_k$ is equivalent to maximizing $\mathcal{L}_{\text{std}}$. Also, note that $-2\ln \mathcal{L}_k$ behaves asymptotically as a χ^2 function (with degrees of freedom equal to the number of bins minus the number of free parameters in the fit) and therefore can be used to compute p-values and extract goodness-of-fit information.

In practice, we can minimize

$$\chi_k^2 = -2\ln\mathcal{L}_k = \sum_i \left[2\mu_{ik} - n_{ik}\ln\left(\mu_{ik}^2\right) - 2n_{ik} + n_{ik}\ln\left(n_{ik}^2\right) \right], \tag{A.153}$$

instead of maximizing the likelihood function.

Assuming the absence of a signal above background, the minimal value χ_{min}^2 should be close to χ_{Null}^2, which is the value of χ^2 for which $g_{a\gamma\gamma}^4 = 0$. For this case, tracking and background data can be directly compared, since no photons from conversion are expected. The difference between χ_{min}^2 and χ_{Null}^2 can be used to confirm the absence of signal.

Once all pressure steps have been taken into account separately and individual ML functions per step are obtained, the global likelihood function $\mathcal{L}$ is obtained by multiplying the individual likelihoods

$$\mathcal{L} = \prod_k \mathcal{L}_k, \tag{A.154}$$

with $k = 0, \ldots$, number of pressure steps -1.[5] Before deriving an upper limit, for all pressure settings, the global ML function has to be maximized or, equivalently, its χ^2-function

$$\chi^2 = -2\ln\mathcal{L}, \tag{A.155}$$

---

[5] In practice, only values of k "close" to the m_a being evaluated have a meaningful contribution to the final result at that particular mass, so, in order to speed up computation time, only a few steps need to be actually combined at any time.

needs to be minimized to determine the best fit value for the axion–photon coupling constant $g_{a\gamma\gamma}^4$ at each axion mass m_a.

The confidence interval for the l-th axion mass can be estimated using

$$\left[\ln\mathcal{L}(g_{a\gamma\gamma}^4)\right]_l = \left[\ln\mathcal{L}_{max}(g_{a\gamma\gamma}^4)\right]_l - \frac{\sigma^2}{2}, \tag{A.156}$$

with $\left[\ln\mathcal{L}_{max}(g_{a\gamma\gamma}^4)\right]_l$ being the maximal value at the l-th mass. The statistical error of the best fit value for $g_{a\gamma\gamma}^4$, i.e., $g_{a\gamma\gamma,min}^4$, can therefore be obtained as

$$\left[\chi^2(g_{a\gamma\gamma}^4)\right]_l = \left[\chi_{min}^2(g_{a\gamma\gamma}^4)\right]_l + \sigma^2. \tag{A.157}$$

Here, $\left[\chi_{min}^2(g_{a\gamma\gamma}^4)\right]_l$ represents the minimal χ^2 at the l-th axion mass. Note that the χ^2-distribution does not necessarily have to be symmetric, and therefore the statistical error can be asymmetric as well.

If the best fit value and its error are compatible with absence of signal, we usually like to express our result as an upper limit to $g_{a\gamma\gamma}$, above which we consider a signal to be rejected by the available data at a given confidence level (CL), like, e.g., 95% CL. There are several methods in the statistics literature to compute upper limits. A conceptually simple one is offered by Bayesian statistics. The Bayesian framework allows for building the probability function of unknown theoretical parameters (like, in this case, $g_{a\gamma\gamma}$). Such a probability is to be viewed as the distribution of our state of knowledge (or degree of belief) on where the true value of $g_{a\gamma\gamma}$ is (and not as the distribution of infinite outcomes of a variable, as is the case in frequentist statistics). This is done by invoking Bayes' theorem, by which the posterior probability of a theoretical parameter θ, after having obtained certain measurements x, is

$$P(\theta, x) = \frac{\pi(\theta)\mathcal{L}(x, \theta)}{\int \pi(\theta')\mathcal{L}(x, \theta')d\theta'}, \tag{A.158}$$

where $\pi(\theta)$ is the prior probability of θ (that is, the state of knowledge about θ that we had before carrying out the observations), and $\mathcal{L}$ is the likelihood function. Note that the denominator is just a normalization to make the numerator a proper probability (its total integral over θ is unity).

Coming back to our case, we identify θ with $g_{a\gamma\gamma}^4$ and assume a flat prior for positive values of this variable, while zero for negative ones, reflecting our prior knowledge that the expected signal must be positive, but otherwise being quite uninformative.[6] Therefore, the posterior probability on $g_{a\gamma\gamma}^4$ is

---

[6] The choice of $g_{a\gamma\gamma}^4$ versus other function of $g_{a\gamma\gamma}$ seems justified as the strength of the signal in a helioscope is proportional to $g_{a\gamma\gamma}^4$, although admittedly the choice of a prior has always some degree of subjectivity that is typical in Bayesian methods.

$$P(g_{a\gamma\gamma}^4) = \frac{\mathcal{L}(g_{a\gamma\gamma}^4)}{\mathcal{L}_0}, \tag{A.159}$$

where $\mathcal{L}_0 = \int_0^\infty \mathcal{L}(g_{a\gamma\gamma}^4) dg_{a\gamma\gamma}^4$ is a normalization factor.

The upper limit at 95% CL, $g_{a\gamma\gamma}^4(95\%)$, is then simply computed by integrating the posterior probability from zero to the value that encompasses 95% of the area:

$$\int_0^{g_{a\gamma\gamma}^4(95\%)} P(g_{a\gamma\gamma}^4) dg_{a\gamma\gamma}^4 = 0.95. \tag{A.160}$$

Repeating this step for different values of m_a, one obtains an exclusion line in the $(g_{a\gamma\gamma}, m_a)$ plane. Note that it is possible in principle to calculate the upper limit for each single pressure setting in this way, but since neighboring pressure settings contribute to the same masses, one loses information in comparison to a combined limit.

In order to combine multiple detectors, one can proceed equivalently, i.e., multiplying the global likelihoods of each detector

$$\mathcal{L}_{\text{Total Detectors } ^4\text{He Phase}} = \mathcal{L}_{\text{Detector 1}} \cdot \mathcal{L}_{\text{Detector 2}} \cdots \cdot \mathcal{L}_{\text{Detector n}}, \tag{A.161}$$

to obtain a global helioscope likelihood function and derive an exclusion plot for a buffer-gas phase (such as CAST's Phase II with ^{4}He in the magnet bores). This result can then also be combined with the achievements of a vacuum data run as well by multiplying the global likelihoods of different experimental phases. Of course, this procedure can also be directly applied to other helioscope experiments like BabyIAXO or IAXO. To find better justification and context on the statistical methods used here, we strongly recommend the interested reader to consult a textbook on statistical methods in particle physics (e.g., see Ref. [5] for a modern one).

Problem 6.1: Magnetic Field Produced by a Spherical Sample

(a) From Eq. (6.21),

$$\mathbf{M} \approx \frac{N\hbar^2\gamma^2 B_0}{2k_B T}\mathbf{I}. \tag{A.162}$$

Plugging in the numerical values from the statement of the problem yields approximately 4.922×10^{-5} A/m.

(b) Multiplying the magnetization by the volume of the sample, the magnetic moment is 2.577×10^{-11} A m^2. The magnetic field along the z-axis is

$$\mathbf{B}(z) = \frac{\mu_0 |\mathbf{m}|}{2\pi z^3} \hat{\mathbf{z}}, \tag{A.163}$$

so, plugging in the numbers, the field at a distance of 1 cm along the magnetization axis is about 5.154 pT.

The flux through a coil enclosing a surface $\mathbf{S}$ is defined as

$$\Phi = \oiint_S \mathbf{B} \cdot d\mathbf{S}, \tag{A.164}$$

where $\mathbf{B}$ is the magnetic field. Using Stokes' theorem, this can be rewritten as

$$\Phi = \oiint_S \nabla \times \mathbf{A} \cdot d\mathbf{S} = \oint_C \mathbf{A} \cdot d\mathbf{C}, \tag{A.165}$$

where $\mathbf{A}$ is the magnetic vector potential and $\mathbf{C}$ is the bounding curve of surface $\mathbf{S}$ (in our case, the coil itself). The vector potential of a magnetic dipole with magnetic moment $\mathbf{m}$ is

$$\mathbf{A}(\mathbf{r}) = \frac{\mu_0}{4\pi} \frac{\mathbf{m} \times \mathbf{r}}{r^3}. \tag{A.166}$$

We parametrize the curve $\mathbf{C}$ as

$$\mathbf{C}(\theta) = \begin{pmatrix} 5 \times 10^{-3} m \cos\theta \\ 5 \times 10^{-3} m \sin\theta \\ 10^{-2} m \end{pmatrix}, \tag{A.167}$$

such that the derivative is

$$\mathbf{C}'(\theta) = \begin{pmatrix} -5 \times 10^{-3} m \sin\theta \\ 5 \times 10^{-3} m \cos\theta \\ 0 \end{pmatrix}. \tag{A.168}$$

Then, the flux is

$$\Phi = \int_0^{2\pi} \mathbf{A}(\mathbf{C}(\theta)) \cdot \mathbf{C}'(\theta) d\theta = 4\mu_0 \sqrt{5} |\mathbf{m}|/m, \tag{A.169}$$

which works out to 2.897×10^{-16} Wb or 0.14 magnetic flux quanta.

(c) The polarization at 1 T and 170 K is $\frac{\hbar \gamma B_0}{2 k_B T} \approx 1.67 \times 10^{-6}$. A nuclear spin polarization of 10% provides an enhancement of nearly 60,000. The resulting flux would be 1.733×10^{-11} Wb or 8.38×10^3 flux quanta.

Problem 6.2: Spin Noise

The spin noise limit is determined by the spin density, the sample volume, and the sensitivity of the detector to magnetic fields. It is independent of the degree of polarization of the ensemble. However, the sensitivity to ALP coupling does scale linearly with the degree of polarization, which is why it is a crucial quantity.

Consider a sample with volume V, containing $N = n_{Xe} V$ atoms, each with magnetic moment μ_{Xe}. Its spin-noise magnetization is given by

$$M_{SPN} = \frac{\mu_{Xe} \sqrt{N}}{V} = \mu_{Xe} \sqrt{\frac{n_{Xe}}{V}}. \tag{A.170}$$

For a sample of spherical shape, the magnetic field immediately outside is given by (see, for example, [1])

$$B_{SPN} = \frac{\mu_0 M_{SPN}}{3} = \frac{\mu_0 \mu_{Xe}}{3} \sqrt{\frac{n_{Xe}}{V}},$$

where μ_0 is the vacuum permeability. In order to find the volume necessary for the measurement to be limited by spin projection noise, this field is equated to the magnetic detector sensitivity: $B_{SPN} = B_{det}$. The result for the sample volume is

$$V_{SPN} = n_{Xe} \left(\frac{\mu_0 \mu_{Xe}}{3 B_{det}} \right)^2.$$

Substituting the values given in the problem gives the numerical result: $V_{SPN} \approx 10^{-6}\,\mathrm{m}^3 = 1\,\mathrm{ml}$.

Problem 7.1: Interaction Basis

Beginning from the expressions for $\bar{A}_\mu$ and $\bar{X}_\mu$ [Eqs. (7.3) and (7.4)], we find A_μ and X_μ in terms of these interaction basis potentials:

$$\bar{A}_\mu - \kappa \bar{X}_\mu = (A_\mu + \kappa X_\mu) - \kappa (X_\mu - \kappa A_\mu), \tag{A.171}$$

$$= A_\mu + \kappa^2 A_\mu \approx A_\mu, \tag{A.172}$$

and

$$\bar{X}_\mu + \kappa \bar{A}_\mu = (X_\mu - \kappa A_\mu) + \kappa (A_\mu + \kappa X_\mu), \tag{A.173}$$

$$= X_\mu + \kappa^2 X_\mu \approx X_\mu, \tag{A.174}$$

where in our approximation we neglect terms of order κ^2.

Note that the electromagnetic field strength tensor $F_{\mu\nu}$ and the hidden photon field strength tensor $\mathcal{F}_{\mu\nu}$ are related to the gauge potentials through

$$F_{\mu\nu} = \partial_\mu A_\nu - \partial_\nu A_\mu , \tag{A.175}$$

$$\mathcal{F}_{\mu\nu} = \partial_\mu X_\nu - \partial_\nu X_\mu , \tag{A.176}$$

from which one can show that

$$F_{\mu\nu} = \bar{F}_{\mu\nu} - \kappa \bar{\mathcal{F}}_{\mu\nu} , \tag{A.177}$$

$$\mathcal{F}_{\mu\nu} = \bar{\mathcal{F}}_{\mu\nu} + \kappa \bar{F}_{\mu\nu} . \tag{A.178}$$

Thus, again neglecting terms of order κ^2, the term in the Lagrangian involving the field strength tensors is unchanged in form going from the mass basis to the interaction basis:

$$F_{\mu\nu} F^{\mu\nu} + \mathcal{F}_{\mu\nu} \mathcal{F}^{\mu\nu} \approx \bar{F}_{\mu\nu} \bar{F}^{\mu\nu} + \bar{\mathcal{F}}_{\mu\nu} \bar{\mathcal{F}}^{\mu\nu} . \tag{A.179}$$

Next, we consider the term from the mass basis proportional to $X_\mu X^\mu$:

$$X_\mu X^\mu = \left(\bar{X}_\mu + \kappa \bar{A}_\mu \right) \left(\bar{X}^\mu + \kappa \bar{A}^\mu \right) , \tag{A.180}$$

$$\approx \bar{X}_\mu \bar{X}^\mu + 2\kappa \bar{A}_\mu \bar{X}^\mu . \tag{A.181}$$

Substituting Eqs. (7.3), (A.179), and (A.181) into the Lagrangian for the mass basis [Eq. (7.1)] yields Eq. (7.5) as desired.

Problem 7.2: Oscillation Frequency of Hidden Electromagnetic Fields

The wave equation for the hidden photon gauge potential is given by

$$\left[\frac{1}{c^2} \frac{\partial^2}{\partial t^2} - \nabla^2 + \left(\frac{m_{\gamma'} c}{\hbar} \right)^2 \right] \bar{X}_\mu = 0 . \tag{A.182}$$

Assuming a plane wave solution for a given mode of the hidden photon field, such that $\bar{X}_\mu \propto e^{i(\mathbf{k}\cdot\mathbf{r} - \omega t)}$, we find that

$$\left[-\frac{\omega^2}{c^2} + k^2 + \left(\frac{m_{\gamma'}c}{\hbar}\right)^2\right]\bar{X}_\mu = 0, \qquad (A.183)$$

which implies that

$$-\frac{\omega^2}{c^2} + k^2 + \left(\frac{m_{\gamma'}c}{\hbar}\right)^2 = 0. \qquad (A.184)$$

Since $k \ll \omega/c$ due to the fact the hidden photons are nonrelativistic,

$$\omega \approx \frac{m_{\gamma'}c^2}{\hbar}, \qquad (A.185)$$

and thus the field oscillates at the hidden photon Compton frequency. Hidden photons with high mode occupation will interfere with one another producing a classical field that has properties similar to thermal light, with characteristic coherence length and time determined by the velocity distribution.

Problem 7.3: DM Energy Density and the Magnetic Field Within Shields

The energy density in the hidden photon field can be related to the amplitude of the hidden photon vector potential based on the free space (vacuum) form result [Eq. (7.12)],

$$\rho_{\text{dm}} = \frac{1}{8\pi}\left(\mathcal{E}'\right)^2 \approx \frac{1}{8\pi}\frac{X_0^2}{\lambda_{\gamma'}^2}, \qquad (A.186)$$

where we have used the fact that in vacuum the hidden electric field is given by

$$\mathcal{E}' = -\partial_\mu \bar{X}^\mu \approx i\frac{\omega}{c}X \approx \frac{iX}{\lambda_{\gamma'}}. \qquad (A.187)$$

While the effect of $\mathcal{E}'$ on the charges in the shield generates a compensating field that largely cancels $\mathcal{E}'$ within the shield, the estimate of X_0 based on ρ_{dm} in Eq. (A.186) is still valid since it is derived solely from the hidden photon field and does not include the vector potential $\mathbf{A}$ that describes the real field generated by the charges in the shield. Thus, from Eq. (A.186), we have

$$X_0 \approx \lambda_{\gamma'}\sqrt{8\pi\rho_{\text{dm}}}. \qquad (A.188)$$

Substituting into Eq. (7.35), we obtain

$$\mathbf{B}(\mathbf{r}, t) = 8\pi\kappa\sqrt{2\pi\rho_{\mathrm{dm}}}\frac{r}{\lambda_{\gamma'}}e^{-i\omega t}\hat{\boldsymbol{\phi}} \, . \tag{A.189}$$

In terms of G, the magnetic field can be estimated by converting the dark matter density into cgs units, finding $\rho_{\mathrm{dm}} \approx 6.4 \times 10^{-4} \mathrm{\ erg/cm^3}$, from which we have

$$B \approx (0.04 \mathrm{\ G}) \times \frac{\kappa R}{\lambda_{\gamma'}} \, . \tag{A.190}$$

To get a rough sense of the $R/\lambda_{\gamma'}$ suppression, let us estimate $R \sim 100$ cm for the shield and assume $m_{\gamma'}c^2 \approx 10^{-12}$ eV, which corresponds to $\lambda_{\gamma'} \approx 2 \times 10^7$ cm. In this case, $B \approx \kappa \times (2 \times 10^{-7} \mathrm{\ G})$.

Problem 7.4: Inductance

The hollow superconducting sheath consists of two concentric cylinders of height h and radii r_1 and r_2. For the purposes of this calculation, we assume the common axis of the two cylinders is along z. The inductance is a measure of the ratio of magnetic flux Φ to the current I generating the flux for a particular conductor geometry and can be expressed as

$$L = \frac{1}{c}\frac{\Phi}{I} \, . \tag{A.191}$$

Assuming a current I flowing in the $+\hat{\mathbf{z}}$ direction on the inner cylinder, the flux between the cylinders can be calculated from Ampère's law:

$$\oint \mathbf{B} \cdot d\boldsymbol{\ell} = \frac{4\pi}{c}I \, . \tag{A.192}$$

Choosing an Ampèrian loop of radius r, where $r_1 < r < r_2$, we find that

$$\mathbf{B} = \frac{2I}{cr}\hat{\boldsymbol{\phi}} \, . \tag{A.193}$$

The flux through a cross section of the sheath can then be found through integration:

$$\Phi = \mathbf{B} \cdot d\mathbf{A} \, , \tag{A.194}$$

$$= \frac{2I}{c}\int_{r=r_1}^{r_2}\int_{z=0}^{h}\left(\frac{1}{r}\hat{\boldsymbol{\phi}}\right) \cdot \left(dr\,dz\,\hat{\boldsymbol{\phi}}\right) \, , \tag{A.195}$$

$$= \frac{2Ih}{c}\ln\left(\frac{r_2}{r_1}\right) \, . \tag{A.196}$$

Inserting the above expression for the flux into Eq. (A.191), we find

$$L = \frac{2h}{c^2} \ln \left(\frac{r_2}{r_1} \right) . \tag{A.197}$$

A toroidal solenoid of the same dimensions, with radius $r = (r_1 + r_2)/2$ and cross-sectional area $A = h(r_2 - r_1)$, having N turns, has inductance

$$L_{ts} = \frac{4}{c^2} \frac{N^2 A}{2r} , \tag{A.198}$$

$$= \frac{2}{c^2} \frac{N^2 h(r_2 - r_1)}{r_1 + r_2} . \tag{A.199}$$

Equating this expression for L_{ts} with our expression for the inductance of the cylindrical sheath [Eq. (A.197)], we find

$$N^2 = \left(\frac{r_1 + r_2}{r_2 - r_1} \right) \ln \left(\frac{r_2}{r_1} \right) . \tag{A.200}$$

Problem 7.5: DM Radio EMF

The magnetic flux through a cross-sectional area of the toroidal solenoid (height $= h$, inner radius $= r_1$, and outer radius $= r_2$) that acts as the "antenna" of the dark matter radio is given by

$$\Phi = \int \mathbf{B} \cdot d\mathbf{A} , \tag{A.201}$$

$$= 8\pi \kappa \sqrt{2\pi \rho_{dm}} \frac{h}{\lambda_{\gamma'}} e^{-i\omega t} \int_{r_1}^{r_2} r \, dr , \tag{A.202}$$

$$= 4\pi \kappa \sqrt{2\pi \rho_{dm}} \frac{h}{\lambda_{\gamma'}} \left(r_2^2 - r_1^2 \right) e^{-i\omega t} . \tag{A.203}$$

Noting that the volume contained within the concentric cylindrical sheath is $V = \pi h(r_2^2 - r_1^2)$, we have

$$\Phi = 4\kappa \sqrt{2\pi \rho_{dm}} \frac{V}{\lambda_{\gamma'}} e^{-i\omega t} . \tag{A.204}$$

The induced EMF is given by

$$\mathcal{V}_{\gamma'} = -\frac{N}{c}\frac{\partial \Phi}{\partial t} \,, \tag{A.205}$$

$$= 4i\kappa\sqrt{2\pi\rho_{dm}}\frac{NV}{\lambda_{\gamma'}^2}e^{-i\omega t} \,, \tag{A.206}$$

where we have used the fact that $c/\omega \approx \lambda_{\gamma'}$ and accounted for the fact that the total flux through the solenoid is $N\Phi$.

Problem 7.6: DM Radio Q-Factor

To understand the signal enhancement, consider the current flowing through the RLC circuit in Fig. 7.4, given by

$$I = \frac{\mathcal{V}_{\gamma'}}{Z} = \frac{\mathcal{V}_{\gamma'}}{\sqrt{R^2 + (\omega L - 1/(\omega C))^2}} \,, \tag{A.207}$$

where Z is the circuit impedance. On resonance $\omega = \omega_0 = 1/\sqrt{LC}$, the current is $I = \mathcal{V}_{\gamma'}/R$, and so the magnetic flux Φ_L in the inductor is given by

$$\Phi_L = cLI \tag{A.208}$$

$$= \frac{L\mathcal{V}_{\gamma'}c}{R} \,, \tag{A.209}$$

$$= \frac{L}{R}4i\kappa\sqrt{2\pi\rho_{dm}}\frac{NVc}{\lambda_{\gamma'}^2}e^{-i\omega t} \,, \tag{A.210}$$

$$= \frac{\omega_0 L}{R}4i\kappa\sqrt{2\pi\rho_{dm}}\frac{NV}{\lambda_{\gamma'}}e^{-i\omega t} \,, \tag{A.211}$$

$$= Q4i\kappa\sqrt{2\pi\rho_{dm}}\frac{NV}{\lambda_{\gamma'}}e^{-i\omega t} \,, \tag{A.212}$$

$$= QN\Phi \,. \tag{A.213}$$

Thus, the flux in the inductor is enhanced by the Q-factor as compared to the flux from the hidden photon field alone.

Problem 8.1: Yukawa Potential in the Monopole–Monopole Interaction

In the nonrelativistic limit, the interaction between two fermions at the tree level can be written from the inverse Born approximation (momentum space to coordinate space Fourier transformation),

$$V(r) = \int \frac{d^3q}{(2\pi)^3} \frac{(\text{vertex 1})(\text{vertex 2})}{|q|^2 + m_b^2} e^{iq \cdot r} , \qquad (A.214)$$

where the q is the transferred momentum and m_b is the mass of the mediating boson. The simplest form of integration can be obtained from the interaction between two monopoles $(g_{s,1}g_{s,2})$:

$$V_{ss}(r) = -g_{s,1}g_{s,2}\mathcal{V}_{ss}(r) = -g_{s,1}g_{s,2} \int \frac{d^3q}{(2\pi)^3} \frac{1}{|q|^2 + m_b^2} e^{iq \cdot r} . \qquad (A.215)$$

This integration can be done in spherical coordinates as follows:

$$\mathcal{V}_{ss}(r) = \int_0^\infty dq \int_0^{2\pi} d\phi \int_{-1}^{1} d\cos\theta \frac{q^2}{(2\pi)^3} \frac{1}{q^2 + m_b^2} e^{iqr\cos\theta} , \quad (A.216)$$

$$= \frac{1}{(2\pi)^2} \int_0^\infty dq \frac{q}{q^2 + m_b^2} \left(\frac{e^{iqr} - e^{-iqr}}{ir} \right) ,$$

$$= \frac{1}{ir} \frac{1}{(2\pi)^2} \int_{-\infty}^\infty dq \frac{q}{q^2 + m_b^2} e^{iqr} .$$

From the Cauchy integral theorem, the complex integration yields

$$\int_{-\infty}^\infty dq \frac{q}{q^2 + m_b^2} e^{iqr} = 2\pi i \left(\frac{im_b}{2im_b} \right) e^{-m_b r} = i\pi e^{-m_b r}, \qquad (A.217)$$

which is the Yukawa-type potential. The potential describing the scalar–scalar interaction becomes

$$V_{ss}(r) = - \left(\frac{g_{s,1}g_{s,2}}{4\pi r} \right) e^{-m_b r}. \qquad (A.218)$$

Problem 8.2: Spin-Dependent Interaction via Spin-0 Boson Exchange: Monopole–Dipole Interaction

The potential describing a monopole (g_s) and dipole $(ig_p \boldsymbol{\sigma} \cdot \boldsymbol{q}/2m)$ interaction between two fermions can be expressed as follows:

$$V_{ps}(r) = -\frac{g_{p,1}g_{s,2}}{2im_1}\mathcal{V}_{ps}(r) = -\frac{g_{p,1}g_{s,2}}{2im_1}\int \frac{d^3q}{(2\pi)^3}\frac{\boldsymbol{\sigma}_1 \cdot \boldsymbol{q}}{|\boldsymbol{q}|^2 + m_b^2}e^{i\boldsymbol{q}\cdot\boldsymbol{r}} , \qquad (A.219)$$

This integration can be obtained by applying the inner product between the spin vector and the gradient of the $\mathcal{V}_{ss}(r)$ from Eq. (A.218),

$$\mathcal{V}_{ps}(r) = -i\boldsymbol{\sigma}_1 \cdot \nabla\mathcal{V}_{ss}(r) . \qquad (A.220)$$

In spherical coordinates, the calculation is straightforward,

$$\mathcal{V}_{ps}(r) = -i\boldsymbol{\sigma}_1 \cdot \nabla\mathcal{V}_{ss}(r) = i\boldsymbol{\sigma}_1 \cdot \hat{\boldsymbol{r}}\left(\frac{m_b}{r} + \frac{1}{r^2}\right)\frac{e^{-m_b r}}{4\pi} . \qquad (A.221)$$

Therefore,

$$V_{ps}(r) = -\frac{g_{p,1}g_{s,2}}{2im_1}\mathcal{V}_{ps}(r) = -\frac{g_{p,1}g_{s,2}}{2m_1}\boldsymbol{\sigma}_1 \cdot \hat{\boldsymbol{r}}\left(\frac{m_b}{r} + \frac{1}{r^2}\right)\frac{e^{-m_b r}}{4\pi} , \qquad (A.222)$$

which can be rewritten in terms of an "effective" or "pseudo-" magnetic field $\mathbf{B}_{\text{eff}}$ as suggested in the statement of the problem:

$$\begin{aligned} V_{ps}(r) &= -\frac{g_{p,1}g_{s,2}}{8\pi m_1}\left(\boldsymbol{\sigma}_1 \cdot \hat{\boldsymbol{r}}\right)\left(\frac{m_b}{r} + \frac{1}{r^2}\right)e^{-m_b r} , \\ &= -\left(\frac{g_{p,1}g_{s,2}}{8\pi m_1}\right)\nabla\left(\frac{1}{r}e^{-m_b r}\right)\cdot\boldsymbol{\sigma} , \\ &= -\nabla U(r)\cdot\boldsymbol{\sigma} , \\ &= -\nabla U(r)\left(\frac{2}{\hbar\gamma_f}\right)\cdot\boldsymbol{\sigma}\left(\frac{\hbar\gamma_f}{2}\right) , \\ &= -\mathbf{B}_{\text{eff}}\cdot\boldsymbol{\mu}_f . \end{aligned} \qquad (A.223)$$

Equation (A.223) shows that indeed the interaction potential acts on a nearby fermion as an "effective" magnetic field

$$\mathbf{B}_{\text{eff}} = \frac{2}{\hbar\gamma_f}\nabla U(r), \qquad (A.224)$$

where γ_f is the gyromagnetic ratio of the fermion. This effective field is different from an ordinary magnetic field. Since it couples to the spin of the particle rather than electric charge or ordinary angular momentum, this field is crucially not subject to Maxwell's equations and therefore cannot be screened by magnetic shielding.

Problem 8.3: Spin-Dependent Interaction via Spin-0 Boson Exchange: Dipole–Dipole Interaction

In the case of pseudoscalar vertices on both sides, the dipole–dipole interaction has following integral form:

$$V_{pp}(r) = \frac{g_{p,1}g_{p,2}}{4m_1m_2}\mathcal{V}_{pp}(r) = \frac{g_{p,1}g_{p,2}}{4m_1m_2} \int \frac{d^3q}{(2\pi)^3} \frac{(\boldsymbol{\sigma}_1 \cdot \boldsymbol{q})(\boldsymbol{\sigma}_2 \cdot \boldsymbol{q})}{|\boldsymbol{q}|^2 + m_b^2} e^{i\boldsymbol{q}\cdot\boldsymbol{r}},$$

(A.225)

where labels 1 and 2 indicate the two fermions. One can expand the inner product with the summation over all possible spin states,

$$\mathcal{V}_{pp}(r) = \sum_a \sum_b \int \frac{d^3q}{(2\pi)^3} \frac{\sigma_{1,a}\sigma_{2,b}q_aq_b}{|\boldsymbol{q}|^2 + m_b^2} e^{i\boldsymbol{q}\cdot\boldsymbol{r}},$$

(A.226)

where a and b are the possible spin state of each fermion ($a, b = 1, 2, 3$). Eq. (A.226) can be expressed in the following way:

$$\mathcal{V}_{pp}(r) = -\sum_a \sum_b \sigma_{1,a}\sigma_{2,b}\partial_a\partial_b\mathcal{V}_{ss},$$

(A.227)

where $\mathcal{V}_{ss}$ is already defined from Eq. (A.218). The partial derivatives of the monopole interaction $\partial_a\partial_b\mathcal{V}_{ss}$ become

$$4\pi \partial_a\partial_b\mathcal{V}_{ss} = \left(\partial_a\partial_b e^{-m_br}\right)\frac{1}{r} + 2\left(\partial_a e^{-m_br}\right)\left(\partial_b\frac{1}{r}\right) + e^{-m_br}\left(\partial_a\partial_b\frac{1}{r}\right).$$

(A.228)

The evaluation of each term in Eq. (A.228) yields

$$\left(\partial_a \partial_b e^{-m_b r}\right)\frac{1}{r} = \partial_a \left(-m_b e^{-m_b r}\frac{r_b}{r}\right)\frac{1}{r}$$

$$= \left(m_b^2 \frac{r_a r_b}{r^2} - \frac{m_b \delta_{a,b}}{r} + \frac{m_b r_a r_b}{r^3}\right)\frac{e^{-m_b r}}{r},$$

$$\left(\partial_a e^{-m_b r}\right)\left(\partial_b \frac{1}{r}\right) = m_b \frac{r_a r_b}{r^4},$$

$$e^{-m_b r}\left(\partial_a \partial_b \frac{1}{r}\right) = \left(-\frac{\delta_{a,b}}{r^3} + \frac{3 r_a r_b}{r^5} - \frac{4\pi}{3}\delta_{a,b}\delta^3(r)\right)e^{-m_b r}.$$

(A.229)

The Dirac delta function in Eq. (A.229) comes from the Laplacian of the $1/r$ term since $\nabla^2(1/r) = -4\pi\delta^3(r)$, and the factor of $1/3$ comes from the normalization of the $\sum_a \sum_b \delta_{a,b}$ running over the indices $1 \to 3$. Using the relationships

$$\sum_{a,b} \sigma_{1,a}\sigma_{2,b}\delta_{a,b} = \boldsymbol{\sigma}_1 \cdot \boldsymbol{\sigma}_2 \qquad (A.230)$$

and

$$\sum_{a,b} \sigma_{1,a}\sigma_{2,b}r_a r_b = (\boldsymbol{\sigma}_1 \cdot \hat{\boldsymbol{r}})(\boldsymbol{\sigma}_2 \cdot \hat{\boldsymbol{r}})r^2 , \qquad (A.231)$$

one can obtain

$$\mathcal{V}_{pp}(r) = \frac{\boldsymbol{\sigma}_1 \cdot \boldsymbol{\sigma}_2}{4\pi}\left(\frac{1}{r^3} + \frac{m_b}{r^2} + \frac{4\pi}{3}\delta(r)\right)e^{-m_b r}$$

$$- \frac{(\boldsymbol{\sigma}_1 \cdot \hat{\boldsymbol{r}})(\boldsymbol{\sigma}_2 \cdot \hat{\boldsymbol{r}})}{4\pi}\left(\frac{m_b^2}{r} + \frac{3 m_b}{r^2} + \frac{3}{r^3}\right)e^{-m_b r} . \qquad (A.232)$$

Problem 8.4: Magnetic Field "Amplification Factor" for a Magnetized NMR Sample Subject to an Effective Axion-Induced "Magnetic Field"

The time-varying magnetic field B_{SQUID} that would be detected by a SQUID pickup loop at a distance of $r = 2$ mm from the center of a 1-mm radius spherical sample can be determined from the dipole approximation by integrating the induced transverse magnetization M_x over the volume V of the sample.

$$B_{SQUID} \approx \frac{\mu_0}{4\pi}\frac{M_x V}{r^3}.$$

Driven for a duration of T_2 by an axion with an effective field B_{eff}, the expected magnetization is

$$M_x \approx \frac{1}{2} \frac{N_s}{V} p \mu_N \gamma T_2 B_{\text{eff}},$$

where N_s is the number of nuclear spins in the sample and p is the polarization fraction.

The numerical value of the "amplification factor" $(B_{\text{SQUID}}/B_{\text{eff}})$ with $T_2 = 1000$ seconds, a spin density of 10^{21} spins per cubic centimeter, and unity polarization $p = 1$ is about 2.7×10^4.

Problem 9.1 Measuring the Mass of the UBDM Field

The experiment will not produce any UBDM field if $|F(qL)| = 0$. This will occur when qL is some greater than zero integer multiple of 2π. We can use the approximation in Eq. (9.5) with the index of refraction equal to one to find the mass:

$$q \approx \frac{m_\varphi^2}{2\omega}. \tag{A.233}$$

The lowest mass that the experiment is insensitive to is therefore

$$m_\varphi = \sqrt{\frac{4\pi\omega}{L}}. \tag{A.234}$$

If we then plug that mass back into the equation for qL, but this time with an index of refraction of n, we get

$$qL \approx \omega L(n - 1) + 2\pi. \tag{A.235}$$

Since $n \geq 1$, the form factor, $|F(qL)|$, will have a maximum value when $qL = 2\pi + \pi/2$. Therefore, we can solve for n to get

$$n \approx \frac{\pi}{2\omega L} + 1. \tag{A.236}$$

If we use roughly the ALPS II parameters of $L = 100\,m$ and a photon energy of $1\,\text{eV}$, we see we get an index of refraction $n \approx 1 + 3 \times 10^{-9}$.

If a regenerated photon signal is observed, this could be used to identify the mass of the UBDM field by performing measurements with and without a small pressure of residual gas present in the system and then comparing the difference in the event rates.

Problem 9.2: Maximum Power Build-Up

If we take the derivative of β_P with respect to T_1, we get the following:

$$\frac{\partial \beta_P}{\partial T_1} = \frac{\beta}{T_1} - \frac{2\beta}{T_1 + T_2 + \rho} \,, \tag{A.237}$$

$$\frac{\partial \beta_P}{\partial T_2} = -\frac{2\beta}{T_1 + T_2 + \rho} \,. \tag{A.238}$$

By setting this equation to zero, we can see that the maximum power build-up will occur when

$$T_1 = T_2 + \rho \,, \tag{A.239}$$

while T_1 and T_2 are as low as possible. We should note here that when the cavity meets the condition in Eq. (A.239), it is in what is called an "impedance matched" configuration.

Problem 9.3: Eigenmode Waist Size Versus Length

Since the cavity is configured with mirrors that have identical curvatures, we know that the waist position will be halfway between them. Therefore, we can use Eq. (9.19) to solve for the Rayleigh range with the knowledge $R(z = L/2) = L$:

$$R\left(\frac{L}{2}\right) = L = \frac{L}{2}\left[1 + \left(\frac{2z_r}{L}\right)^2\right] \,. \tag{A.240}$$

Here, we can see that $z_r = L/2$. We can then plug this into Eq. (A.243) to get

$$w_0 = \sqrt{\frac{\lambda L}{2\pi}} \,. \tag{A.241}$$

Problem 9.4: Clipping Losses and Cavity Length

If we integrate the intensity in a Gaussian beam out to a radius r_{ap}, we get the following equation:

$$\int_0^{2\pi} \int_0^{r_{ap}} I(r, z_{ap}) r \, dr \, d\theta = P_{max}\left(1 - \exp\left(-2\frac{r_{ap}^2}{w^2}\right)\right), \quad (A.242)$$

where the second term is equal to the clipping losses. With this, we can derive the following relationship between the clipping losses and the waist radius:

$$w_0 = \sqrt{-2\frac{r_{ap}^2}{\ln(\rho)}}. \quad (A.243)$$

Since we know that the minimum clipping losses will occur when the length of the cavity is the Rayleigh length, we can then find the following relationship between the bore radius r_{ap}, the clipping losses, and the length of the cavity:

$$L = -2\pi \frac{r^2}{\lambda \ln(\rho)}. \quad (A.244)$$

Let us assume $T_1 \gg T_1$. Then, the maximum power build-up factor will occur when $T_1 = \rho$, and using $\beta = 10,000$ necessitates that $\rho = 100\,ppm$. From this, we can see that we can only make the cavity $\sim 400\,m$ before we will incur more than $100\,ppm$ of clipping losses.

Problem 9.5: Black-Body Pile-Ups

In a given rise time, probability of 2 photons from this energy band hitting the detector is

$$\mathcal{P}_{2\gamma} = \frac{(\tau_{rise}\gamma_r)e^{-\tau_{rise}\gamma_r}}{2!} \approx 5 \times 10^{-9}, \quad (A.245)$$

where γ_r is the background rate for this energy band. Over the course of a $10^6\,s$ measurement 10^{12}, "rise times" will occur. Therefore, the expected number of unresolvable pile-ups will be 500.

Problem 10.1: Dark Matter-Induced Variation of Fundamental Constants

We focus on the linear portal (10.1) (derivations for the quadratic portal are identical)

$$\mathcal{L}_{\text{clk}}^{(1)} = \left(-\Gamma_f m_0 c^2 \bar{\psi}\psi + \frac{\Gamma_\alpha}{4} F_{\mu\nu} F^{\mu\nu}\right) \sqrt{\hbar c}\, \phi\,.$$

Here, we streamlined notation in Eq. (10.1). Recall that the SM Lagrangian for a fermion coupled to an electromagnetic field reads

$$\mathcal{L}_f^{\text{SM}} = i(\hbar c)\bar{\psi}\gamma_\mu \partial^\mu \psi - m_0 c^2 \bar{\psi}\psi - q_0 \bar{\psi}\gamma^\mu \psi A_\mu - \frac{1}{4} F_{\mu\nu} F^{\mu\nu}\,,$$

with the Faraday tensor $F_{\mu\nu} = \partial_\mu A_\nu - \partial_\nu A_\mu$ expressed in terms of the electromagnetic 4-potential A_μ and q_0 being the fermion's electric charge. Comparing the mass terms in $\mathcal{L}_{\text{clk}}^{(1)}$ and $\mathcal{L}_f^{\text{SM}}$ leads to a combination $m_0 c^2 \bar{\psi}\psi\left(1 + \Gamma_f \sqrt{\hbar c}\,\phi\right)$. Thereby, the factor $m_0 \times \left(1 + \Gamma_f \sqrt{\hbar c}\,\phi\right)$ is the effective, DM field-dressed mass (10.3).

The proof of DM-induced variation of α, Eq. (10.4), is more involved. We start by combining the Faraday tensor contributions in $\mathcal{L}_{\text{clk}}^{(1)}$ and $\mathcal{L}_f^{\text{SM}}$: $\frac{1}{4} F_{\mu\nu} F^{\mu\nu}\left(1 - \Gamma_\alpha \sqrt{\hbar c}\,\phi\right)$. Next, we rescale the Faraday tensor

$$F_{\mu\nu} = F'_{\mu\nu}\left(1 - \Gamma_\alpha \sqrt{\hbar c}\,\phi\right)^{-1/2}$$

to bring the resulting Lagrangian contribution to the canonical $-F'_{\mu\nu} F'^{\mu\nu}/4$ form. This rescaling is equivalent to the redefinition of the 4-potential,

$$A_\mu = A'_\mu\left(1 - \Gamma_\alpha \sqrt{\hbar c}\,\phi\right)^{-1/2}\,,$$

in the assumption that the field ϕ does not vary appreciably over the atomic sample size or over atomic time-scales. This assumption is well justified for conventional atomic clocks and ultralight DM fields. The rescaling of the 4-potential brings the gauge interaction in $\mathcal{L}_f^{\text{SM}}$ into

$$-q_0 \bar{\psi}\gamma^\mu \psi A'_\mu\left(1 - \Gamma_\alpha \sqrt{\hbar c}\,\phi\right)^{-1/2}\,.$$

In other words, the electric charge is dressed by the DM field

$$q(r, t) = q_0 \times \left(1 - \Gamma_\alpha \sqrt{\hbar c}\,\phi\right)^{-1/2}\,.$$

Since the fine structure constant is $\alpha = q^2/\hbar c$, this leads to Eq. (10.4), where we used $\Gamma_\alpha \sqrt{\hbar c}\, \phi \ll 1$.

Problem 10.2: Dark Matter-Induced Pseudo-Magnetic Field

The relevant contribution to the Dirac Hamiltonian can be computed as

$$\mathcal{H}_{\text{int}}\psi = -\gamma_0 \left[\frac{\partial \mathcal{L}_{\text{int}}}{\partial \bar{\psi}} - \partial_\mu \left(\frac{\partial \mathcal{L}_{\text{int}}}{\partial \left(\partial_\mu \bar{\psi} \right)} \right) \right], \tag{A.246}$$

with $\mathcal{L}_{\text{int}} \equiv \mathcal{L}_{\text{mag}}^{(1)}$. Explicit evaluation leads to

$$\mathcal{H}_{\text{mag}}^{(1)} = -\frac{1}{f_l}\left(\gamma_5 \frac{\partial}{c\partial t}\phi + \Sigma \cdot \nabla\phi \right).$$

Here, we used the identities $\gamma_0\gamma_0 = 1$ and $\gamma_0\gamma^i\gamma_5 = \Sigma^i$ with the spin matrix

$$\Sigma = \begin{pmatrix} \sigma & 0 \\ 0 & \sigma \end{pmatrix}. \tag{A.247}$$

In the nonrelativistic limit for atomic electrons, we arrive at the effective spin-dependent interaction

$$\mathcal{H}_{\text{mag}}^{(1)} \approx -\frac{2(\hbar c)^{3/2}}{f_l} S \cdot \nabla\phi.$$

The terms containing time derivatives of the ϕ field are neglected in the nonrelativistic limit for atomic electrons or nucleons as the γ_5 matrix mixes large and small components of the Dirac bi-spinors. This term is additionally suppressed due to cold DM being nonrelativistic, as $\frac{\partial}{c\partial t}\phi \sim \frac{v_g}{c}\phi$.

By comparing $\mathcal{H}_{\text{mag}}^{(1)}$ to the Zeeman interaction $\mathcal{H}_{\text{Zeeman}} = -\gamma S \cdot B_{\text{dm}}$, one can immediately read out the DM-induced pseudo-magnetic field B_{dm}. Here, γ is the gyromagnetic ratio for the magnetometer's atoms or nuclei.

Problem 10.3: Atomic Projection Noise Limit on Magnetometric Sensitivity

Consider a fully polarized ensemble of N_{at} atoms, each of which has a total angular momentum F. A magnetic field B, perpendicular to the spin orientation, induces atomic magnetic moment precession with the Larmor frequency ν_L. The phase φ acquired by a freely processing atom before their depolarization is given by

$$\varphi = 2\pi \nu_L T_2 = \gamma T_2 B,$$

where T_2 is the transverse relaxation time and γ is the gyromagnetic ratio of the atom. In atomic magnetometers, determination of the net phase φ acquired by the atoms due to the field provides information about the magnetic field. Thereby, the phase uncertainty $\delta\varphi$ determines the uncertainty δB of magnetic field determination

$$\delta B = \frac{1}{\gamma T_2}\delta\varphi.$$

At the most fundamental level, the phase uncertainty $\delta\varphi$ is given by the uncertainty of the spin orientation $\delta\varphi = \delta F_i/F$, where δF_i is the uncertainty of spin orientation in i-th direction. From the Heisenberg uncertainty principle, one gets

$$\delta F_i^2 \delta F_j^2 \geq \frac{|[F_i, F_j]|^2}{4} = \frac{\hbar^2 F_k^2}{4}.$$

For a coherent state, this relation equates, and based on the symmetry of the problem, we can write

$$\delta F_i = \sqrt{\frac{\hbar F_k}{2}}.$$

Introducing this relation into equation for $\delta\varphi$, we can write

$$\delta\varphi = \frac{\delta F_i}{N_{at} F} = \sqrt{\frac{2\hbar}{N_{at} F}},$$

which gives the magnetometric limit in a single measurement

$$\delta B = \frac{1}{\gamma} \frac{\sqrt{2\hbar}}{T_2 \sqrt{N_{at} F}}.$$

For the total measurement time τ_{meas}, the measurement can be repeated τ_{meas}/T_2 times. Since this is random white noise, this reduces the magnetic field uncertainty $\sqrt{\tau_{\text{meas}}/T_2}$ times

$$\delta B = \frac{1}{\gamma} \frac{\sqrt{2\hbar}}{\sqrt{N_{at} F T_2 \tau_{\text{meas}}}}.$$

Assuming that $\sqrt{2\hbar/F}$ is on the order of unity and replacing γ by $g\mu_B/\hbar$, with g being the Landé factor and μ_B being the Bohr magneton, we get the spin-projection-noise limit on the sensitivity of atomic magnetometer

$$\delta B \approx \frac{\hbar}{g\mu_B}\sqrt{\frac{1}{N_{\text{at}}T_2\tau_{\text{meas}}}}.$$

Problem 10.4: Noise Suppression of False Positive Events with a Sensor Network

Let us start with a discussion of the spatiotemporal pattern of signals of the network encountering a two-dimensional DM topological defect (domain wall). To reconstruct the pattern, we notice that the timing of DM-induced signals recorded by specific sensors is determined by the velocity $v_\perp$, with which Earth moves along the normal to the wall ($v_\perp = \boldsymbol{v} \cdot \boldsymbol{n}$, where $\boldsymbol{v}$ is the wall-Earth relative velocity). The same velocity component also determines the duration τ of the transient signal measured with a single sensor (assuming that the bandwidth of the sensor is larger than the pulse spectral width)

$$\tau = \frac{d}{v_\perp}, \tag{A.248}$$

where d is the thickness of the wall. The time delay between DM-induced transient signals in two sensors i and j separated by the distance L_{ij} can be calculated based on

$$t_i - t_j = \boldsymbol{L}_{ij} \cdot \boldsymbol{n}/v_\perp, \tag{A.249}$$

where the vector connecting the two sites is projected onto the normal to the wall. From Eq. (A.249), we can show that in order to fully determine the velocity $v_\perp$ we need to have four sensors.

Let us now estimate the rate of false positive detection by the network. If the network consists of identical sensors recording signals with random noise spikes appearing with an average rate of $1/\tau_n$, the average number N_{1234} of false positive events due to these random noise spikes within the measurement campaign lasting for time $\mathcal{T}$ is given by

$$N_{1234} \sim \frac{\mathcal{T}}{\tau_n}\left(\frac{t_{\text{tran}}}{\tau_n}\right)^3, \tag{A.250}$$

where t_{tran} is the transit time of the wall through the entire network and we assumed that $\mathcal{T} \gg \tau_n \gg t_{tran}$ [if the left side of Eq. (A.250) is larger than one we associated it with a high likelihood of detecting at least one false positive event during the campaign]. Equation (A.250) follows from the fact that the number of noise spikes measured by one sensor throughout the campaign is $\sim \mathcal{T}/\tau_n$, and in order for a false positive, there should be at least one event recorded in three other sensors within the transit time t_{tran}, which reduces the number of events by $\sim (t_{tran}/\tau_n)^3$.

Based on the timing of signals in specific sensors, one can precisely determine the time t_5 of appearance of the signal in additional sensor. Recalling that the signal lasts for a time τ, one can estimate the average number of false positives with a network consisting of five sensors

$$N_{12345} \approx N_{1234} \left(\frac{\tau}{\tau_n} \right). \tag{A.251}$$

This demonstrates that each additional sensor further reduces the probability of false positives by $\sim \tau/\tau_n$, which clearly shows the advantage of introducing additional sensors to the network.

We can now put it in perspective using experimental parameters of the GNOME. For that, we assume a 4-month long campaign with 9 active sensors, domain-wall transit time through a single detector of 10 ms, and the entire network passage time of 30 s. The noise spikes' characteristic time τ_n strongly depends on a signal-to-noise ratio and a confidence level one wants to achieve (this determines the threshold level). Herein, we assume $\tau_n = 100$ s, which comes from feasible parameters. Putting all these numbers into Eq. (A.250), one gets

$$N_{1234} \approx 2800, \tag{A.252}$$

which roughly corresponds to a false positive detection every hour. Introduction of five additional sensors gives an average number of false positive detections

$$N_{1-9} \approx 10^{-17}, \tag{A.253}$$

meaning that observation of a false positive event is nearly impossible.

References

1. J.D. Jackson, *Classical Electrodynamics*, 2nd edn. (Wiley, New York, NY, 1975)
2. M.S. Turner, Phys. Rev. D **28**, 1243 (1983)
3. E. Masso, F. Rota, G. Zsembinszki, Phys. Rev. D **72**, 084007 (2005)
4. P.A. Zyla, et al., *The Review of Particle Physics*, vol. 2020 (Oxford University Press, Oxford, 2020)
5. L. Lista, *Statistical Methods for Data Analysis in Particle Physics*, vol. 941 (Springer, New York, 2017)

Index

Printed in the United States
by Baker & Taylor Publisher Services